NON-LINEAR PARTIAL DIFFERENTIAL EQUATIONS

AN ALGEBRAIC VIEW OF GENERALIZED SOLUTIONS

NORTH-HOLLAND MATHEMATICS STUDIES 164

(Continuation of the Notas de Matemática)

NORTH-HOLLAND – AMSTERDAM • NEW YORK • OXFORD • TOKYO

NON-LINEAR PARTIAL DIFFERENTIAL EQUATIONS

AN ALGEBRAIC VIEW OF GENERALIZED SOLUTIONS

Elemér E. ROSINGER

Department of Mathematics
University of Pretoria
Pretoria, South Africa

1990

NORTH-HOLLAND – AMSTERDAM • NEW YORK • OXFORD • TOKYO

ELSEVIER SCIENCE PUBLISHERS B.V.
Sara Burgerhartstraat 25
P.O. Box 211, 1000 AE Amsterdam, The Netherlands

Distributors for the U.S.A. and Canada:

ELSEVIER SCIENCE PUBLISHING COMPANY, INC.
655 Avenue of the Americas
New York, N.Y. 10010, U.S.A.

Library of Congress Cataloging-in-Publication Data

Rosinger, Elemer E.
Non-linear partial differential equations : an algebraic view of generalized solutions / Elemér E. Rosinger.
p. cm. -- (North-Holland mathematics studies ; 164)
Includes bibliographical references.
ISBN 0-444-88700-8
1. Differential equations, Partial. 2. Differential equations, Nonlinear. I. Title. II. Series.
QA377.R68 1990
515'.353--dc20 90-47848
CIP

ISBN: 0 444 88700 8

Printed in the Netherlands

DEDICATED TO MY DAUGHTER

MYRA-SHARON

FOREWORD

A massive transition of interest from solving linear partial differential equation to solving nonlinear ones has taken place during the last two or three decades.

The availability of better digital computers often made numerical experimentations progress faster than the theoretical understanding of nonlinear partial differential equations.

The three most important nonlinear phenomena observed so far both experimentally and numerically, and studied theoretically in connection with such equations have been the solitons, shock waves and turbulence or chaotical processes. In many ways, these phenomena have presented increasing difficulties in the mentioned order. In particular, the latter two phenomena necessarily lead to *nonclassical* or generalized solutions for nonlinear partial differential equations.

While L. Schwartz's 1950 linear theory of distributions or generalized functions has proved to be of significant value in the theoretical understanding of linear, especially constant coefficient partial differential equations, sufficiently general and comprehensive nonlinear theories of generalized functions which may conveniently handle shock waves or turbulence have been late to appear.

Curiously, the insufficiency of L. Schwartz's linear theory and therefore the need for going beyond it was pointed out quite early. Indeed, in 1957, H. Lewy showed that most simple linear, variable coefficient, first order partial differential equations cannot have solutions within the L. Schwartz distributions. Unfortunately, that early warning has been disregarded for quite a while. One of the more important reasons for that seems to be the misunderstanding of L. Schwarz's so called impossibility result of 1954, which has often been wrongly interpreted as proving that no convenient nonlinear theory of generalized functions could be possible, Hörmander [1].

Nevertheless, various ad-hoc weak solution methods have been used in order to obtain nonclassical, generalized solutions for certain classes of nonlinear partial differential equations, such as for instance presented in Lions [1,2], without however developing any systematic and wide ranging *nonlinear* theory of generalized functions.

As a consequence, the attempts in extending weak solution methods from the linear to the nonlinear case, have often overlooked essentially nonlinear phenomena. In this way, the resulting weak solution methods used in the case of nonlinear partial differential equations proved to be insufficiently founded. Indeed, in the case of weak solutions obtained by various compactness arguments for instance, one remains open to nonlinear stability paradoxes, such as the existence of *weak* and *strong* solutions for the nonlinear system

$$u = 0$$

$$u^2 = 1$$

which would of course mean that we have somehow managed to prove the equality

$$0^2 = 1$$

within the real numbers $\mathbb{R}$, see for details Chapter 1, Section 8 and Appendix 6.

Lately, there appears to be an awareness about the fact that nonlinear operations - such as those involved in nonlinear partial differential equations - often fail to be weakly continuous, Dacorogna. As a consequence, certain particular and limited solution method have been developed, such as for instance those based on compensated compactness and the Young measure associated with weakly convergent sequences of functions subjected to differential constraints on algebraic manifolds, Ball, Murat, Tartar, Di Perna, Rauch & Reed, Slemrod. The fact remains however that with these methods only special, for instance conservation type nonlinear partial differential equations can be dealt with, since the basic philosophy of these methods is to get around the nonlinear failures of weak convergence by *imposing further restrictions* both on the nonlinearities and the weakly convergent sequences considered, Dacorogna. Therefore, with such methods there is no attempt to develop a comprehensive nonlinear theory of generalized functions, which may be capable of handling large enough classes of nonlinear partial differential equations. These methods only try to *avoid* the difficulties by *particularizing* the problems considered. Thus very little is done in order to better understand the deeper nature of these difficulties, nature which, as shown in this volume, see also Rosinger [1,2,3], is rather *algebraic* then topological.

The conceptual difficulties which so often arise when trying to extend linear methods to essentially nonlinear situations cannot and should not be overlooked or disregarded. However, as the above kind of nonlinear stability paradoxes show it, the transition from linear to nonlinear methods is not always done in a proper way. For a better glimpse into some of the more important such transitions, one can consult for instance the excellent historical survey in Zabuski.

For the sake of completeness, and in order to further stress the critical importance of the care for rigour when trying to extend linear ideas and methods to essentially nonlinear situations, one should perhaps mention as well the following. The transition of methods and concepts from linear to nonlinear partial differential equations has in fact produced two sets of paradoxes. The one above is connected with exact solutions. A second one concerns the numerical convergence paradox implied by the Lax equivalence result, and it is presented in detail in Rosinger [4,5,6].

Since the late seventies, two systematic attempts have been made in order to remedy the mentioned inadequate situation concerning weak and generalized solutions of sufficiently large classes of nonlinear partial differential equations. The main publications have been Rosinger [1,2], Colombeau [1,2] and Rosinger [3], a first presentation of some of the basic ideas involved being given earlier in Rosinger [7,8].

Colombeau's nonlinear theory of generalized functions, although developed in the early eighties, has started in a rather independent manner. However, it proves to be a particular case of the more general nonlinear theory of generalized functions in Rosinger [1,2], see for details Rosinger [3,pp. 300-306].

In fact, the two theories in Rosinger [1,2,3] and Colombeau [1,2] have so far been somewhat complementary to each other, as they approach the field of generalized solutions for nonlinear partial differential equations from rather opposite points of view. Indeed, both theories aim to construct *differential algebras* A of generalized functions which extend the L. Schwartz distriburtions, that is, admit embeddings

$$\mathcal{D}'(\Omega) \subset A, \quad \text{with} \quad \Omega \subset \mathbb{R}^n \quad \text{open.}$$

Given then linear or nonlinear partial differential operators $T(x,D)$ on Ω, one can extend them easily, so that they may act for instance as mappings

$$T(x,D) : A \longrightarrow A$$

In that case the respective linear or nonlinear partial differential equations

$$T(x,D)U(x) = f(x), \quad x \in \Omega$$

with $f \in A$ given, may have generalized solutions $U \in A$, which under customary conditions may prove to be unique, regular, etc.

And in view of H. Lewy's mentioned impossibility result, extensions of the L. Schwartz distributions given by embeddings

$$\mathcal{D}'(\Omega) \subset A$$

of the above or similar type prove to be *necessary* even when solving linear variable coefficient partial differential equations.

Now, Colombeau's nonlinear theory develops what appears to be the most *natural* and *central* class of differential algebras A which contain the $\mathcal{D}'(\Omega)$ distributions, see for details Chapter 8, as well as Rosinger [3,pp. 115-123]. The power of that approach is quite impressive as it leads to existence, uniqueness and regularity results concerning solutions of large classes of linear and nonlinear partial differential equations, equations which earlier were not solved, or were even proved to be unsolvable within the distributions or hyperfunctions, see for details Rosinger [3, pp. 145-192]. In addition, Colombeau's nonlinear theory has important applications in the numerical solution of nonlinear and nonconservative shocks for instance, see for details Biagioni [2].

On the other hand, the earlier and more general nonlinear theory in Rosinger [1,2,3], has started with the clarification of the *algebraic* and *differential* foundations of what may conveniently be considered as *all possible* nonlinear theories of generalized functions. That approach leads to the characterization and construction of a very large class of differential algebras A which contain the $\mathcal{D}'$ distributions, and which can be used in order to give the solution of most general nonlinear partial differential equations. In that context, in addition to the usual problems of existence, uniqueness and regularity of solutions, a first and fundamental role is played by the problems of *stability*, *generality* and *exactness* of such solutions, see for details Chapter 1, Sections 8-12.

This general approach yields several results which are a *first* in the literature.

For instance, one obtains *global* generalized solutions for *all analytic nonlinear* partial differential equations. These solutions are *analytic* on the whole of the domain of analyticity of the respective equations, except for closed, nowhere dense subsets, which can be chosen to have zero Lebesque measure, see Chapter 2 and Rosinger [3].

A second result gives an *algebraic* characterization for the existence of generalized solutions for all polynomial nonlinear partial differential equations with continuous coefficients, see Chapter 3 and Rosinger [3]. This algebraic characterization happens to be given by a version of the so called *neutrix* or *off diagonality* condition, see (1.6.11) in Chapter 1.

A third type of results concerns the *characterization* of a very large class of differential algebras containing the distributions. One of this characterizations is given by the mentioned neutrix or off diagonality condition on differential algebras of generalized functions constructed as quotient algebras

$$A = \mathcal{A}/\mathcal{I}$$

where $\mathcal{A} = (\mathcal{C}^0(\Omega))^{\mathbb{N}}$ and $\mathcal{I}$ is an ideal in $\mathcal{A}$, see Chapter 6, as well as Rosinger [1,2,3]. Within a more general framework of quotient algebras

$$A = \mathcal{A}/\mathcal{I}$$

where $\mathcal{A}$ is a subalgebra in $(\mathcal{C}^0(\Omega))^{\mathbb{N}}$ and $\mathcal{I}$ is an ideal in $\mathcal{A}$, a further characterization of the structure of these algebras is given. Indeed, it is shown that the algebraic type neutrix or off diagonality condition is equivalent to a *topological* type condition of *dense vanishing*, see Chapter 3.

The above three results use the full generality of the nonlinear theory developed in Rosinger [1,2,3], and it is an open question whether similar results may be obtainable within the particular nonlinear theory in Colombeau [1,2].

Several other results which so far could only be obtained within the framework of the nonlinear theory in Rosinger [1,2,3] are presented shortly in Chapters 6 and 7. More detailed accounts, including additional such results can be found in Rosinger [1,2,3].

At this stage it may be important to point out the *utility* of considering the problem of generalized solution for nonlinear partial differential equations within sufficiently large frameworks. Indeed, as H. Lewy's 1957 example shows it, the framework $\mathcal{D}'(\mathbb{R}^n)$ of the L. Schwartz distribution is too restrictive even for linear, variable coefficient partial differential equations. Colombeau's particular nonlinear theory, owing to its natural, central position proves to be unusually powerful, both in generalized and numerical solutions for wide classes of linear and nonlinear partial differential equations. However, results such as in Chapters 2 and 3 for instance, find their natural framework within the general nonlinear theory introduced in Rosinger [1,2,3], and so far could not be reproduced within the framework in Colombeau [1,2].

What to us seems however less than surprising is that this is not yet the end of the story. Indeed, as seen in the results in Chapter 4, contributed recently by M. Oberguggenberger, further extensions of the general framework in Rosinger [1,2,3] are particularly useful.

All this development seems to create the feeling that, inspite of the rather extended framework presented in this volume, the nonlinear theories of generalized functions may still be at their beginnings.

And now a few words about the point of view and approach pursued in this volume.

At least since Sobolev [1,2], the main, in fact nearly exclusive approach in the study of weak and generalized solutions for linear and nonlinear partial differential equations has been that of functional analysis, used most often in infinite dimensional vector spaces. That includes as well the way Colombeau's nonlinear theory of generalized functions was started in Colombeau [1].

The difficulties in such a functional analytic approach in the case of solving *nonlinear* partial differential equations are well known. And they come mainly from the fact that the strength of present day functional analysis is rather in the linear than the nonlinear realm.

In addition, an exaggerated preference for a functional analytic point of view can have the unfortunate tendency to fail to see simple but fundamental facts for what they really are, and instead, to notice them only through some of their more sophisticated consequences, as they may emerge when translated into the functional analytic language.

The effect may be an unnecessary obfuscation, and hence, misunderstanding, as happened for instance with L. Schwartz's so called impossibility result. Indeed, the point this result tries to emphasize in its original formulation is that in a differential algebra which contains just a few continuous functions, the multiplication of these functions cannot be the usual function multiplication, unless we are ready to accept certain apparently unpleasant consequences, see for details Proposition 1 in Chapter 1, Section 2. However, those few continuous functions are *not* $\mathcal{C}^\infty$-smooth. In fact, their first or at most second order derivatives happen to be *discon-*

tinuous. In this way, at a deeper level, the difficulty which the so called Schwartz impossibility result is trying to tell us inspite of all misunderstandings, is that there exists a certain *conflict* between *discontinuity*, *multiplication* and *differentiation*.

And as seen in Chapter 1, Section 1 and Appendix 1, this conflict is of a *most simple algebraic* nature, which already happens to occur within a rock-bottom, very general framework, far from being in any way restricted or specific to the L. Schwartz distributions.

And then, what can be done?

Well, we can remember that one way to see modern mathematics is as being a multilayered theory in which successive layers are built upon and include earlier, more fundamental ones. For instance, one may list some of them as follows, according to the way successive layers depend on previous ones:

- set theory
- binary relations, order
- algebra
- topology
- functional analysis
- etc.

In this way, it may appear useful to try to identify the roots of a problem or difficulty at the deeper relevant layers. Such an approach will bring the so called L. Schwartz impossibility result, and in general, the problem of distribution multiplication to the *algebra* level of the basic *conflict* between discontinuity, multiplication and differentiation, mentioned above.

This is then in short the essence and the novelty of the 'algebra first' approach pursued in the present volume.

The reader who may wonder about the possible effectiveness of such a *desescalation* from involved and sophisticated functional analysis to basic mathematical structures, may perhaps first - and equally - wonder about the rather incisive insight of the celebrated seventeenth century Dutch philosopher Spinoza, according to whom the ultimate aim of science is to reduce the whole world to a tautology.

In mathematics, a good part of this dynamics is expressed in the well known adage that, old theorems never die: they just become definitions!

Indeed, it is obvious that a lot of knowledge, understanding, experience and hopefully *simplification* is needed in order to set up an appropriate mathematical structure, in particular axioms or definitions. In this way, the knowledge in 'old theorems' becomes *explicit* in the very mathematical structure itself. A good illustration for that, in particular for *simplification*, is the transition from Colombeau [1] to Colombeau [2].

Certainly, in a deduction $A \Rightarrow B$, B *cannot* be more than A, that is, it cannot contain more information than A, and the nearer B is to A, the more the *information* which was gotten through the deduction is near to 100%. Of course, we are not interested in a theory which mainly has 100%

efficient deductions $A \Rightarrow A$. So, we should keep somewhat away from tautology. On the other hand, the more the amount of near tautological deductions in a theory, the greater our understanding of what is going on: after all, the best *analogy* is a tautology and the best *explicit* knowledge is an analogy. Isn't it that a proper key is better than a skeleton key precisely to the extent that it is more analogous with the lock, containing more explicit knowledge in its very structure?

In this respect the *presence* of 'hard theorems' - which are hard owing to their far from tautological proofs - is a sign of insufficient insight on the level of the structure of the theory as a *whole*. Let us just remember how the so called 'Fundamental Theorem of Algebra' according to which an algebraic equation has at least one complex root, lost its 'hard' status from the time of D'Alembert to the time of Cauchy, owing to the emergence of complex function theory.

Now, as if to give some much desired comfort to those who may nevertheless feel that, within a good mathematical theory one should, at least here and there, have some 'hard theorems', the mentioned desescalation proves to leave room for such theorems. Indeed, let us mention just some of them, present also within the part of the general theory contained in this volume. In Sections 4 and 6 in Chapter 2, one uses a transfinite inductive exhaustion process for open sets in Euclidean spaces and, respectively, a rather involved topological and measure theoretical argument in Euclidean spaces. In Section 4 of Chapter 3, a similarly involved, twice iterated use of the Baire category argument in Euclidean spaces is employed. Further, in Section 5 of Chapter 3, a deep property of upper semicontinuous functions is used in a critical manner. In Section 4 of Chapter 4, functional analytic methods are employed. And we can also mention the cardinality arguments on sets of continuous functions on Euclidean spaces, which are fundamental in the results presented in Chapter 6.

It should be mentioned that the presentations in Rosinger [1,2,3] and Colombeau [2], have also pursued an 'algebra first' approach, although in a different, more obvious manner, which is directly inspired by the classical weak solution method. Indeed, it is well known that $C^\infty(\Omega)$ for instance, is weakly sequentially dense in $\mathcal{D}'(\Omega)$. Therefore, in some sense to be defined precisely, we have an 'inclusion'

$$\mathcal{D}'(\Omega) \text{ '}\subset\text{' } (\mathcal{C}^\infty(\Omega))^{\mathbb{N}}$$

Moreover

$$A = (\mathcal{C}^\infty(\Omega))^{\mathbb{N}}$$

is obviously an associative and commutative differential algebra, when considered with the usual term wise operations on functions. In this way, all what a nonlinear theory of generalized functions needs to do is to give a precise meaning to the above 'inclusion'.

In the present volume, this second algebraic approach follows after the earlier mentioned one, namely, that dealing with the conflict between discontinuity, multiplication and differentiation.

The outcome of such an approach is that functional analytic methods need not be brought into play for quite a long time. In fact, just as in Rosinger [1,2,3] and Colombeau [2], such methods are not used at all, except for Chapter 4. The mathematics which is used consists of basic algebra of rings of functions, calculus and some topology, all in Euclidean spaces alone. The connection with partial differential equations is made through certain asymptotic interpretations in the spirit of the 'neutrix calculus' in Van der Corput, see Chapter 1, Section 6 and Appendix 4.

The fact that functional analysis is not needed from the very beginning should not come as a surprise. Indeed, in the customary approaches to partial differential equations there are *three reasons* for the use of functional analysis, namely, first, in order to define partial derivatives for generalized functions, then, in order to approximate generalized solutions by regular functions, and finally, in order to define the generalized functions as elements in the completion of certain spaces of regular functions. But, by constructing embeddings

$$\mathcal{D}'(\Omega) \subset A$$

into quotient algebras

$$A = \mathcal{A}/\mathcal{I}, \ \mathcal{I} \subset \mathcal{A} \subset (\mathcal{C}^\infty(\Omega))^{\mathbb{N}}$$

one can avoid functional analysis to a good extent. Indeed, the partial derivatives of generalized functions $T \in A = \mathcal{A}/\mathcal{I}$, can be *reduced* to the usual partial derivatives of the smooth functions in the sequences representing T. Further, an algebraic study of *exact* solutions given by generalized functions need not involve approximations from the very beginning. Finally, the respective algebras A of generalized functions - used as 'reservoirs' of solutions - can easily be kept large enough by simply using suitably chosen subalgebras $\mathcal{A}$ and ideals $\mathcal{I}$ in the construction of the quotient algebras $A = \mathcal{A}/\mathcal{I}$.

The ease in such an approach and the extent to which it works can be seen in this volume, as well as in the cited main publications Rosinger [1,2,3] and Colombeau [1,2]. Further developments have been contributed in a number of papers and two research monographs, published or due to appear, by M. Adamczewski, J. Aragona, H.A. Biagioni, J.J. Cauret, J.F. Colombeau, J.E. Galé, F. Lafon, M. Langlais, A.Y. Le Roux, A. Noussair, M. Oberguggenberger, B. Perot and T.D. Todorov, see the References.

As before, a close contact with J.F. Colombeau and M. Oberguggenberger has offered the author a particularly useful help, not least owing to the exchange of different views on what appears to be a fast emerging nonlinear theory of generalized functions.

A special mention is due to M. Oberguggenberger's recent cycle of research on Semilinear Wave Equations with rough initial data, which is shortly presented in Chapter 4. In addition to its obvious importance as a powerful contribution to a clearer understanding of the propogation of singularities in nonlinear wave phenomena, its impact on the emerging nonlinear theory of generalized functions is uniquely important at this stage. Indeed, the method used in Oberguggenberger's research is so simple and powerful precisely due to the fact that it is sufficiently general, in fact so general that it goes beyond the framework of Rosinger [1,2,3] as well. A detailed presentation of recent results on the propagation of singularities in nonlinear wave phenomena, as well as a thorough analysis of the connection between the emerging nonlinear theory of generalized functions and earlier, partial attempts at distribution multiplication are presented in the research monograph Oberguggenberger [15].

Another special mention is due to H.A. Biagioni's research monograph, Biagioni [2], which presents a most important development of Colombeau's nonlinear theory, namely, in the field of numerical solutions for nonlinear, nonconservative partial differential equations. This line of research is still very much at the beginning of reaching its full potential, and owing to the extensive space it takes to present the results already obtained, it could not be included in this volume.

The author owes a deep and warm gratitude to Professor J. Swart, the head of the mathematics department, and to Professor N. Sauer, the dean of the science faculty at the University of Pretoria for the unique academic and research conditions created and for the friendly and kind support over the years.

As on so many previous occasions, the outstanding work of careful typing of the manuscript was done by Mrs. A.E. Van Rensburg. It is hard to ever truly appreciate the contribution of such help.

As on earlier occasions, the author is particularly grateful to Drs. A Sevenster of the Elsevier Science Publishers for a truly supportive approach.

This is the author's third research monograph in ten years in Prof. L. Nachbin's series of the North-Holland Mathematics Studies. In our era, when 'Big Science' so often tries to dwarf us into negligible and disposable entities, subjecting us to the 'Big Industry', conveyor belt type management by 'Publish or Perish!, one should perhaps better not think about how the world may look without editors like Prof. L. Nachbin, who are still ready to offer us a most outstanding encouragement and support.

And what may in fact be wrong with 'Big Science'?

Well, was it Henry Ford, of the 'History is bunk' fame, who found it necessary to insist that:

'Big Organizations can never be humane'?

Yet, after WW II, to the more traditional Big Organizations of Army, Priesthood, Bureaucracy and Industry, we have been so busy adding that of Big Science ...

E.E. Rosinger

Pretoria, May 1990

TABLE OF CONTENTS

CHAPTER 1

CONFLICT BETWEEN DISCONTINUITY, MULTIPLICATION AND DIFFERENTIATION

§1. A BASIC CONFLICT

There exist basic algebraic - in particular, ring theoretic - aspects involved in the problem of finding generalized solutions for nonlinear partial differential equations.

Why generalized solutions?

The answer is well known and a short, first account of it is given in Chapter 5, as well as in the literature mentioned there.

But then equally, if not even more so, one may ask: why algebra, and why precisely in the realms of *nonlinear* partial differential equations?

Fortunately, the answer to this second question is much simpler, and it can be presented here, without the need for any special introduction.

In fact, it is our aim to show for the first time in the known literature, that the issue of generalized solutions for nonlinear partial differential equations can be approached in a relevant and useful way by first considering the algebraic problems involved in the *basic trio* of:

- discontinuity
- multiplication
- differentiation.

The interests in such an 'algebra first' approach can be multiple.

First of all, the ring theoretic type of algebra involved belongs to a more fundamental kind of mathematics than the usual calculus, functional analytic or topological methods which are customary in the study of partial differential equations. And by using such more fundamental mathematics, one can hope for a better and deeper understanding of the issues involved, as well as for easier solutions. Fortunately, such advantages happen to materialize to a good extent.

Another reason for pursuing 'algebra first' is in trying to draw the attention of mathematicians working in fields far removed from analysis or functional analysis - such as for instance algebra, or rings of functions - upon the possibility of significant applications of their methods and results in the field of solving nonlinear partial differential equations.

Finally, there is an interest in showing to many analysts and functional analysts working in the field of nonlinear partial differential equations, that the road can lead not only from theories and methods which are already quite complicated towards other ones, yet more complicated. On the contrary, and at least as a temporary detour, the 'algebra first' road can lead to quite a few simplifications and clarifications. However, in view of the results already obtained along that road, Rosinger [1,2,3], Colombeau [1,2], one may as well see it as being much more than a temporary affair. Indeed, one of such results - a first in the literature - is a *globalized* version of the classical Cauchy-Kovalevskaia theorem concerning the existence of solutions for *arbitrary analytic nonlinear* partial differential equations, Rosinger [3, pp. 259-266]. By using algebraic, ring theoretic methods, one can prove the existence of generalized solutions on the whole of the domain of analyticity of the respective nonlinear equations. Furthermore, these generalized solutions are analytic on the whole of the respective domains, except for subsets of zero Lebesque measure. Details are presented in Chapter 2. Similar ring theoretic methods can lead to another first in the literature, Rosinger [3, pp. 233-247], namely, an algebraic characterization for the solvability of a very large class of nonlinear partial differential equations, reviewed in Chapter 3. Important first results in the literature are obtained in the particular theory in Colombeau [1,2] as well. For instance, solutions are found for large classes of nonlinear partial differential equations which earlier were unsolved or proved to be unsolvable in distributions. In addition, solutions are constructed for the first time for arbitrary systems of linear partial differential equations with smooth coefficients, thus going beyond the celebrated impossibility result of Lewy, as well as its various extensions, Hörmander.

Let us now turn to the motivation of the basic algebraic setting presented later in its basic form in Section 3. For that purpose we shall give here a most simple example of the kind of conflicts we can expect when dealing with the mentioned trio of discontinuity, multiplication and differentiation. Fortunately, an attentive study of the conflict involved in this simple, one dimensional example can lead a long way towards the clarification and solution of most important problems concerning the solution of wide classes of important nonlinear partial differential equations.

As is known, see also Chapter 5, the classical solutions of linear or nonlinear partial differential equations are given by sufficiently smooth functions

$$(1.1.1) \qquad U : \Omega \longrightarrow \mathbb{R}$$

where $\Omega \subset \mathbb{R}^n$ is a certain domain. For our purposes, we can restrict the setting to the simplest, one dimensional case, when $n = 1$ and $\Omega = \mathbb{R}$.

As also seen in Chapter 5, nonlinear partial differential equations have important nonclassical, that is, generalized solutions. In particular, such nonclassical solutions may be given by nonsmooth or discontinuous functions U in (1.1.1)

For our purposes, it will be sufficient at first to consider a most simple discontinuous function, such as the well known Heaviside function

$$(1.1.2) \qquad H : \mathbb{R} \longrightarrow \mathbb{R}$$

defined by

$$(1.1.3) \qquad H(x) = \begin{cases} 0 & \text{if } x \leq 0 \\ 1 & \text{if } x > 0 \end{cases}$$

Now, when appearing in generalized solutions of nonlinear partial differential equations, a discontinuous function such as H, will be involved in multiplication and differentiation.

Therefore, it appears that the basic setting we are looking for should be given by a ring of functions

$$(1.1.4) \qquad A \subset \mathbb{R} \rightarrow \mathbb{R}$$

such that

$$(1.1.5) \qquad H \in A$$

and A has a derivative operator

$$(1.1.6) \qquad D : A \rightarrow A$$

that is, an operator D which is linear and also satisfies the Leibnitz rule of product derivative

$$(1.1.7) \qquad D(f \cdot g) = (Df) \cdot g + f \cdot (Dg), \quad f,g \in A$$

Unfortunately, the problem already starts right here.

Indeed, no matter how intuitive and natural is the above setting as an extension of the classical, smooth case, the relations (1.1.2)-(1.1.7) have rather inconvenient consequences!

For that, first we note that (1.1.4), (1.1.5) yield the relations

$$(1.1.8) \qquad H^m = H, \quad m \in \mathbb{N}, \quad m \geq 1$$

Further, in view of (1.1.4), A is associative and commutative. Hence (1.1.8), (1.1.7) give the relations

$$(1.1.9) \qquad mH \cdot (DH) = DH, \quad m \in \mathbb{N}, \quad m \geq 2$$

Now, if $p,q \in \mathbb{N}$, $p,q \geq 2$ and $p \neq q$, then (1.1.9) implies

$$\frac{1}{p}DH = \frac{1}{q}DH = H \cdot (DH)$$

or

$$(\frac{1}{p} - \frac{1}{q})DH = 0 \in A$$

which results in

(1.1.10) $DH = 0 \in A$

However, as seen in Chapter 5, there exist particularly strong reasons to expect that the derivative operator D in (1.1.6), (1.1.7) is such that

(1.1.11) $DH = \delta$

where δ is the well known Dirac delta function. In this case, the relations (1.1.10), (1.1.11) would imply

(1.1.12) $\delta = 0 \in A$

which is false, since the Dirac delta function is not the identically zero function.

To recapitulate, the discontinuous function H has the multiplication property (1.1.8), which by differentiation gives the relation (1.1.10). Then, assuming the natural relation in (1.1.11), one obtains the incorrect result in (1.1.12).

The way out is obviously by trying to relax some of the assumptions involved.

What we have to try to hold to, in view of reasons such as those presented in Chapter 5, is the discontinuous function H and the relation $DH = \delta$ describing its derivative.

But we are more *free* in *two* other respects, namely, in choosing the algebra A and the derivative operator D.

Indeed, while A should contain functions such as $H : \mathbb{R} \longrightarrow \mathbb{R}$, it need not be an algebra of functions from $\mathbb{R}$ to $\mathbb{R}$. In other words, A may as well contain more general elements, and the multiplication in A need not be so closely related to the multiplication of functions. In particular, (1.1.8) need not necessarily hold.

Concerning the derivative operator D, a most important point to note is the particularly restrictive nature of the assumption in (1.1.6). Indeed, this assumption implies that the elements of A are *indefinitely derivable*, that is

(1.1.13) $D^m a$ exists, for all $a \in A$ and $m \in \mathbb{N}$, $m \geq 1$

This would of course happen if we had

(1.1.14) $A \subset \mathcal{C}^\infty(\mathbb{R})$

which is however *not possible* in view of (1.1.5). It follows, that we should keep open the possibility when the derivative operator D is defined as follows

$$(1.1.15) \qquad D : A \longrightarrow \bar{A}$$

where $\bar{A}$ is *another* algebra of generalized functions. In this case, in order to preserve the Leibnitz rule of product derivative, we can assume the existence of an algebra homomorphism

$$(1.1.16) \qquad \begin{array}{ccc} A & \longrightarrow & \bar{A} \\ a & \longmapsto & \bar{a} \end{array}$$

and then rewrite (1.1.7) as follows

$$(1.1.17) \qquad D(f\cdot g) = (Df)\cdot\bar{g} + \bar{f}\cdot(Dg), \quad f,g \in A$$

It will be shown in Chapters 2 and 3, that the above two kind of relaxations, namely, on the algebra A and the derivative operator D, are more then sufficient in order to find generalized solutions for wide classes of nonlinear partial differential equations. In particular one can obtain the mentioned globalized version for the Cauchy-Kovalevskaia theorem, in which one can prove the existence of generalized solutions, on the whole of the domain of analyticity, of arbitrary analytic nonlinear partial differential equation. Furthermore, one can obtain an algebraic characterization for the solvability of a very large class of nonlinear partial differential equations.

§2. A FEW REMARKS

It is *particularly important* to note that the argument leading to (1.1.10) does *not* use calculus and it is purely *algebraic*, more precisely it only uses the algebra structure of A and the fact that D is linear and it satisfies the Leibnitz rule of product derivative. Two, more abstract variants of this argument are presented in Appendix 1.

The further results in this chapter on the conflict within the trio of discontinuity, multiplication and differentiation will also be of a similar purely algebraic nature. This is precisely the reason why the 'algebra first' approach is useful, and should be systematically pursued.

Now, let us recall the essential role played by the property, see (1.1.8)

$$(1.2.1) \qquad H^m = H, \quad m \in \mathbb{N}, \quad m \geq 1$$

in obtaining the undesirable relation (1.1.10).

We note that the infinite family of equations

$$(1.2.2) \qquad x^m = x, \quad x \in \mathbb{R}, \quad m \in \mathbb{N}, \quad m \geq 1$$

only has two solutions, namely

(1.2.3) $x = 0$ or $x = 1$

Therefore, the relation (1.2.1) determines uniquely the Heaviside function H among all functions from $\mathbb{R}$ to $\mathbb{R}$, which are discontinuous only at $x = 0 \in \mathbb{R}$, and are nondecreasing.

Let us denote by 0 and 1 the functions defined on $\mathbb{R}$ which take everywhere the value 0, respectively 1. Then, within the framework of (1.1.4)-(1.1.7), we obviously have $0 \in A$. Let us further assume that

(1.2.4) $1 \in A$

then from (1.1.7), we obtain easily the relations

(1.2.5) $D0 = D1 = 0 \in A$

which are to be expected, since the functions 0 and 1 are constant.

The unexpected fact is that, although H is not a constant function, (1.1.4)-(1.1.7) will nevertheless imply, see (1.1.10)

(1.2.6) $DH = 0 \in A$

In view of (1.2.3), the functions 0 and 1 are the only continuous functions which satisfy (1.2.1). It follows that the framework in (1.1.4)-(1.1.7) and (1.2.4) is *too coarse* in order to *distinguish* between the derivatives of continuous and discontinuous functions.

The above conflict between discontinuity, multiplication and derivative appears already at that rather simple and fundamental level. Needless to say, it has many further, more involved implications, such as for instance, the long misunderstood, so called L. Schwartz impossibility result, see below.

However, a proper treatment of that conflict can only benefit from its identification at its most basic and also simple levels. Otherwise, the complications involved may lead to misunderstandings, as happened with the mentioned result of L. Schwartz.

Suffice it here to say that one of the most important consequences of the conflict between discontinuity, multiplication and derivative is the following. Above a certain level of irregular, discontinuous or nonsmooth functions, multiplication *can no longer* be made in a unique, canonical or best way. This is the price we have to pay if we nevertheless want to bring together into one mathematical structure both discontinuity and multiplication, as well as derivative. In this way, the message of L. Schwartz's mentioned result is not that 'one cannot multiply distributions', Hörmander [p. 9], but on the contrary, that one can, and inevitably has to face the fact that they can be multiplied in many different ways.

By the way, for the sake of rigour, let us specify that the citation from Hörmander mentioned above reads as follows: 'It has been proved by Schwartz ... that an associative multiplication of two arbitrary distributions cannot be defined'.

The reader less familiar with the linear theory of the L. Schwartz distributions can omit the rest of this section and go straight to the next one, that is, to Section 3. A few useful details on distribution theory are presented in Appendix 1 to Chapter 5. After consulting them, as well as the basic issues concerning generalized solutions for nonlinear partial differential equations presented in the mentioned chapter, the reader may return to the rest of this section.

The second remark concerns the longstanding misunderstanding connected with the so called impossibility result of L. Schwartz, established in 1954, Schwartz [2].

This result has often been overstated by claiming that it proves the impossibility of conveniently multiplying distributions.

In fact, L. Schwartz's mentioned result only shows that - similar to the situation following from (1.1.4)-(1.1.7) above - an insufficiently careful choice of an algebraic framework for generalized functions can lead to undesirable consequences. In addition, its set up is more complicated than the one in (1.1.4)-(1.1.7), therefore, to an extent it hides the simplicity of the fundamental conflict in the trio of discontinuity, multiplication and differentiation. For convenience, we recall here L. Schwartz's mentioned result, a detailed proof of which can be found for instance in Rosinger [3, pp. 27-30].

<u>Proposition 1</u> (Schwartz [2])

Suppose A is an associative algebra with a derivative operator $D : A \longrightarrow A$, that is, a linear mapping which satisfies the Leibnitz rule of product derivative, see (1.1.7).

Further suppose that

(1.2.7) the following four $\mathcal{C}^0$-smooth functions 1, x, $x(\ell n\ |x| - 1)$ and $x^2(\ell n\ |x| - 1)$ belong to A, where for $x = 0$, the latter two functions are assumed to vanish,

(1.2.8) the function 1 is the unit in A

(1.2.9) the multiplication in A is such that

$$(x)\cdot(x(\ell n\ |x| - 1)) = x^2(\ell n\ |x| - 1)$$

(1.2.10) the derivative operator $D : A \longrightarrow A$ applied to the following three $\mathcal{C}^1$-smooth functions
$1, x, x^2(\ell n\ |x| - 1)$
is the usual derivative of functions.

Then, there exists no $\delta \in A$, $\delta \neq 0 \in A$, such that

(1.2.11) $x\delta = 0 \in A$ □

The usual interpretation of Proposition 1 goes along the following lines. If δ is the Dirac delta function, or more precisely distribution, then it is well known that

(1.2.12) $\delta \in \mathcal{D}'(\mathbb{R})$, $\delta \neq 0 \in \mathcal{D}'(\mathbb{R})$

where $\mathcal{D}'(\mathbb{R})$ is the set of the Schwartz distributions on $\mathbb{R}$. Furthermore, we can multiply each distribution $T \in \mathcal{D}'(\mathbb{R})$ with each function $\psi \in \mathcal{C}^\infty(\mathbb{R})$ and obtain $\psi T \in \mathcal{D}'(\mathbb{R})$, see Appendix 1 to Chapter 5. In particular, we have

(1.2.13) $\psi\delta = \psi(0) \in \mathcal{D}'(\mathbb{R})$

therefore

(1.2.14) $x\delta = 0 \in \mathcal{D}'(\mathbb{R})$

It should be recalled here that

(1.2.15) $\mathcal{C}^\infty(\mathbb{R}) \subsetneqq \mathcal{D}'(\mathbb{R})$

in particular

(1.2.16) $\delta \in \mathcal{D}'(\mathbb{R}) \backslash \mathcal{C}^\infty(\mathbb{R})$

Now, if we want to multiply two arbitrary distributions T and S from $\mathcal{D}'(\mathbb{R})$, the above mentioned procedure where the multiplication of any $T \in \mathcal{D}'(\mathbb{R})$ with any $\psi \in \mathcal{C}^\infty(\mathbb{R})$ gives $\psi T \in \mathcal{D}'(\mathbb{R})$ will in general not work, since both distributions T and S may fail to belong to $\mathcal{C}^\infty(\mathbb{R})$. For instance, in view of (1.2.16), we cannot compute $\delta^2 = \delta\delta$ by the above procedure.

One way to multiply arbitrary distributions from $\mathcal{D}'(\mathbb{R})$ is by finding an embedding

(1.2.17) $\mathcal{D}'(\mathbb{R}) \subset A$

where A is an algebra, and then performing the multiplication in A. Now, one notes that, in view of the well known inclusion $\mathcal{C}^0(\mathbb{R}) \subsetneqq \mathcal{D}'(\mathbb{R})$, the four functions in (1.2.7) must belong to A. It follows that the conditions (1.2.8)-(1.2.10) on the algebra A required in L. Schwartz's above result are rather *natural and minimal*. Nevertheless, they lead to (1.2.11), which is in conflict with the customary distributional properties of the Dirac delta funtion presented in the relations (1.2.12), (1.2.14).

And here, the usual misinterpretation occurs, according to which it is stated that there cannot exist convenient algebras A such as in (1.2.17) in which to embed the distributions. Sometime it is also concluded that in view of (1.2.11) any algebra A which satisfies the natural and minimal conditions in (1.2.7)-(1.2.10), cannot contain the Dirac delta function. One can as well encounter the general conclusion that the multiplication of arbitrary distributions is not possible, Hörmander.

Let us now return to Proposition 1 above and try to assess its correct meaning.

First we note that in view of (1.2.12), we shall necessarily have

(1.2.18) $\delta \in A, \quad \delta \neq 0 \in A$

for any algebra A containing the distributions, such as in (1.2.17).

Now, an algebra A as in Proposition 1, may or may not contain the Dirac delta function δ. But if A is large enough, such as for instance in (1.2.17), then, in view of (1.2.18), it will contain δ as a nonzero element.

What is then the message in (1.2.11)?

Well, it is simply the following. The relation (1.2.14) is valid for the usual multiplication between $\mathcal{C}^\infty$-smooth functions and distributions, see Appendix 1 to Chapter 5. In other words, x and δ are *zero divisors* in that multiplication. Or, from an analysis point of view, the *singularity* of δ at the point $0 \in \mathbb{R}$ is of *lower order* than that of the function 1/x.

In the same time, in view of (1.2.11), the multiplication in any algebra A such as in Proposition 1, and which contains δ, will by necessity give

(1.2.19) $x\delta \neq 0 \in A$

Hence, x and δ are *no* longer zero divisors in A. Which in the language of analysis means that, when seen in A, the *singularity* of δ at the point $0 \in \mathbb{R}$ is *not* of *lower order* than that of the function 1/x.

§3. AN ALGEBRA SETTING FOR GENERALIZED FUNCTIONS

As mentioned, it will be sufficient to deal alone with the one dimensional case of generalized functions on $\mathbb{R}$, since extensions to arbitrary higher dimensions follow easily. Further, we have seen that we may have to go as far as to construct embeddings

$$(1.3.1) \qquad \mathcal{D}'(\mathbb{R}) \subset A \longrightarrow \bar{A}$$

where A and $\bar{A}$ are associative algebras, with a given algebra homomorphism

$$(1.3.2) \qquad \begin{array}{ccc} A & \longrightarrow & \bar{A} \\ a & \longmapsto & \bar{a} \end{array}$$

and a derivative operator

$$(1.3.3) \qquad D : A \longrightarrow \bar{A}$$

which is a linear mapping such that that Leibnitz rule of product derivative is satisfied

$$(1.3.4) \qquad D(f\cdot g) = (Df)\cdot\bar{g} + \bar{f}\cdot(Dg), \quad f,g \in A$$

Now, the first problem in setting up a framework such as in (1.3.1)-(1.3.4) arises from the fact that unrestricted differentiation on, and certain multiplications with elements of $\mathcal{D}'(\mathbb{R})$ can be performed within the classical linear theory of distributions, see Appendix 1 Chapter 5. Let us recall a few relevant details. We have the well known inclusions

$$(1.3.5) \qquad \mathcal{C}^{\infty}(\mathbb{R}) \subsetneqq \mathcal{C}^{0}(\mathbb{R}) \subsetneqq \mathcal{L}^{\infty}_{\ell oc}(\mathbb{R}) \subsetneqq \mathcal{D}'(\mathbb{R})$$

where the three spaces, except for $\mathcal{D}'(\mathbb{R})$, are algebras of functions. Further, we have the multiplication, see in particular (1.2.13)

$$(1.3.6) \qquad \mathcal{C}^{\infty}(\mathbb{R}) \times \mathcal{D}'(\mathbb{R}) \ni (\psi,T) \rightarrow \psi T \in \mathcal{D}'(\mathbb{R})$$

which for $T \in \mathcal{L}^{\infty}_{\ell oc}(\mathbb{R})$ reduces to the multiplication in $\mathcal{L}^{\infty}_{\ell oc}(\mathbb{R})$. Finally, we have the distributional derivative

$$(1.3.7) \qquad D : \mathcal{D}'(\mathbb{R}) \rightarrow \mathcal{D}'(\mathbb{R})$$

which coincides with the classical derivative of functions, when restricted to $\mathcal{C}^1(\mathbb{R})$, and satisfies the following version of the Leibnitz rule of product derivative

(1.3.8) $D(\psi T) = (D\psi)T + \psi(DT), \ \psi \in \mathcal{C}^\infty(\mathbb{R}), \ T \in \mathcal{D}'(\mathbb{R})$

It follows that the *first problem* is to see to what extent the structures of *algebra and differentiation* on A can be *compatible* with the respective classical structures in (1.3.5)-(1.3.8).

In this respect, we have already noted a few facts, which within (1.3.5)-(1.3.8), can be formulated as follows

(1.3.9) $H \in \mathcal{L}^\infty_{loc}(\mathbb{R}), \ \delta \in \mathcal{D}'(\mathbb{R}), \ H \neq 0, \ \delta \neq 0$

then, with the multiplication in $\mathcal{L}^\infty_{loc}(\mathbb{R})$ we have

(1.3.10) $H^m = H, \ m \in \mathbb{N}, \ m \geq 1$

while the multiplication in (1.3.6) gives

(1.3.11) $x\delta = 0 \in \mathcal{D}'(\mathbb{R})$

and finally

(1.3.12) $DH = \delta$

Let us see to what extent the above few basic properties in (1.3.10)-(1.3.12) can be preserved within the algebra and differential structure of A.

An argument similar to that used in proving (1.1.10) shows that (1.3.1)-(1.3.3) together with (1.3.10) and (1.3.12) yield

(1..3.13) $DH = \delta = 0 \in \bar{A}$

In case

(1.3.14) $\bar{A} = A$

and (1.3.2) is the identity mapping, (1.3.13) contradicts the fact that $\delta \neq 0$ in A.

It follows that the algebra and differential structure on A can only to a *limited extent be compatible* with the classical multiplication and differentiation on $\mathcal{D}'(\mathbb{R})$, summarized in (1.3.5)-(1.3.8).

§4. LIMITS TO COMPATIBILITY

It will be useful to further investigate the *limitations upon the compatibility* between the algebra and differential structure of the extensions (1.3.1)-(1.3.4), called for short

EAD,

and on the other hand, the classical structures of multiplication and differentiation in (1.3.5)-(1.3.8), denoted by

CMD.

For convenience, we shall denote by

EAD_0

the particular case of EAD, when (1.3.14) holds.

L. Schwartz's so called impossibility result in Proposition 1, Section 2, is one example of such a limitation, and in Chapter 6, we shall present several related results.

First however, a few simpler and more basic results on such limitations on the compatibility between EAD and CMD will be mentioned. These results concern a *yet more general trio*, namely that of:

- insufficient smoothness
- multiplication
- differentiation.

Let us define the continuous functions

(1.4.1) $$x_+, x_- \in \mathcal{C}^0(\mathbb{R})$$

by

(1.4.2) $$x_+ = \begin{vmatrix} x & \text{if} & x \geq 0 \\ 0 & \text{if} & x \leq 0 \end{vmatrix}, \quad x_- = \begin{vmatrix} x & \text{if} & x \leq 0 \\ 0 & \text{if} & x \geq 0 \end{vmatrix}$$

Then, within the algebra of functions $\mathcal{C}^0(\mathbb{R})$, we have the relations

(1..4.3) $$x_+ + x_- = x$$

(1.4.4) $$xx_+ = (x_+)^2, \quad xx_- = (x_-)^2$$

while obviously

(1.4.5) $$(x_+)^2, \ (x_-)^2 \in \mathcal{C}^1(\mathbb{R})$$

and with the classical derivative of functions, we have

(1.4.6) $Dx = 1,\ D(x_+)^2 = 2x_+,\ D(x_-)^2 = 2x_-$

Further, with the distributional derivative (1.3.7), we obtain

(1.4.7) $Dx_+ = H,\quad D^2x_+ = DH = \delta$

We shall start with some rather general differential algebras A, which need *not* contain all the distributions in $\mathcal{D}'(\mathbb{R})$. The interest in such results is that they show the mentioned kind of limits on compatibility, even if the algebras A contain only a *few* of the classical functions or distributions.

We shall name by

DA

any associative and commutative algebra A, together with a derivative operator $D : A \longrightarrow A$ which is a linear mapping and it satisfies the Leibnitz rule of product derivative (1.1.7).

Suppose now that our DA satisfies the conditions

(1.4.8) $1, x, x_+, x_- \in A$

(1.4.9) 1 is the unit element in A

further, that the relations (1.4.3), (1.4.4) hold in A, and finally, the derivative D on A satisfies the relations in (1.4.6).

<u>Proposition 2</u> (Rosinger [3], p. 318)

The following relations hold in A

(1.4.10) $xDx_+ = x_+\ ,\quad xDx_- = x_-$

(1.4.11) $x_+Dx_+ = x_+\ ,\quad x_-Dx_- = x_-$

(1.4.12) $x_+Dx_- = x_-Dx_+ = 0$

(1.4.13) $xD^2x_+ = xD^2x_- = 0$

(1.4.14) $Dx_+Dx_- = x_+D^2x_- = x_-D^2x_+ = x_+D^2x_+ = x_-D^2x_- = 0$

(1.4.15) $(Dx_+)^2 = Dx_+, \ (Dx_-)^2 = Dx_-$

(1.4.16) $D^2x_+ = D^2x_- = 0$

Remark 1

(1) The relations (1.4.7) and (1.4.16) show that in case A satisfies the minimal compatibility conditions (1.4.3), (1.4.4), (1.4.6), (1.4.8) and (1.4.9), the derivative D on A *cannot be compatible* with the distributional derivative, even for the *continuous* function x_+, since $\delta \neq 0$.

(2) The interest in Propositon 2 comes from the fact that none of the conditions on the algebra A involve discontinuous functions. Furthermore, both the results and as seen next, their proof are *purely algebraic*, in the sense mentioned at the beginning of Section 2.

Proof of Proposition 2.

For convenience, let us denote

(1.4.17) $a = x_+, \ b = x_-$

In view of (1.4.4) we have

(1.4.18) $x \cdot a = a^2, \ x \cdot b = b^2$

hence by derivation and owing to (1.4.6) we obtain

$$a + x \cdot Da = 2 \cdot a, \quad b + x \cdot Db = 2 \cdot b$$

which yield (1.4.10). Now, a derivation of (1.4.10) gives through (1.4.6)

$$Da + x \cdot D^2 a = Da, \quad Db + x \cdot D^2 b = Db$$

and thus (1.4.13). A derivation of (1.4.18) gives directly

$$a + x \cdot Da = 2 \cdot a \cdot Da, \quad b + x \cdot Db = 2 \cdot b \cdot Db$$

which in view of (1.4.10) yields (1.4.11). In view of (1.4.3) we have

$$a \cdot Db = (x - b) \cdot Db = x \cdot Db - b \cdot Db = 0$$

the last equality being implied by (1.4.10) and (1.4.11). Similarly

$$b \cdot Da = (x - a) \cdot Da = x \cdot Da - a \cdot Da = 0$$

and (1.4.12) is proved. From (1.4.11) by derivation, we obtain

$$Da \cdot Db + a \cdot D^2 b = 0, \quad Da \cdot Db + b \cdot D^2 a = 0$$

hence

(1.4.19) $$Da \cdot Db = -a \cdot D^2 b = -b \cdot D^2 a$$

But (1.4.3), (1.4.13) yield

(1.4.20) $$\begin{aligned} a \cdot D^2 b &= (x - b) \cdot D^2 b = xD^2 b - b \cdot D^2 b = -b \cdot D^2 b \\ b \cdot D^2 a &= (x - a) \cdot D^2 a = x \cdot D^2 a - a \cdot D^2 a = -a \cdot D^2 a \end{aligned}$$

while (1.4.3), (1.4.6) give by derivation

(1.4.21) $$Da + Db = 1$$

hence

(1.4.22) $$\begin{aligned} Da \cdot Db &= Da \cdot (1 - Da) = Da - (Da)^2 \\ Da \cdot Db &= (1 - Db) \cdot Db = Db - (Db)^2 \end{aligned}$$

Now (1.4.19)-(1.4.22) give

(1.4.23) $$\begin{aligned} Da \cdot Db &= -a \cdot D^2 b = -b \cdot D^2 a = a \cdot D^2 a = b \cdot D^2 b = \\ &= Da - (Da)^2 = Db - (Db)^2 = c \end{aligned}$$

for a certain $c \in A$. We shall show that

(1.4.24) $$c = 0$$

Indeed applying twice the derivative to (1.4.12) we obtain successively

$$Da \cdot Db + a \cdot D^2 b = 0$$
$$D^2 a \cdot Db + Da \cdot D^2 b + Da \cdot D^2 b + a \cdot D^3 b = 0$$

and multiplying the last relation by x, we obtain in view of (1.4.13) the relation

(1.4.25) $$a \cdot (x \cdot D^3 b) = 0$$

But a derivation of (1.4.13) yields

$$D^2 b + x \cdot D^3 b = 0$$

which if multiplied by a, gives together with (1.4.25) the relation

$$a \cdot D^2 b = 0$$

hence in view of (1.4.23) we obtain (1.4.24). Now obviously (1.4.23) and (1.4.24) imply (1.4.14), as well as (1.4.15).

In view of (1.4.14) it follows that

$$(Da)^p = Da, (Db)^p = Db, \quad p \in \mathbb{N}, \quad p \geq 1$$

hence by derivation

$$p \cdot (Da)^{p-1} \cdot D^2 a = D^2 a, \quad p \in \mathbb{N}, \quad p \geq 2$$

and then again by (1.4.15), we have

$$p \cdot Da \cdot D^2 a = D^2 a, \quad p \in \mathbb{N}, \quad p \geq 2$$

or

$$\frac{1}{p} \cdot D^2 a = Da \cdot D^2 a, \quad p \in \mathbb{N}, \quad p \geq 2$$

hence

$$(\frac{1}{p} - \frac{1}{q}) \cdot D^2 a = 0, \quad p,q \in \mathbb{N}, \quad p,q \geq 2$$

which obviously yields

$$D^2 a = 0$$

Since we can similarly obtain

$$D^2 b = 0$$

the proof of (1.4.16) is completed. □

We turn now to the more particular trio of:

- singular functions, such as δ
- multiplication
- differentiation.

Suppose given a DA such that

(1.4.26) $\qquad \delta \in A, \quad \delta \neq 0 \in A$

Before we go further, let us recall, see (1.2.14), that with the multiplication in (1.3.6) available in $\mathcal{D}'(\mathbb{R})$, we have the relation

(1.4.27) $\qquad x\delta = 0 \in \mathcal{D}'(\mathbb{R})$

which means that at the point $0 \in \mathbb{R}$, δ has a singularity of an order less than that of the function $1/x$.

On the other hand, as seen at the end of Section 2, in particular in (1.2.19), the *order of singularity* of δ at the point $0 \in \mathbb{R}$ may be quite high, if not even *infinite*, if the multiplication is considered in a DA which satisfies (1.4.26).

In Propositions 3 and 4 next, results detailing these opposite possibilities are presented.

Suppose that in addition to (1.4.27), the algebra A satisfies the following conditions

(1.4.28) $x^m \in A, \; m \in \mathbb{N}$

(1.4.29) the multiplication in A induces on the monomials in (1.4.28) the usual multiplication

(1.4.30) 1 is the unit element in A

(1.4.31) D applied to monomials in (1.4.28) coincides with the classical derivative of functions

Proposition 3 (Rosinger [3], p. 35)

The following relations hold within the algebra A

(1.4.32) $x^p \cdot D^q \delta = 0 \in A, \quad p,q \in \mathbb{N}, \quad p > q$

(1.4.33) $(p + 1) \cdot D^p \delta + x^{p+1} \delta = 0 \in A, \quad p \in \mathbb{N}$

(1.4.34) $x^p \cdot (D^p \delta)^q = 0 \in A, \quad p,q \in \mathbb{N}, \quad q \geq 2$

(1.4.35) $(\delta)^2 = \delta \cdot D\delta = 0 \in A$

Proof

In view of (1.4.28)-(1.4.31) D applied to (1.4.27) yields

(1.4.36) $\delta + x \cdot D\delta = 0 \in A$

which multiplied by x, and in view of (1.4.29) and (1.4.27), yields

$$x^2 \cdot D\delta = 0 \in A$$

If we apply D to the latter relation and then multiply by x, we have in the same way

$$x^3 \cdot D^2 \delta = 0 \in A$$

hence, by repeating this procedure, (1.4.32) is obtained.

In view of (1.4.30) and (1.4.31), a repeated application of D to (1.4.36) will yield (1.4.31).

Further, if we multiply (1.83) by x^p, then, in view of (1.4.28)-(1.4.31), we obtain

$$(p + 1) \cdot x^p D^p \delta + x^{p+1} D^{p+1} \delta = 0 \in A$$

Multiplying this latter relation by $(D^p \delta)^{q-1}$ and taking into account (1.4.32), we obtain (1.4.34).

Finally, for $p = 0$ and $q = 2$, (1.4.34) yields

$$(\delta)^2 = 0 \in A$$

and applying D to this latter relation, in view of (1.4.30), we obtain (1.4.35). □

Remark 2

The above *degeneracy* result in (1.4.35) is *not* in agreement with various other results encountered and used in the literature, see for instance Mikusinski [2], Braunss & Liese. In particular, in various distribution multiplication theories, as those for instance given by differential algebras containing the distributions, it follows that $(\delta)^2 \notin \mathcal{D}'$, hence $(\delta)^2 \neq 0$, see for instance Rosinger [1, p. 11], Rosinger [2, p. 66], Colombeau [1, p. 69], Colombeau [2, p. 38].

It follows that, in case we embed $\mathcal{D}'(\mathbb{R})$, or some of its subsets into a *single* differential algebra A, certain products involving the Dirac delta distribution or its derivatives, may vanish as in Proposition 3 above. However, in view of (1.2.19), in such algebras, we may have to expect that

(1.4.37) $\qquad x\delta \neq 0 \in A$

The next result in Proposition 4, shows that, under rather general conditions, we necessarily have in such algebras A the relations

(1.4.38) $\qquad x^m \delta \neq 0 \in A, \quad m \in \mathbb{N}$

which means that with the multiplication in such algebras, the Dirac delta distribution has an *infinite order singularity* at the point $0 \in \mathbb{R}$.

It is *particularly important* to mention that the above fact - which can be seen in *purely algebraic* terms as well - conditions much of the way the L. Schwartz distributions can be embedded in differential algebras. Details on the most general forms of possible embeddings are presented in Chapter 6.

Suppose given a DA which satisfies (1.4.26), as well as

(1.4.39) $x^m \in A, \quad m \in \mathbb{N}$

(1.4.40) 1 is the unit element in A

(1.4.41) $Dx^{m+1} = (m+1)x^m, \quad m \in \mathbb{N}$

Proposition 4 (Rosinger [3], p. 323)

If

(1.4.42) $\delta, \delta^2, \delta^3, \ldots \neq 0 \in A$

then we have for $m \in \mathbb{N}$

(1.4.43) $x^m \cdot \delta \neq 0 \in A$

Proof

Assume that for a certain $m \in \mathbb{N}$, we have

(1.4.44) $x^{m+1} \cdot \delta = 0 \in A$

Then we shall have

(1.4.45) $x^m \cdot \delta^2 = x^m \cdot \delta^3 = \ldots = 0 \in A$

Indeed, if $p \in \mathbb{N}$, $p \geq 2$, then (1.4.44) yields

$$x^{m+1} \cdot \delta^p = 0 \in A$$

hence by differentiation

$$(m+1) \cdot x^m \cdot \delta^p + p \cdot x^{m+1} \cdot \delta^{p-1} \cdot D\delta = 0 \in A$$

but $p-1 \geq 1$, thus (1.4.44) yields

$$(m+1) \cdot x^m \cdot \delta^p = 0 \in A$$

and the proof of (1.4.45) is completed.

Starting with

$$x^m \cdot \delta^2 = 0 \in A$$

obtained in (1.4.45), in a similar way we obtain

$$x^{m-1} \cdot \delta^3 = x^{m-1} \cdot \delta^4 = \ldots = 0 \in A$$

Continuing the argument, we end up with

$$\delta^{m+2} = 0 \in A$$

which contradicts (1.4.42). □

The conclusion which emerges from Propositions 1-4 and Remarks 1-2 above is the following. The EAD setting in (1.3.1)-(1.3.4) contains a large variety of rather different, if not even conflicting possiblities concerning the ways to settle the *purely algebraic conflict* between insufficient smoothness, multiplication and differentiation. And the outcomes of some of these ways may be unacceptable in certain circumstances. Indeed, under rather general conditions we can have the *degeneracy* property (1.4.35) of δ, namely

$$\delta^2 = 0$$

On the other hand, under similarly general conditions δ may prove to be *infinitely singular*, namely

$$x^m \cdot \delta \neq 0, \quad m \in \mathbb{N}$$

as follows for instance from (1.4.43).

It is therefore desirable if we can find a sufficiently natural and general way the mentioned conflict can be settled.

In Section 5 next, we present such a way, which is inspired by most elementary considerations concerning rings of functions. And we are led to these rings of functions in a most direct way whenever we consider the concept of weak solution introduced in, and massively used since Sobolev [1,2].

The framework in Section 5 is the basis for the general nonlinear theory in Rosinger [1,2,3], as well as for its particular case in Colombeau [1,2].

The insistence on the 'algebra first' approach proves to lead to a rather powerful tool, although at present we seem to be at the beginning of its fuller use and understanding, in spite of results such as those in Chapters 2 and 3.

§5. CONSTRUCTION OF ALGEBRAS CONTAINING THE DISTRIBUTIONS

So far, one of the major interest in sufficiently systematic nonlinear theories of generalized functions has come from the fact that, as seen in Chapter 5, important classes of nonlinear partial differential equations have solutions of applicative interest which are no longer given by sufficiently smooth functions, therefore, they fail to be classical solutions. To the extent that L. Schwartz's linear theory of distributions or generalized functions does not allow *within itself* for the unrestricted performance of nonlinear operations, in particular multiplication, see Appendix 1 to Chapter 5, one should try to embed the distributions in larger algebras of generalized functions. Fortunately, a wide range of nonlinear operations - beyond multiplication - will equally be available within these algebras.

For algebraic convenience, let us start with the class of *polynomial nonlinear* partial differential equations, having the form

(1.5.1) $$\sum_{1\leq i\leq h} c_i(x) \prod_{1\leq j\leq k_i} D^{p_{ij}} U(x) = f(x), \quad x \in \Omega$$

where $\Omega \subset \mathbb{R}^n$ is nonvoid, open, $p_{ij} \in \mathbb{N}^n$ and $c_i, f \in \mathcal{C}^0(\Omega)$ are given, while

(1.5.2) $$U : \Omega \longrightarrow \mathbb{R}$$

is the unknown function. The order of the equation (1.5.1) is by definition

(1.5.3) $$m = \max\{|p_{ij}| \,|\, 1 \leq i \leq h, \quad 1 \leq j \leq k_i\}$$

assuming that none of the c_i vanishes on the whole of Ω. Let us denote by

(1.5.4) $$T(D) = \sum_{1\leq i\leq h} c_i(x) \prod_{1\leq j\leq k_j} D^{p_{ij}}, \quad x \in \Omega$$

the partial differential operator generated by the left hand term in (1.5.1). Then obviously we have the mapping

(1.5.5) $$T(D) : \mathcal{C}^m(\Omega) \longrightarrow \mathcal{C}^0(\Omega)$$

and a *classical solution* (1.5.2) of (1.5.3) is any *smooth* function

(1.5.6) $$U \in \mathcal{C}^m(\Omega)$$

such that, for the mapping in (1.5.5), we have

(1.5.7) $T(D)U = f \quad \text{in} \quad \mathcal{C}^0(\Omega)$

Unfortunately, the important case of *nonclassical solutions* of (1.5.1), where

(1.5.8) $U \notin \mathcal{C}^m(\Omega)$

cannot be obtained through (1.5.5)-(1.5.7).

Therefore, the mapping $T(D)$ in (1.5.5) has to be *extended* beyond $\mathcal{C}^m(\Omega)$, to suitable spaces of generalized functions.

The various ways the mapping $T(D)$ in (1.5.5) has usually been extended beyond $\mathcal{C}^m(\Omega)$, have been *biased* by an improper balance in the handling of our often conflicting interests concerning *representation* and *interpretation* in mathematical models and theories, for further details, see Section 7, as well as Appendices 2 and 5 at the end of this chapter.

Indeed, especially since Sobolev [1,2], there has been a natural intuitive tendency to conceive of the needed extensions of $\mathcal{C}^m(\Omega)$ as being given by an embedding

(1.5.9) $\mathcal{C}^m(\Omega) \subset E$

where E is a suitable topological vector space of generalized functions, obtained *solely* based on an *interpretation* of *approximation*. That is, each generalized function

(1.5.10) $U \in E \backslash \mathcal{C}^m(\Omega)$

is supposed to be *some kind of limit*

(1.5.11) $U = \lim_{\nu\to\infty} \psi_\nu$

of classical functions $\psi_\nu \in \mathcal{C}^m(\Omega)$, with $\nu \in \mathbb{N}$.

In this way, the excessive and early - in fact, a priori and exclusive - stress on the mentioned kind of *approximation type interpretation*, has inevitably led to the situation where *topology alone* is supposed to give us extensions E, such as in (1.5.9).

However, as seen in Rosinger [1,2,3] and Colombeau [1,2], it proves to be particularly useful to avoid such an early and exclusive stress on any approximation interpretation, stress which can easily lead to the usual 'topology first', if not even 'topology only' approach.

The alternative is based on the realization that in mathematics, it may often be useful to allow representation *to go farther than* interpretation.

This may also mean that we should start with a *rather general representation*, and then later, become concerned with interpretation. Fortunately from the point of view of simplicity and clarity, as seen next, such a course will lead us to an 'algebra first' approach.

In order to have a better understanding of the kind of general representation we may need in order to construct E in (1.5.9), let us first have a look at the usual way generalized solutions (1.5.8) are constructed for equations (1.5.7). This way, which can be called the *sequential method*, has its systematic origin in Sobolev [1,2], and proceeds as follows.

Based on specific features of the partial differential equation (1.5.7) an *infinite* sequence of so called approximating equations or systems of equations

(1.5.12) $T_\nu(D)\psi_\nu(x) = 0, \quad x \in \Omega, \ \nu \in \mathbb{N}$

are constructed, with $T_\nu(D)$ being ordinary or partial differential operators. The essential point in the construction of the approximating equations (1.5.12) is that they have sufficiently smooth classical solutions

(1.5.13) $\psi_\nu \in \mathcal{F}, \quad v \in \mathbb{N}$

where

(1.5.14) $\mathcal{F} \subset \mathcal{C}^m(\Omega)$

is a suitable topological vector space of funtions. Now, from (1.5.13) one can often extract a Cauchy sequence in the given uniform topology on $\mathcal{F}$, by using compactness, monotonicity, fixed points or other arguments. Thus one obtains a Cauchy sequence of sufficiently smooth function

(1.5.15) $s = (\psi_\nu | \nu \in \mathbb{N}) \in \mathcal{F}^{\mathbb{N}}$

labeled for convenience with the same indices as in (1.5.13).

And now one is supposed to take a 'leap of faith' and declare the element, which often is not a usual function (1.5.2), but a generalized function

(1.5.16) $U = \lim_{\nu \to \infty} \psi_\nu \in \bar{\mathcal{F}}$

to be the *generalized* solution of the partial differential equation (1.5.7), that is, one *declares* that U in (1.5.16) does solve the equation

(1.5.17) $T(D)U = f, \quad U \in \bar{\mathcal{F}}$

where $\bar{\mathcal{F}}$ is the *completion* of $\mathcal{F}$ in its given uniform topology.

In this way, we are suggested to choose in (1.5.9)

(1.5.18) $\quad E = \bar{\mathcal{F}}$

Let us now review from a more abstract point of view the representational structure above, and do so for the time being without too early and limiting attempts at interpretation.

As is well known in general topology, in case the uniform topology on $\mathcal{F}$ is metrizable, its completion $\bar{\mathcal{F}}$ can be obtained as follows

(1.5.19) $\quad \bar{\mathcal{F}} = \mathcal{S}/\mathcal{V}$

where

(1.5.20) $\quad \mathcal{V} \subset \mathcal{S} \subset \mathcal{F}^{\mathbb{N}}$

and $\mathcal{S}$ is the set of all Cauchy sequences in $\mathcal{F}$, while $\mathcal{V}$ is the set of all sequences convergent to zero in $\mathcal{F}$, see also Appendices 2 and 5.

It follows that the general form of representation for spaces of generalized functions E in (1.5.9) can be expected to be given by *quotient spaces* of the form

(1.5.21) $\quad E = \mathcal{S}/\mathcal{V}$

where

(1.5.22) $\quad \mathcal{V} \subset \mathcal{S} \subset \mathcal{F}^{\Lambda}$

Here Λ is a given *infinite* index set, while

(1.5.23) $\quad \mathcal{F} \subset \mathcal{C}^{m}(\Omega)$

are suitable vector subspaces. Then, with the term wise operation on sequences, $\mathcal{F}^{\Lambda}$ is in a natural way a vector space. Finally, $\mathcal{V}$ and $\mathcal{S}$ are appropriately chosen vector subspaces in $\mathcal{F}^{\Lambda}$. Obviously, within (1.5.21)-(1.5.23), E will again be a vector space.

In the case of solving *nonlinear* partial differential equations (1.5.1), we shall be interested in extensions (1.5.9) with E replaced by an *algebra* of generalized functions. The above construction in (1.5.21)-(1.5.23) can easily be particularized for that purpose. Indeed, we can choose a suitable subalgebra of smooth functions

(1.5.24) $\quad \mathcal{U} \subset \mathcal{C}^{m}(\Omega)$

and

(1.5.25) $\mathcal{I} \subset \mathcal{A} \subset \mathcal{X}^{\Lambda}$

where $\mathcal{A}$ is a subalgebra in $\mathcal{X}^{\Lambda}$, while $\mathcal{I}$ is an ideal in $\mathcal{A}$. Then the *quotient algebra*

(1.5.26) $A = \mathcal{A}/\mathcal{I}$

can offer the *representation* for our algebra of generalized functions. We should note that $\mathcal{X}$ in (1.5.24) is automatically an associative and commutative algebra. Therefore, with the term wise operation on sequences, $\mathcal{X}^{\Lambda}$ is in a natural way an associative and commutative algebra. It follows that the algebra A of generalized functions in (1.5.26) will also be associative and commutative.

At first sight, it may appear that there exists too much arbitrariness with constructions of algebras of generalized functions such as in (1.5.24)-(1.5.26). Such concerns will be addressed and clarified to a good extent in several stages in the sequel.

First, the so called *neutrix condition* will particularize the above framework in (1.5.24)-(1.5.26). This purely algebraic condition, dealt with in the next section, characterizes the natural requirement that

(1.5.27) $\mathcal{X} \subset A$

yet it proves to be surprisingly powerful.

A second kind of particularization will come from the obvious requirement that the nonlinear partial differential operators (1.5.5) should be extendable to the algebras of generalized functions A. This issue is dealt with in Section 9.

Finally, a third necessity for particularization, presented in Section 10, concerns the important concepts of *stability, generality and exactness*, which are naturally associated with the concept of generalized solutions of nonlinear partial differential equations.

For further comment on the necessary degree of generality in (1.5.21)-(1.5.23), or in particular in (1.5.24)-(1.5.26), see Appendices 3, 5, 6 and 7.

§6. **THE NEUTRIX CONDITION**

Given in general a vector space $\mathcal{F}$ of classical functions $\psi:\Omega \longrightarrow \mathbb{R}$, we have taken vector subspaces $\mathcal{V} \subset \mathcal{S} \subset \mathcal{F}^{\Lambda}$ and have defined $E = \mathcal{S}/\mathcal{V}$ as a space whose elements $U \in E$ generalize the classical functions $\psi \in \mathcal{F}$. It follows that we should have a *vector space embedding*

(1.6.1) $\mathcal{F} \subset E = \mathcal{S}/\mathcal{V}$

defined by the *linear injection*

(1.6.2) $\mathcal{F} \ni \psi \longrightarrow u(\psi) + \mathcal{V} \in E$

where

(1.6.3) $u(\psi) = (\psi_\lambda | \lambda \in \Lambda) \in \mathcal{S}$

is the *constant* sequence with the terms $\psi_\lambda = \psi$, for $\lambda \in \Lambda$.

We reformulate (1.6.1)-(1.6.3) in a more convenient form. Let us denote by $\mathcal{O}$ the null vector subspace in $\mathcal{F}^\Lambda$ and let

(1.6.4) $\mathcal{U}_{\mathcal{F},\Lambda} = \{u(\psi) | \psi \in \mathcal{F}\}$

be the vector subspace in $\mathcal{F}^\Lambda$ of all the *constant* sequences of functions in $\mathcal{F}$, that is, the *diagonal* in the cartesian product $\mathcal{F}^\Lambda$. Then it is easy to see that (1.6.1)-(1.6.3) are equivalent with the *inclusion diagram*

(1.6.5)
$$\begin{array}{ccccc} \mathcal{V} & \longrightarrow & \mathcal{S} & \longrightarrow & \mathcal{F}^\Lambda \\ \uparrow & & \uparrow & & \\ \mathcal{O} & \text{———} & \mathcal{U}_{\mathcal{F},\Lambda} & & \end{array}$$

together with the *off diagonality* condition

(1.6.6) $\mathcal{V} \cap \mathcal{U}_{\mathcal{F},\Lambda} = \mathcal{O}$

which we shall call in the sequel the *neutrix condition.*

The name is suggested by similar ideas introduced earlier in Van der Corput, within a so called 'neutrix calculus' developed in order to simplify and systematize methods in *asymptotic analysis*, see Appendix 4. With the terminology in Van der Corput, the sequences of functions $v = (\chi_\lambda | \lambda \in \Lambda) \in \mathcal{V}$ are called '$\mathcal{V}$-negligible'. In this sense, given two functions $\alpha, \beta \in \mathcal{F}$, their difference $\alpha - \beta \in \mathcal{F}$ is '$\mathcal{V}$-neglibible' if and only if $u(\alpha) - u(\beta) = u(\alpha - \beta) \in \mathcal{V}$, which in view of (1.6.6) is equivalent to $\alpha = \beta$. In other words, (1.6.6) means that the quotient structure $E = \mathcal{S}/\mathcal{V}$ does *distinguish* between classical functions in $\mathcal{F}$.

Let us denote by

(1.6.7) $VS_{\mathcal{F},\Lambda}$

the set of all the *quotient vector spaces* $E = \mathcal{S}/\mathcal{V}$ which satisfy (1.6.5) and (1.6.6).

Since we are interested in generalized solutions for *nonlinear* partial differential equations, it is useful to consider the particular cases of the above quotient structures given by quotient algebras. For that purpose we proceed as follows.

Suppose $\mathcal{X}$ is an *algebra* of classical functions $\psi:\Omega \longrightarrow \mathbb{R}$, for instance $\mathcal{X}$ can be a subalgebra in the algebra

$$(1.6.8) \qquad \mathcal{M}(\Omega)$$

of all the measurable and a.e. finite functions $\psi:\Omega \to \mathbb{R}$. In that case, with the term wise operations on sequences of functions, $\mathcal{X}^\Lambda$ will be not only a vector space but also an algebra. Let us denote by

$$(1.6.9) \qquad \mathrm{AL}_{\mathcal{X},\Lambda}$$

the set of all the *quotient algebras* $A = \mathcal{A}/\mathcal{I}$, where $\mathcal{A} \subset \mathcal{X}^\Lambda$ is a subalgebra and $\mathcal{I}$ is an ideal in $\mathcal{A}$ such that the *inclusion diagram*

$$(1.6.10) \qquad \begin{array}{ccccc} \mathcal{I} & \longrightarrow & \mathcal{A} & \longrightarrow & \mathcal{X}^\Lambda \\ \uparrow & & \uparrow & & \\ \mathcal{O} & \text{———} & \mathcal{U}_{\mathcal{X},\Lambda} & & \end{array}$$

satisfies the *neutrix* or *off diagonality condition*

$$(1.6.11) \qquad \mathcal{I} \cap \mathcal{U}_{\mathcal{X},\Lambda} = \mathcal{O}$$

Obviously, if $\mathcal{F}$ in (1.6.7) is an algebra, then

$$(1.6.12) \qquad \mathrm{AL}_{\mathcal{F},\Lambda} \subset \mathrm{VS}_{\mathcal{F},\Lambda}$$

In general, it follows that similar to (1.6.1), (1.6.2), for every *quotient algebra* $A = \mathcal{A}/\mathcal{I} \in \mathrm{AL}_{\mathcal{X},\Lambda}$ we have the *algebra embedding*

$$(1.6.13) \qquad \mathcal{X} \subset A = \mathcal{A}/\mathcal{I}$$

given by the *injective algebra homomorphism*

$$(1.6.14) \qquad \mathcal{X} \ni \psi \longmapsto u(\psi) + \mathcal{I} \in A$$

Again, the conditions (1.6.10) and (1.6.11) are necessary and sufficient for (1.6.13) and (1.6.14). In this way, one obtains an answer to (1.5.7).

Here we should like to draw the attention upon a *most important* fact. In the case of *quotient algebras* $A = \mathcal{A}/\mathcal{I} \in AL_{\mathcal{X},\Lambda}$, the purely algebraic *neutrix* or *off diagonality condition* (1.6.11) is particularly *powerful*, although it appears to be simple and trivial. Indeed, variants of the neutrix condition characterize the *existence* and *structure* of a large class of *chains of differential algebras* of generalized functions, as seen in Rosinger [2,3]. In particular, the neutrix condition determines to a good extent the *structure* of ideals $\mathcal{I}$ which play a crucial role in the stability, generality and exactness properties of the algebras of generalized functions, see Section 11 as well as Chapters 3 and 6.

§7. REPRESENTATION VERSUS INTERPRETATION

Let us start with an example which can illustrate the fact that in mathematical modelling it may be useful to allow representation to go *further* than interpretation. Indeed, in the case of the infinite set $\mathbb{N}$ of natural numbers, the Peano axioms give a rigorous *representation* for all $n \in \mathbb{N}$. However, when it comes to interpret various $n \in \mathbb{N}$, we *cannot* go too far. For instance, if we take

$$(1.7.1) \qquad n = 10^{10^{10}}$$

it is hard to find for it an interpretation which may be satisfactory to the extent that, let us say, it could distinguish meaningfully between n above, and $n + 1$. And yet, n in (1.7.1) can be rigorously represented by using six digits only!

On the other hand, an integer $n \in \mathbb{N}$ which is not too large can to a good extent be its own interpretation.

The point to note here is that we first represent rigorously *all* $n \in \mathbb{N}$, and then, later, we only interpret *some* of the $n \in \mathbb{N}$!

Certainly, rigorous computations can only be done based on rigorous representations. And representations which are far reaching and convenient will allow for generality and ease in computations.

Let us now consider shortly a second example which is nearer to the quotient algebras of generalized functions considered in Sections 5 and 6, and used extensively in the sequel. This example concerns one of the ways nonstandard numbers can be constructed, and as such, it is an extension of the classical Cauchy-Bolzano construction of the real numbers, presented in Appendix 2. For us, the interest in this second example is in the fact that the way the representation of nonstandard numbers is constructed is to a good extent *free* of numerical, approximation, or in general, metric topological interpretations. Yet, as is well known, the nonstandard numbers thus obtained prove to be particularly useful.

The construction of the *nonstandard set* $^*\mathbf{Q}$ of real numbers is given by the quotient field

$$(1.7.2) \qquad {}^*\mathbf{Q} = {}^*\mathcal{A}/{}^*\mathcal{I}$$

where $^*\mathcal{A} = \mathbf{Q}^{\mathbb{N}}$ and $^*\mathcal{I}$ is an ideal in $^*\mathcal{A}$, while $\mathbf{Q}$ is the set of the usual rational numbers, see Schmieden & Laugwitz, or Stroyan & Luxemburg [pp. 7-9]. It follows easily that $^*\mathbf{Q}$ in (1.7.2) is a *quotient algebra* in the sense of (1.6.9), that is

$$(1.7.3) \qquad {}^*\mathbf{Q} = {}^*\mathcal{A}/{}^*\mathcal{I} \in AL_{\mathbf{Q},\mathbb{N}}$$

Thus in particular, we have satisfied the *neutrix condition*, see (1.6.11)

$$(1.7.4) \qquad {}^*\mathcal{I} \cap \mathcal{U}_{\mathbf{Q},\mathbb{N}} = \mathcal{O}$$

and we have the *embedding of algebras*, in fact *fields*, given by

$$(1.7.5) \qquad \mathbf{Q} \ni r \longmapsto {}^*r = u(t) + {}^*\mathcal{I} \in {}^*\mathbf{Q} = {}^*\mathcal{A}/{}^*\mathcal{I}$$

The *important* fact to note concerning (1.7.2)-(1.7.5) is that, unlike with the quotient algebra construction in Appendix 2, the *quotient representation* of elements of $^*\mathbf{Q}$

$$(1.7.6) \qquad {}^*x = s + {}^*\mathcal{I} \in {}^*\mathbf{Q} = {}^*\mathcal{A}/{}^*\mathcal{I}, \quad s \in {}^*\mathcal{A} = \mathbf{Q}^{\mathbb{N}}$$

cannot have metric, in particular approximation or numerical interpretations. Indeed, $^*\mathbf{Q}$ contains *nonzero infinitesimals*, whose metric interpretation would of course have to be zero. Then $^*\mathbf{Q}$ contains many different *infinite* numbers, whose metric interpretation could only be $\pm\infty$, since $^*\mathbf{Q}$ is ordered. What can however be done is the following: one can define a mapping

$$(1.7.7) \qquad {}^*\mathbf{Q} \supset {}^*\mathbf{Q}_0 \xrightarrow{\ st\ } \mathbb{R}$$

on the strict subset $^*\mathbf{Q}_0$ of $^*\mathbf{Q}$, such that $st(x^*) \in \mathbb{R}$ will be the *standard part* of the nonstandard number $^*x \in {}^*\mathbf{Q}_0$. But there will be nonstandard numbers $^*x \in {}^*\mathbf{Q} \setminus {}^*\mathbf{Q}_0$ which do *not* have standard part.

Further examples and comments on the possible relationship between representation and interpretation can be seen in Appendix 5.

By the way, it should be pointed out that the nonlinear theory of generalized functions as developed in Rosinger [1,2,3] and Colombeau [1,2], has strong connections with nonstandard methods, although the algebras of generalized functions constructed by the mentioned nonlinear theory are much more large than the fields of nonstandard numbers. For details see Oberguggenberger [6] and the literature cited there.

We may perhaps conclude that, since Cantor's set theory, one of the most important powers of mathematics has come from its capability to rigorously represent. And to the extent that such representations could be so general and far reaching, one of our weaknesses as mathematicians has been in interpreting the powerful representations available. This however need not lead to a situation where our limitations in interpretation would come to dictate the way we may use the representational power of mathematics.

Mathematical research is for the researcher a learning process as well. And one way to learn may as well come from trying to deal with the gap between representation and interpretation, without collapsing the former within the given limits of the latter.

§8. NONLINEAR STABILITY PARADOXES, OR HOW TO PROVE THAT $0^2 = 1$ IN $\mathbb{R}$

The nonlinear stability paradoxes mentioned in this section point to long existent basic *conceptual* deficiencies of the customary weak solution method for nonlinear partial differential equations, as developed for instance in Lions [1,2], and used in a wide range of later applications.

Indeed, as shown next, according to that customary method, one can prove the existence of weak - and strong - solutions U for the *nonlinear system* of equations

$$(1.8.1) \qquad \begin{aligned} U &= 0 \\ U^2 &= 1 \end{aligned}$$

therefore admittedly proving that in $\mathbb{R}$, we have

$$(1.8.2) \qquad 0^2 = 1$$

For that, let us now present in some detail the mentioned customary weak solution method, sketched shortly in Section 5, see (1.5.12)-(1.5.18).

Suppose given a nonlinear partial differential equation, see (1.5.1), (1.5.7)

$$(1.8.3) \qquad T(D)U(x) = f(x), \quad x \in \Omega \subset \mathbb{R}^n$$

Then, one constructs an infinite sequence of approximating equations

$$(1.8.4) \qquad T_\nu(D)\psi_\nu(x) = 0, \quad x \in \Omega, \quad \nu \in \mathbb{N}$$

which admit classical solutions

(1.8.5) $\psi_\nu \in \mathcal{F}, \quad \nu \in \mathbb{N}$

in other words $\mathcal{F}$ is a vector space of sufficiently smooth functions, such as for instance in (1.5.14).

Now, with a suitable choice of (1.8.4), (1.8.5) and of a metrizable topology on $\mathcal{F}$, one can often find a Cauchy subsequence in (1.8.5), that is

(1.8.6) $s = (\psi_\nu | \nu \in \mathbb{N}) \in \mathcal{F}^{\mathbb{N}}$

which for convenience was labeled with the same indices as in (1.8.5).

Usually, this subsequence s is obtained from a compactness, monotonicity or fixed point argument.

The final and *most objectionable* step comes now, when the limit

(1.8.7) $U = \lim_{\nu \to \infty} \psi_\nu \in \bar{\mathcal{F}}$

is declared to be a solution of (1.8.3), where $\bar{\mathcal{F}}$ is the completion of $\mathcal{F}$ in its given topology.

It should further be pointed out that, owing to the usual difficulties involved in the steps (1.8.4)-(1.8.6), especially when initial and/or boundary value problems are associated with the nonlinear partial differential equation (1.8.3), one ends up with one single or with *very few* Cauchy sequences s in (1.8.6).

What happened can be recapitulated as follows:
We have a sequence s in (1.8.6) such that

s converges to U, and

T(D)s converges to f

Then, based alone on that, we feel entitled to *define* U as being a generalized solution of the nonlinear partial differential equation

$T(D)U = f$

Now let us go back to the framework in (1.8.3)-(1.8.7) and see what the 'leap of faith' in *declaring* (1.8.7) to be a solution of (1.8.3) amounts to.

It is obvious that the above is *equivalent* to saying that the *nonlinear* mapping (1.5.5) has been *extended* to a nonlinear mapping

(1.8.8) $T(D):(\{U\} \cup \mathcal{C}^m(\Omega)) \to \mathcal{X}$

in such a way that this extension satisfies

(1.8.9) $T(D)U = f \in \mathcal{X}$

where $\mathcal{X}$ is a suitable topological vector space of functions or generalized functions on Ω.

In this way, the above customary method for finding generalized solutions to nonlinear partial differential equations amounts to nothing else but *ad-hoc point-wise* extensions of the type (1.8.8), (1.8.9) for *nonlinear* mappings (1.5.5).

Now the *deficiency* of this solution method is obvious. Indeed, the nonlinear extension (1.8.8), (1.8.9) was made based *not* upon *all* the Cauchy sequences t in $\mathcal{F}$ which converge to the same given $U \in \mathcal{F}$, but only upon the very few, quite often one single sequence s in (1.8.6). Not to mention that the sequences s in (1.8.6) and thus U in (1.8.7), are usually obtained by a rather arbitrary subsequence selection from (1.8.5), such as for instance, compactness arguments.

However one *critical point* is often overlooked. Namely, that in general, the *nonlinear* mapping (1.5.5) is *not compatible* with the vector space topologies on $\mathcal{F}$ and $\mathcal{X}$, which means that for sequences $t \in \mathcal{F}^{\mathbb{N}}$ we have in general

(1.8.10) t Cauchy in $\mathcal{F} \not\Longrightarrow T(D)t$ Cauchy in $\mathcal{X}$

And worse yet: it may happen that, even if $T(D)t_1$ and $T(D)t_2$ are Cauchy in $\mathcal{X}$, for two given Cauchy sequences t_1 and t_2 in $\mathcal{F}$, nevertheless

(1.8.11) $\lim_{\mathcal{F}} t_1 = \lim_{\mathcal{F}} t_2 \not\Longrightarrow \lim_{\mathcal{X}} T(D)t_1 = \lim_{\mathcal{X}} T(D)t_2$

A most simple example in this connection is given by the zero order *nonlinear* partial differential operator

(1.8.12) $T : \mathcal{C}^0(\mathbb{R}) \longrightarrow \mathcal{C}^0(\mathbb{R}), \quad TU = U^2$

connected with the stability paradoxes (1.8.1), (1.8.2). Indeed, if we take $\mathcal{F} = \mathcal{C}^\infty(\mathbb{R})$ with the topology induced by $\mathcal{D}'(\mathbb{R})$, and we take $\mathcal{X} = \mathcal{D}'(\mathbb{R})$, then the sequence

(1.8.13) $v = (\chi_\nu \mid \nu \in \mathbb{N}), \quad \chi_\nu(x) = \sqrt{2} \cos \nu x, \quad x \in \mathbb{R}, \quad v \in \mathbb{N}$

is Cauchy in $\mathcal{F}$ and

(1.8.14) $\lim_{\mathcal{F}} v = U = 0$

both weakly and strongly in $\mathcal{F} = \mathcal{D}'(\mathbb{R})$. Yet, we shall also have

(1.8.15) $$\lim_{\mathcal{X}} Tv = \lim_{\mathcal{X}} v^2 = 1$$

both weakly and strongly in $\mathcal{X} = \mathcal{D}'(\mathbb{R})$. Therefore, according to the customary method (1.8.3)-(1.8.7), the sequence v in (1.8.13) defines *both a weak and strong solution* for the system

(1.8.16) $$\begin{aligned} U &= 0 \\ U^2 &= 1 \end{aligned}$$

For the sake of clarity let us show the *more general* effects stability paradoxes such as in (1.8.16) can have upon the customary sequential approach (1.8.3)-(1.8.7). Suppose $T(D)$ in (1.5.5) contains U^2 as the only nonlinear term, and U^2 has the coefficient 1, that is

(1.8.17) $$T(D)U = L(D)U + U^2$$

where $L(D)$ is a linear partial differential operator. Further, suppose that the system (1.8.16) has a solution in $\mathcal{F}$, that is, there exist a Cauchy sequence $v \in \mathcal{F}^{\mathbb{N}}$ such that

(1.8.18) $$\lim_{\mathcal{F}} v = 0, \quad \lim_{\mathcal{X}} v^2 = 1$$

Let us take s in (1.8.6) which is supposed to define a solution $U \in \mathcal{F}$ of (1.8.3) through (1.8.7). Then, for a given but arbitrary $\lambda \in \mathbb{R}$ let us define

(1.8.19) $$t = s + \lambda v \in \mathcal{F}^{\mathbb{N}}$$

Then obviously, t is a Cauchy sequence in $\mathcal{F}$ and in view of (1.8.7) we have

(1.8.20) $$\lim_{\mathcal{F}} t = \lim_{\mathcal{F}} s = U \in \mathcal{F}$$

hence t defines the *same* generalized function U as given in (1.8.7) by s. If now, as is usual, the generalized meaning of (1.8.9) is taken to be

(1.8.21) $$\lim_{\mathcal{X}} T(D)s = f$$

it follows that

$$(1.8.22) \qquad T(D)t = T(D)s + \lambda L(D)v + 2\lambda sv + \lambda^2 v^2$$

hence (1.8.18), (1.8.21) yield

$$(1.8.23) \qquad \lim_{\mathcal{X}} T(D)t = f + \lambda^2 + 2\lambda \lim_{\mathcal{F}} sv$$

provided that

$$(1.8.24) \qquad sv \in \mathcal{F}^{\mathbb{N}} \text{ is Cauchy}$$

and that also, as it happens with many of the usually encountered topologies on $\mathcal{F}$ and $\mathcal{X}$, we have satisfied

$$(1.8.25) \qquad \lim_{\mathcal{X}} L(D)v = 0$$

which is for instance the case of the topology on $\mathcal{D}'(R^n)$, if the coefficients in $L(D)$ are $\mathcal{C}^\infty$-smooth.

Since $\lambda \in \mathbb{R}$ is arbitrary, (1.8.23) yields

$$(1.8.26) \qquad T(D)U \neq f$$

if in (1.8.20) we use the representation

$$(1.8.27) \qquad U = \lim_{\mathcal{F}} t \in \mathcal{F}$$

In that way, the very same $U \in \mathcal{F}$, once happens to be the generalized solution of the equation $T(D)U = f$, as in (1.8.9), while another time, if $U \in \mathcal{F}$ is given as in (1.8.27), then in view of (1.8.26), it happens to be no longer a solution.

It is obvious from (1.8.23) that in case (1.8.23) does not hold, the above situation in (1.8.26) remains the same.

In case $T(D)$ contains nonlinear terms in U other than in (1.8.17), the overall situation remains the same, since stability paradoxes similar to (1.8.16) can easily be obtained for the respective nonlinear terms. Further examples and clarifications can be found in Appendix 6.

It follows that in the case of *nonlinear* partial differential equations, the extension of the concept of classical solution to that of the concept of generalized solution along the lines (1.8.3)-(1.8.7) is an *improper* generalization of various classical extensions, such as for instance the extension $\mathbb{Q} \subset \mathbb{R}$ of the rational numbers into the real numbers. Indeed, since the usual topology on $\mathbb{Q}$ is compatible with multiplication, non-rational solutions $u \in \mathbb{R}$ of equations such as for instance

(1.8.28) $u^2 = 2$

can be defined by the sequential method

(1.8.29) $u = \lim_{\mathbb{R}} s$

where $s \in \mathbb{Q}^{\mathbb{N}}$ is *any* Cauchy sequence in $\mathbb{Q}$, such that

(1.8.30) $\lim_{\mathbb{R}} s^2 = 2$

Indeed, if $v \in \mathbb{Q}^{\mathbb{N}}$ is such that $\lim_{\mathbb{Q}} v = 0$, the Cauchy sequence $t = s + v \in \mathbb{Q}^{\mathbb{N}}$ will again satisfy both (1.8.29) and (1.8.30). Unfortunately however, as seen above, the usual topologies encoutered on the spaces $\mathcal{F}$ and $\mathcal{X}$ are *not* compatible with multiplication.

One can wonder about the reasons the customary sequential approach to generalized solutions for *nonlinear* partial differential equations has managed to *overlook* the above deficiency. One likely reason is that in the case of *linear* partial differential operators with sufficiently smooth coefficients, the above customary method (1.8.3)-(1.8.7) *happens* to be correct because of what is called in general the phenomenon of *automatic continuity* of certain classes of linear operators. Indeed, let us suppose that $T(D)$ in (1.5.4) has the following particular *linear* form

(1.8.31) $L(D) = \sum_{1 \leq i \leq h} c_i(x) D^{p_i}, \quad x \in \Omega$

with $c_i \in \mathcal{C}^\infty(\Omega)$. Suppose given a Cauchy sequence $s \in \mathcal{F}^{\mathbb{N}}$ and let us assume that

(1.8.32) $\lim_{\mathcal{F}} s = U \in \mathcal{F}$

(1.8.33) $\lim_{\mathcal{X}} L(D)s = f$

Then, even if (1.8.32) and (1.8.33) have been obtained for *one single* sequence s, these two relations can nevertheless be *interpreted* as giving a generalized solution $U \in \mathcal{F}$ of the equation

(1.8.34) $L(D)U = f \quad \text{in} \quad \mathcal{X}$

Indeed, if we take *any* Cauchy sequence $v \in \mathcal{F}^{\mathbb{N}}$ such that

(1.8.35) $$\lim_{\mathcal{F}} v = 0$$

and define the Cauchy sequence in $\mathcal{F}$ given by

(1.8.36) $$t = s + v$$

then (1.8.35), (1.8.36) yield

(1.8.37) $$\lim_{\mathcal{F}} t = U \in \mathcal{F}$$

while the *linearity* of $L(D)$ gives

(1.8.38) $$L(D)t = L(D)s + L(D)v$$

Now in view of the smoothness of the coefficients in (1.8.31), we usually have the *automatic continuity* property

(1.8.39) $$\lim_{\mathcal{F}} v = 0 \Rightarrow \lim_{\mathcal{X}} L(D)v = 0$$

In this way (1.8.33), (1.8.35), (1.8.38) and (1.8.39) yield

(1.8.40) $$\lim_{\mathcal{X}} L(D)t = f$$

The relations (1.8.38)-(1.8.40) prove the validity of the interpretation in (1.8.34) *irrespective* of the sequences s in (1.8.32) and (1.8.33).

It is now obvious that, if the *linear* partial differential operator $L(D)$ in (1.8.31) is replaced by a *nonlinear* partial differential operator as in (1.5.4), then *both* crucial steps in (1.8.38) and (1.8.39) will in general break down.

It is precisely this *double break down* which has usually been *overlooked* when going from solution methods for linear partial differential equations to solution methods for nonlinear ones. In general, the extension of linear methods to nonlinear ones can often involve difficulties which require essentially new ways of thinking. A particularly illuminating survey of several such well known extensions and of the mentioned kind of difficulties can be found in Zabuski.

Let us now have a better look at the above nonlinear stability paradoxes and do it in a way which *avoids* the usual exclusive use of approximation or topological interpretations. In particular, let us consider these paradoxes in the terms of the *representations* in Sections 5 and 6, terms which are *purely algebraic*.

For convenience, let us consider the nonlinear partial differential equation (1.5.1) in the following particular case

$$(1.8.41)\qquad \sum_{1\le i\le h} c_i(x) \prod_{1\le j\le k_i} D^{p_{ij}}U(x) = f(x),\ x \in \Omega = \mathbb{R}^n$$

with $c_i, f \in \mathcal{C}^\infty(\mathbb{R}^n)$, and let us denote by

$$(1.8.42)\qquad P(D) = \sum_{1\le i\le h} c_i(x) \prod_{1\le j\le k_i} D^{p_{ij}},\quad x \in \Omega = \mathbb{R}^n$$

the corresponding nonlinear partial differential operator. Further, let us denote by, see Appendix 1, Chapter 5,

$$(1.8.43)\qquad \mathcal{S}^\infty,\ \mathcal{V}^\infty$$

the set of all sequences of $\mathcal{C}^\infty$-smooth functions on $\mathbb{R}^n$ which converge weakly in $\mathcal{D}'(\mathbb{R}^n)$ to a distribution, respectively to zero. Then $\mathcal{V}^\infty$ and $\mathcal{S}^\infty$ are vector subspaces in $(\mathcal{C}^\infty(\mathbb{R}^n))^{\mathbb{N}}$ and the mapping

$$(1.8.44)\qquad \mathcal{S}^\infty/\mathcal{V}^\infty \ni s + \mathcal{V}^\infty \longmapsto U \in \mathcal{D}'(\mathbb{R}^n)$$

where, for $s = (\psi_\nu | \nu \in \mathbb{N}) \in \mathcal{S}^\infty$, we define

$$(1.8.45)\qquad U(\alpha) = \lim_{\nu\to\infty} \int_{\mathbb{R}^n} \psi_\nu(x)\alpha(x)dx,\quad \alpha \in \mathcal{D}(\mathbb{R}^n)$$

is a *vector space isomorphism.*

Now, the mentioned customary method for constructing weak solutions for nonlinear partial differential equations can often be reformulated as follows.

One tries to find a solution $U \in \mathcal{D}'(\mathbb{R}^n)$ of the equation (1.8.41) by constructing a sequence $s = (\psi_\nu | \nu \in \mathbb{N}) \in \mathcal{S}^\infty$ such that

$$(1.8.46)\qquad P(D)s = (P(D)\psi_\nu | \nu \in \mathbb{N}) \in \mathcal{V}^\infty$$

while, in the sense of the mapping (1.8.44), we obtain in the same time

$$(1.8.47)\qquad s + \mathcal{V}^\infty \longmapsto U$$

In view of (1.8.42), it is obvious that $P(D)$ generates a mapping

$$(1.8.48)\qquad P(D): (\mathcal{C}^\infty(\mathbb{R}^n))^{\mathbb{N}} \to (\mathcal{C}^\infty(\mathbb{R}^n))^{\mathbb{N}}$$

if applied term by term to sequences of $\mathcal{C}^\infty$-smooth functions.

For a moment, let us assume that P(D) in (1.8.42) is linear, i.e., it has the form

$$(1.8.49) \qquad P(D)U = \sum_{1\leq i\leq h} c_i(x)D^{p_i}U$$

then it follows easily, see Appendix 1, Chapter 5, that P(D) is *compatible* with the quotient structure $\mathcal{S}^\infty/\mathcal{V}^\infty$, that is

$$(1.8.50) \qquad P(D)\mathcal{S}^\infty \subset \mathcal{S}^\infty, \; P(D)\mathcal{V}^\infty \subset \mathcal{V}^\infty$$

thus, one can define the linear mapping

$$(1.8.51) \qquad P(D) : \mathcal{S}^\infty/\mathcal{V}^\infty \longrightarrow \mathcal{S}^\infty/\mathcal{V}^\infty$$

by

$$(1.8.52) \qquad P(D)(s + \mathcal{V}^\infty) = P(D)s + \mathcal{V}^\infty, \quad s \in \mathcal{S}^\infty$$

It follows that in the linear case (1.8.49), it *sufficies* to construct *one single* sequence $s \in \mathcal{S}^\infty$ with the property (1.8.46), since for any other sequence $t \in \mathcal{S}^\infty$ with $t-s \in \mathcal{V}^\infty$, the linearity of P(D) and the relations (1.8.46), (1.8.50) will yield

$$(1.8.53) \qquad P(D)t = P(D)s + P(D)(t - s) \in \mathcal{V}^\infty + \mathcal{V}^\infty \subset \mathcal{V}^\infty$$

However, the general *nonlinear* mapping (1.8.42) is *not* necessarily compatible with the quotient structure $\mathcal{S}^\infty/\mathcal{V}^\infty$, since instead of (1.8.50), we may have

$$(1.8.54) \qquad P(D)\mathcal{S}^\infty \not\subset \mathcal{S}^\infty$$

and also

$$(1.8.55) \qquad P(D)\mathcal{V}^\infty \not\subset \mathcal{V}^\infty$$

Indeed, as seen with the weak and strong solution (1.8.13)-(1.8.15) of the nonlinear system (1.8.16), it follows that

$$(1.8.56) \qquad v \in \mathcal{V}^\infty, \; v^2 \in (\mathcal{V}^\infty\cdot\mathcal{V}^\infty) \cap \mathcal{S}^\infty, \quad v^2 \notin \mathcal{V}^\infty$$

therefore

$$(1.8.57) \qquad (\mathcal{V}^\infty\cdot\mathcal{V}^\infty) \cap \mathcal{S}^\infty \not\subset \mathcal{V}^\infty$$

in particular

(1.8.58) $\mathcal{V}^\infty \cdot \mathcal{V}^\infty \not\subset \mathcal{V}^\infty$

In this way, (1.8.55) follows from (1.8.58), even in the case of the *zero order nonlinear differential operator*

(1.8.59) $T(D)U = U^2$

The relation (1.8.54) can be obtained in a similar way, using the argument that $\delta^2 \notin \mathcal{D}'$, see Rosinger [1, p. 11], or Rosinger [2, p. 66].

Now, in view of (1.8.54) and (1.8.55), it is obvious that we *cannot* always have (1.8.53) in the case of a general nonlinear operator $P(D)$ in (1.8.42). In this way, the customary weak solution method is invalidated in the general nonlinear case.

As the above simple and general algebraic argument shows it, the mentioned nonlinear stability paradoxes can reach rather deep in the study of weak solutions of nonlinear partial differential equations. Therefore these paradoxes deserve as general and complete a study as possible.

It is easy to see that one of the *basic reasons* for the above nonlinear stability paradoxes are the relations in (1.8.57) and (1.8.58).

Obviously, in order to reestablish the inclusion '$\subset$' in these two relations, we can follow one of the following *three* possibilities:
First, to replace $\mathcal{V}^\infty$ with a *smaller* vector subspace

(1.8.60) $\mathcal{I} \subset \mathcal{V}^\infty$

Secondly, to replace $\mathcal{V}^\infty$ with a *larger* vector subspace

(1.8.61) $\mathcal{V}^\infty \subset \mathcal{I} \subset (\mathcal{C}^\infty(\mathbb{R}^n))^{\mathbb{N}}$

Or thirdly, to replace $\mathcal{V}^\infty$ with any other suitable vector subspace

(1.8.62) $\mathcal{I} \subset (\mathcal{C}^\infty(\mathbb{R}^n))^{\mathbb{N}}$

As seen in Chapter 8, Colombeau's method does in a way replace $\mathcal{V}^\infty$ with one particular, *smaller* $\mathcal{I} \subset \mathcal{V}^\infty$. On the other hand, the method in Rosinger [1,2,3], presented shortly in Chapters 2, 3, 5 and 6 is based on a study of the *more general* third possibility mentioned above in (1.8.62). One of the first results of that study is that $\mathcal{V}^\infty$ is *too large* to be suitable for *nonlinear* theories of generalized functions. This is also suggested by the following simple result.

Lemma 1

Suppose given any sequence of positive numbers a_ν, with $\nu \in \mathbb{N}$, such that $a_\nu \longrightarrow \infty$, when $\nu \longrightarrow \infty$.

Then, there exist sequences $v = (\chi_\nu | \nu \in \mathbb{N}) \in \mathcal{V}^\infty$, such that

$$(1.8.63) \qquad \int_{\mathbb{R}} \chi_\nu^2(x)\alpha(x)dx \geq a_\nu \alpha(0), \quad \alpha \in \mathcal{D}(\mathbb{R}), \quad \alpha \geq 0$$

for $\nu \in \mathbb{N}$, $\nu \geq \mu$, with $\mu \in \mathbb{N}$ suitably chosen, possibly dependent on α.

Proof

Let us take $\beta \in \mathcal{D}(\mathbb{R})$, $\beta \geq 0$, such that

$$(1.8.64) \qquad \int_{\mathbb{R}} \beta(x)dx = 1$$

For $\lambda \in \mathbb{R}$, $\lambda > 0$, let us define $\gamma \in \mathcal{D}(\mathbb{R})$ by

$$(1.8.65) \qquad \gamma(x) = \lambda\beta(\lambda x), \quad x \in \mathbb{R}$$

then obviously $\gamma \geq 0$ and in view of (1.8.64) we have

$$(1.8.66) \qquad \int_{\mathbb{R}} \gamma(x)dx = 1, \quad \int_{\mathbb{R}} \gamma^2(x)dx = \lambda \int_{\mathbb{R}} \beta^2(x)dx$$

thus we can assume that with suitably chosen λ, we have

$$(1.8.67) \qquad \int_{\mathbb{R}} \gamma^2(x)dx > 1$$

Let us take $\rho \in (0,1)$ and define the sequences of positive numbers b_ν, c_ν, with $\nu \in \mathbb{N}$, by

$$(1.8.68) \qquad b_\nu = a^{1/(1-\rho)}, \quad c_\nu = b_\nu^{1+\rho}$$

Finally, let us define $v = (\chi_\nu | \nu \in \mathbb{N}) \in (\mathcal{C}^\infty(\mathbb{R}))^{\mathbb{N}}$

by

(1.8.69) $\chi_\nu(x) = b_\nu \gamma(c_\nu x), \quad x \in \mathbb{R}, \quad \nu \in \mathbb{N}$

If $\alpha \in \mathcal{D}(\mathbb{R})$ then obviously

(1.8.70) $$\int_{\mathbb{R}} \chi_\nu(x)\alpha(x)dx = \frac{b_\nu}{c_\nu} \int_{\mathbb{R}} \gamma(x)\alpha(x/c_\nu)dx, \quad \nu \in \mathbb{N}$$

hence, in view of (1.8.66) and (1.8.68), the integral in the left hand term of (1.8.70) tends to zero, when $\nu \to \infty$. It follows that

(1.8.71) $v = (\chi_\nu \mid \nu \in \mathbb{N}) \in \mathcal{V}^\infty$

Assume now $\alpha \in \mathcal{D}(\mathbb{R})$ and $\alpha \geq 0$, then obviously

$$\int_{\mathbb{R}} \chi_\nu^2(x)\alpha(x)dx = \frac{b_\nu^2}{c_\nu} \int_{\mathbb{R}} \gamma^2(x)\alpha(x/c_\nu)dx, \quad \nu \in \mathbb{N}$$

hence (1.8.63) follows in view of (1.8.67) and (1.8.68). □

The result in (1.8.63) shows that the *square* v^2 of a sequence v which converges weakly to zero, can diverge weakly to infinity *arbitrarily fast*.

In connection with the phenomenon of nonlinear stability paradoxes it is useful to consider the following milder version of Lemma 1 above. Let us define the sequence of $\mathcal{C}^\infty$-smooth functions on $\mathbb{R}$

(1.8.72) $w = (\omega_\nu \mid \nu \in \mathbb{N})$

by

(1.8.73) $\omega_\nu(x) = \sqrt{\nu}\ \alpha(\nu x), \quad x \in \mathbb{R}, \quad \nu \in \mathbb{N}$

where we have chosen

(1.8.74) $\alpha \in \mathcal{D}(\mathbb{R})$, with $\alpha \geq 0$ and $\int_{\mathbb{R}} \alpha^2(x)dx = 1$

Then again, an easy direct computation will give

(1.8.75) $\omega_\nu \longrightarrow 0$ and $\omega_\nu^2 \longrightarrow \delta$, when $\nu \longrightarrow \infty$

in the sense of both the *weak* and *strong* topology in $\mathcal{D}'(\mathbb{R})$.

It follows that (1.8.72)-(1.8.75) gives a *weak and strong* solution for the nonlinear system

$$(1.8.76) \qquad \begin{aligned} U &= 0 \\ U^2 &= \delta \end{aligned}$$

thus admittedly proves that, in addition to the relation (1.8.2), we can have as well

$$(1.8.77) \qquad 0^2 = \delta$$

One of the interesting differences between the above solutions of (1.8.16) and (1.8.76) is the following. The weakly and strongly convergent sequence v in (1.8.13) which solves the system (1.8.16), has highly oscillatory terms χ_ν, when $\nu \rightarrow \infty$. On the contrary, the weakly and strongly convergent sequence w in (1.8.73) has terms ω_ν which do not oscillate more and more frequently, when $\nu \rightarrow \infty$. The same is true for the sequence v in Lemma 1.

As a final remark, we should note that recently, there has been a certain limited awareness in a few specific instances of nonlinear partial differential equations of the possibility of the above kind of nonlinear stability paradoxes associated with weak solution methods, see Ball, Murat, Tartar, Dacorogna, Di Perna, Rauch & Reed, Slemrod. However, the respective methods developed in order to avoid nonlinear stability paradoxes of weak solutions present several important limitations.

Indeed, first of all, these methods are developed within the traditional functional analytic view point, where alone *topological interpretation* is used, with the consequent exclusion of the vast possibilities offered by more fundamental *algebraic representations*. The limiting vision implied by the usual unquestioned and automatic topological interpretation is clearly expressed for instance in Dacorogna, [p. 4]. There, the sequence

$$u^\epsilon(x) = \sin \frac{x}{\epsilon}, \ x \in (0,2\pi), \ \epsilon > 0$$

and the nonlinear continuous function

$$f(u) = -u^2, \quad u \in \mathbb{R}$$

are considered, for which one obviously has the weak convergence properties

$$u^\epsilon \rightarrow 0, \quad f(u^\epsilon) \rightarrow -\frac{1}{2}$$

when $\epsilon \rightarrow 0$. That example, similar to the one in (1.8.13)-(1.8.16) above, is claimed to lead to the conclusion:
'...Therefore in order to obtain weak continuity ... one has to impose some

restrictions on the sequence $\{u^\epsilon\}$ and on the nonlinear function f...'.

In this way, only certain particular types of nonlinear partial differential equations and sequential solutions can be dealt with. Not to mention that there is no attempt to develop a comprehensive nonlinear theory of generalized functions, capable of handling large classes of nonlinear partial differential equations.

The effect of such particular and limited approaches based on topological interpretation - for instance, the Tartar-Murat compensated compactness and the Young measure associated with weakly convergent sequences of functions subjected to differential constraints on an algebraic manifold - has been a distancing from the basic algebraic reasons underlying the nonlinear stability paradoxes. That distancing has contributed to the clouding of what in fact prove to be rather simple ring theoretic phenomena.

§9. EXTENDING NONLINEAR PARTIAL DIFFERENTIAL OPERATORS TO GENERALIZED FUNCTIONS

It has been noted that the basic and rather elementary algebraic conflict between discontinuity, multiplication and differentiation presented in Section 1 leads to the setting in (1.3.1)-(1.3.4), where A and $\bar{A}$ are algebras of generalized functions extending the distributions. The way such algebras can be constructed is shown in (1.5.24)-(1.5.27), and equivalently, in (1.6.9)-(1.6.11).

As mentioned at the end of Section 5, the generality of the above constructions comes to be subjected to several natural particularizations.

The first of them, dealt with in the section, is imposed by the way polynomial nonlinear partial differential operators T(D) in (1.5.5) can be defined as acting between such spaces of generalized functions.

In order not to miss on any of the possibly relevant phenomena involved, we shall approach the problem of the *extension* of T(D) to spaces of generalized functions within the most general framework of the respective spaces. For that, it is easy to observe that one can obtain an extension

(1.9.1) $\quad T(D) : E \longrightarrow A$

for suitable $E = \mathcal{S}/\mathcal{V} \in VS_{\mathcal{F},\Lambda}$ and $A = \mathcal{A}/\mathcal{I} \in AL_{\mathcal{X},\Lambda}$, see (1.6.7), (1.6.9), provided that one can define the following extensions of the usual partial derivatives

(1.9.2) $\quad D^p : E \longrightarrow A, \quad p \in \mathbb{N}^n, \quad |p| \leq m$

where m is the order of T(D), see (1.5.3).

Indeed, in such a case, for given $U \in E$, the nonlinear, that is polynomial operations involved in $T(D)U$ will take place not in the domain E but in the range A of $T(D)$ in (1.9.1). It follows that the *most general* framework for the extension of $T(D)$ can involve *different* spaces for the domain and the range, and *only the range has to be an algebra*. It should be noted however that further extensions of the framework in (1.9.1) are still possible and useful, see Chapter 4.

Let us now see the way extensions (1.9.2) can be obtained. We can easily do that if we make the following natural assumptions

(1.9.3) $\qquad \mathcal{F} \subset \mathcal{C}^m(\Omega), \quad \mathcal{C}^0(\Omega) \subset \mathcal{X}$

(1.9.4) $\qquad D^p\mathcal{V} \subset \mathcal{I}, \quad D^p\mathcal{S} \subset \mathcal{A}, \quad p \in \mathbb{N}^n. \quad |p| \leq m$

Indeed, in this case (1.9.2) can be defined by

(1.9.5) $\qquad D^pU = D^ps + \mathcal{I} \in A = \mathcal{A}/\mathcal{I}, \quad p \in \mathbb{N}^n, \quad |p| \leq m$

for every

(1.9.6) $\qquad U = s + \mathcal{V} \in E = \mathcal{S}/\mathcal{V}, \quad s \in \mathcal{S}$

where for every

(1.9.7) $\qquad s = (\psi_\lambda | \lambda \in \Lambda) \in \mathcal{F}^\Lambda$

we define

(1.9.8) $\qquad D^ps = (D^p\psi_\lambda | \lambda \in \Lambda), \quad p \in \mathbb{N}^n, \quad |p| \leq m$

However, the above method for extension by reduction to representants can be used in the following *less restrictive* manner. Let us suppose that the *vector subspace* $\mathcal{F} \subset \mathcal{M}(\Omega)$ and the *subalgebra* $\mathcal{X} \subset \mathcal{M}(\Omega)$ are such that

(1.9.9) $\qquad D^p(\mathcal{F} \cap \mathcal{C}^m(\Omega)) \subset \mathcal{X}, \quad p \in \mathbb{N}^n, \quad |p| \leq m$

We note that (1.9.9) holds whenever $\mathcal{X} \supset \mathcal{C}^0(\Lambda)$, see (1.9.3).

Suppose now that

(1.9.10) $\quad D^p(\mathcal{V} \cap (\mathcal{C}^m(\Omega))^\Lambda) \subset \mathcal{I}, \quad D^p(\mathcal{S} \cap (\mathcal{C}^m(\Omega))^\Lambda) \subset \mathcal{A}, \; p \in \mathbb{N}^n, \quad |p| \leq m$

We also note that (1.9.10) hold whenever (1.9.4) is satisfied.

It is obvious that in the conditions (1.9.9) and (1.9.10), we can replace m with any $\ell \in \overline{\mathbb{N}} = \mathbb{N} \cup \{\infty\}$. In this general case, it will be convenient in the sequel to denote (1.9.9) and (1.9.10) together under the simpler form

(1.9.11) $\quad E \overset{\ell}{\leq} A$

Finally, let us denote by

(1.9.12) $\quad VS^{\ell}_{\mathcal{F},\Lambda}$

the set of all the quotient vector spaces $E = \mathcal{S}/\mathcal{V} \in VS_{\mathcal{F},\Lambda}$ such that

(1.9.13) $\quad \mathcal{S} \subset \mathcal{V} + (\mathcal{C}^{\ell}(\Omega))^{\Lambda}$

We note that (1.9.13) holds whenever $\mathcal{F} \subset \mathcal{C}^{\ell}(\Omega)$, see (1.9.3).

Similar to (1.9.12), we can also define $AL^{\ell}_{\mathcal{X},\Lambda}$.

Now the above definition (1.9.5)-(1.9.8) of the partial derivatives (1.9.2) can further be extended as follows.

Suppose given

(1.9.14) $\quad E = \mathcal{S}/\mathcal{V} \in VS^{m}_{\mathcal{F},\Lambda}, \quad A = \mathcal{A}/\mathcal{I} \in AL_{\mathcal{X},\Lambda}$

such that

(1.9.15) $\quad E \overset{m}{\leq} A$

Then we can define the *partial derivatives of generalized functions* as being given by the linear operators

(1.9.16) $\quad D^p : E \longrightarrow A, \quad p \in \mathbb{N}^n, \quad |p| \leq m$

as follows: for given

(1.9.17) $\quad U = s + \mathcal{V} \in E = \mathcal{S}/\mathcal{V}, \quad s \in \mathcal{S}$

we define

(1.9.18) $\quad D^pU = D^pt + \mathcal{I} \in A = \mathcal{A}/\mathcal{I}, \quad p \in \mathbb{N}^n, \quad |p| \leq m$

where

(1.9.19) $\quad t \in \mathcal{S} \cap (\mathcal{C}^m(\Omega))^{\Lambda}, \quad t - s \in \mathcal{V}$

It is easy to see that the above definition (1.9.16)-(1.9.19) of partial derivatives of generalized functions is correct and it contains as a particular case the previous definition (1.9.5).

Furthermore, when restricted to classical functions in $\mathcal{F} \cap \mathcal{C}^m(\Omega)$, the partial derivatives in (1.9.16) coincide with the usual ones. Finally, the partial derivatives (1.9.16) are *linear* mappings, and if $\mathcal{F}$ is a subalgebra and $E = \mathcal{S}/\mathcal{V}$ is a quotient algebra, then they satisfy in A the *Leibnitz rule* of product derivatives.

Now we can return to the problem of defining an *extension* (1.9.1).

The framework will of course be the same with the above used in defining partial derivatives for generalized functions. Namely, we suppose given

(1.9.20) $$E = \mathcal{S}/\mathcal{V} \in VS^m_{\mathcal{F},\Lambda}, \quad A = \mathcal{A}/\mathcal{I} \in AL_{\mathcal{X},\Lambda}$$

such that

(1.9.21) $$E \overset{m}{\leq} A$$

where m is the order of $T(D)$. Further, we shall also assume that the coefficients in $T(D)$, see (1.5.4), satisfy

(1.9.22) $$c_i \in \mathcal{X}, \quad 1 \leq i \leq h$$

which obviously holds whenever $\mathcal{C}^0(\Omega) \subset \mathcal{X}$, see (1.9.3).

Now we can define the *extension*

(1.9.23) $$T(D) : E \longrightarrow A$$

as follows: for given

(1.9.24) $$U = s + \mathcal{V} \in E = \mathcal{S}/\mathcal{V}, \quad s \in \mathcal{S}$$

we define

(1.9.25) $$T(D)U = T(D)t + \mathcal{I} \in A = \mathcal{A}/\mathcal{I}$$

where

(1.9.26) $$t \in \mathcal{S} \cap (\mathcal{C}^m(\Omega))^\Lambda, \quad t - s \in \mathcal{V}$$

Let us show that the above definition (1.9.23)-(1.9.26) is correct. First we note that owing to its *polynomial* nonlinearity, $T(D)$ satisfies the following relation for every $z,w \in (\mathcal{C}^m(\Omega))^\Lambda$

(1.9.27) $$T(D)(z + w) = T(D)z + \sum_{\alpha} z_{\alpha} \cdot D^{p_{\alpha}} w$$

where z_{α} are products of c_i, $D^{p_{ij}}z$ and possibly $D^{p_{ij}}w$, while p_{α} are some of the p_{ij}. In view of (1.9.27) we obtain the following succession of implications: if

(1.9.28) $$z \in \mathcal{S} \cap (\mathcal{C}^m(\Omega))^{\Lambda}, \quad w \in \mathcal{V} \cap (\mathcal{C}^m(\Omega))^{\Lambda}$$

then

(1.9.29) $$\sum_{\alpha} z_{\alpha} \cdot D^{p_{\alpha}} w \in \mathcal{I}$$

therefore

(1.9.30) $$T(D)(z + w) - T(D)z \in \mathcal{I}$$

Now let us take $s_1 \in \mathcal{S}$, $s_1 - s \in \mathcal{V}$, $t_1 \in \mathcal{S} \cap (\mathcal{C}^m(\Omega))^{\Lambda}$, $t_1 - s_1 \in \mathcal{V}$ for alternative representations in (1.9.24)-(1.9.26). Then obviously

$$z = t \in \mathcal{S} \cap (\mathcal{C}^m(\Omega))^{\Lambda} \text{ and}$$

$$w = t_1 - t = (t_1 - s_1) - (t - s) + (s_1 - s) \in \mathcal{V}$$

hence (1.9.30) yields $T(D)t_1 - T(D)t \in \mathcal{I}$, which proves that the definition in (1.9.24)-(1.9.26) does not depend on the representants s or t.

It is easy to see that the restriction of the extended $T(D)$ in (1.9.23) to classical functions in $\mathcal{F} \cap \mathcal{C}^m(\Omega)$ acts in the same way with the usual nonlinear partial differential operator in (1.5.5).

§10. NOTIONS OF GENERALIZED SOLUTION

Given the above constructed extension to generalized functions

(1.10.1) $$T(D) : E \longrightarrow A$$

of the polynomial nonlinear partial differential operator $T(D)$ in (1.5.5), it is now a rather simple matter to define a notion of *generalized solution* for the equation

(1.10.2) $$T(D)U = f, \quad f \in A$$

as being any $U \in E$ which will satisfy (1.10.2) with $T(D)$ defined in (1.10.1).

However, we should not miss the fact that there are less simple phenomena involved here. Indeed, if we are given a nonlinear partial differential equation, such as for instance in (1.5.1), that equation is prior to, and therefore independent of the various possible generalized function spaces involved in extensions such as those in (1.10.1). And obviously, there can be a large variety of such generalized function spaces which could appear in these extensions.

The utility of considering an equation (1.5.1) within different extensions (1.10.1) will become obvious in Section 11 in connection with the nonlinear stability, generality and exactness properties of generalized solutions.

Here however we should like to recall the *third* element which, in addition to the partial differential equations and their possible generalized solutions, does in a natural way belong to the picture, and which is constituted from the various specific *solution methods*. Such specific solution methods, which quite often encompass a wealth of mathematical, physical and other insight and information, usually lead to sequences or in general, families

(1.10.3) $$s = (\psi_\lambda \mid \lambda \in \Lambda)$$

of sufficiently smooth functions $\psi_\lambda \in \Omega \longrightarrow \mathbb{R}$, with $\lambda \in \Lambda$, which are supposed to define in certain ways - often, by approximation - classical or generalized solutions U, see for instance (1.8.3)-(1.8.7).

It should be noted that the nonlinear stability paradoxes point out the questionable way generalized solutions U are *associated* with families s in the customary sequential approach for solving nonlinear partial differential equations. What the families s and the methods which lead to them are concerned, they may have their own merits, depending on the particulars of the situation involved.

In view of the above, we shall define now a solution concept which focuses on such families s in (1.10.3). The interest in such a solution concept is in the fact that it eliminates the problem of nonlinear stability paradoxes, since the *association* $s \longrightarrow U$ of a *single* family s with a generalized function U takes place in the framework of an extension (1.10.1), see details in Section 12. Moreover, it allows a deeper study of the stability, generality and exactness properties of generalized solutions for nonlinear partial differential equations, see Section 11.

Suppose T(D) in (1.5.5) has order m and we are given the equation

(1.10.4) $$T(D)U(x) = f(x), \quad x \in \Omega$$

in which, for the sake of generality, we can assume this time that c_i, $f \in \mathcal{M}(\Omega)$. Further, suppose given a family

(1.10.5) $$s = (\psi_\lambda \mid \lambda \in \Lambda)$$

of functions $\psi_\lambda \in \mathcal{M}(\Omega)$, with $\lambda \in \Lambda$.

Then s is called a *sequential solution* for (1.10.4), if and only if there exists a vector subspace $\mathcal{F}$ and a subalgebra $\mathcal{X}$ in $\mathcal{M}(\Omega)$, as well as $E = \mathcal{S}/\mathcal{V} \in VS^m_{\mathcal{F},\Lambda}$ and $A = \mathcal{A}/\mathcal{I} \in AL_{\mathcal{X},\Lambda}$, with $E \overset{m}{\leq} A$, such that

(1.10.6) $\quad c_1,\ldots,c_h, \quad f \in \mathcal{X}$

and for $T(D)$ in (1.10.1) we have

(1.10.7) $\quad T(D)U = f$

where

(1.10.8) $\quad U = s + \mathcal{V} \in E = \mathcal{S}/\mathcal{V}, \quad s \in \mathcal{S}$

When it is useful to mention the spaces E and A of generalized functions in the above definition, we shall say that s is an $E \longrightarrow A$ *sequential solution* for (1.10.4).

In view of (1.9.24)-(1.9.26), it is easy to see that (1.10.7) and (1.10.8) are equivalent with the condition

$$(1.10.9) \qquad \begin{array}{rl} \exists & t \in \mathcal{S} \cap (\mathcal{C}^m(\Omega))^\Lambda : \\ *) & t - s \in \mathcal{V} \\ **) & T(D)t - u(f) \in \mathcal{I} \end{array}$$

which is further equivalent with the following two conditions

(1.10.10) $\quad s \in \mathcal{S}$

and

$$(1.10.11) \qquad \begin{array}{l} \forall \quad t \in \mathcal{S} \cap (\mathcal{C}^m(\Omega))^\Lambda : \\ \quad t - s \in \mathcal{V} \Rightarrow T(D)t - u(f) \in \mathcal{I} \end{array}$$

Remark 3

Usually, it is considered convenient to solve partial differential equations with the respective partial differential operators acting within one single space of generalized functions. With the notation in (1.10.1) and in the case of a *nonlinear* partial differential operator $T(D)$, that would mean the particular situation when

(1.10.12) $\quad T(D) : A \longrightarrow A$

with

(1.10.13) $\quad A = \mathcal{A}/\mathcal{I} \in AL^m_{\mathcal{X},\Lambda}, \quad A \overset{m}{\leq} A$

As seen in Chapter 8, Colombeau's nonlinear theory of generalized functions leads to such a situation, where the linear or nonlinear partial differential operators are acting within the *same* space of generalized functions

(1.10.14) $\quad T(D) : \mathcal{G} \longrightarrow \mathcal{G}$

It should nevertheless be mentioned that, even in the case of *linear* partial differential operators, the utility of *different* domain and range vector spaces of generalized functions is well documented in the literature. Hörmander, Treves [1,2,3].

However, it is *important* to point out that, as shown in Rosinger [1,2,3] and presented in the sequel, see in particular Chapters 2, 6 and 7, a proper handling of such difficulties as the nonlinear stability paradoxes and the so called Schwartz impossibility result is facilitated if we consider that the nonlinear partial derivative operators $T(D)$ act within the following particular case of (1.10.1), given by

(1.10.15) $\quad T(D) : A \longrightarrow A'$

where

(1.10.16) $\quad A = \mathcal{A}/\mathcal{I} \in AL^m_{\mathcal{X},\Lambda}, \quad A' = \mathcal{A}'/\mathcal{I}' \in AL_{\mathcal{X}',\Lambda}$

and

(1.10.17) $\quad A \overset{m}{\leq} A'$

which is more general than (1.10.2) or (1.10.14). It should be mentioned that the utility of considering different algebras A and A' in (1.10.15) does ultimately lead to the consideration of infinite *chains* of such algebras of generalized functions, see Chapter 6.

§11. NONLINEAR STABILITY, GENERALITY AND EXACTNESS

Now we come to the *three basic properties* which lead to the necessary structure of any nonlinear theory of generalized functions based on the sequential approach initiated earlier in Section 5. The associated notions of stability, generality and exactness have first been introduced in Rosinger [2], where further details can be found. These three properties which relate to generalized solutions as well as the respective spaces of generalized functions are *essential* for a proper handling of the problems which arise from the nonlinear stability paradoxes and the so called Schwartz impossibility result.

Suppose given the framework in (1.9.20)-(1.9.22), in which case we can define, see (1.9.23), the mapping

(1.11.1) $$E = \mathcal{S}/\mathcal{V} \xrightarrow{T(D)} A = \mathcal{A}/\mathcal{I}$$

Then the *generalized solutions* of the nonlinear partial differential equation

(1.11.2) $$T(D)U = f, \quad f \in \mathcal{X}$$

have the form

(1.11.3) $$U = s + \mathcal{V} \in E = \mathcal{S}/\mathcal{V}, \quad s \in \mathcal{S}$$

Let us have a better look at the relationship between U and s in (1.11.3). In view of (1.10.11), for the same U kept fixed, we can replace s with any t which satisfies the conditions

(1.11.4) $$\begin{aligned} & t \in \mathcal{S} \cap (\mathcal{C}^m(\Omega))^\Lambda \\ & t - s \in \mathcal{V} \end{aligned}$$

Therefore, it is obvious that the *maximal stability* of U means

(1.11.5) $$\text{maximal } \mathcal{V}$$

Remark 4

Here it is *most important* to point out that owing to the *neutrix condition* (1.6.6) which has to be satisfied by the spaces of generalized functions, it is obvious that one cannot speak about the *largest* $\mathcal{V}$, since the *off diagonality* condition (1.6.6) means that $\mathcal{V}$ has to be contained in some vetor subspace which is complementary to $\mathcal{U}_{\mathcal{F},\Lambda}$. Similarly, one cannot speak about the *largest* ideals $\mathcal{I}$ which satisfy the respective neutrix, or *off diagonality* condition (1.6.11), see for details Appendix 8. This fact alone is *sufficient reason* to expect that a proper approach to generalized solutions of nonlinear partial differential equations requires the consideration of *various* spaces of generalized functions, since a *canonical* space of generalized functions does not appear in a natural way.

Now, if we go further, we note that the neutrix condition (1.6.6) appears in connection with the embedding (1.6.1) which expresses the requirement that classical functions should be particular cases of generalized functions. Or in other words, generalized functions should be *general* enough in order to include classical functions.

Owing to the well established role of the L. Schwartz distributions $\mathcal{D}'$ in the study of generalized functions, we could ask the following stronger version of the embedding condition (1.6.1), namely

$$(1.11.6) \qquad \mathcal{D}'(\Omega) \subset E = \mathcal{S}/\mathcal{V}$$

which is also satisfied by Colombeau's generalized functions, see details in Chapter 8.

It should be pointed out that, owing to the inexistence of solutions of certain partial differential equations within particular spaces of generalized functions, such as for instance $\mathcal{D}'$, see Rosinger [3, Part 1, Chapter 3, Section 1, or Part 2, Chapter 2, Section 5], there exists an interest in *large* spaces of generalized functions $E = \mathcal{S}/\mathcal{V}$ which can offer a satisfactory 'reservoir' for the existence of generalized solutions U. In this way we are led to a second quality of the spaces of generalized functions $E = \mathcal{S}/\mathcal{V}$ called in the sequel *generality*, and meaning

$$(1.11.7) \qquad \text{large } E = \mathcal{S}/\mathcal{V}$$

or equivalently

$$(1.11.8) \qquad \text{large } \mathcal{S} \text{ and small } \mathcal{V}$$

Remark 5

(1) In view of (1.11.5) and (1.11.8), it is obvious that *stability* and *generality* are *conflicting*. It follows in particular that there is *no* interest in *maximal* generality, unless one is ready to sacrifice stability.

(2) In order to obtain a generality for a quotient space $E = \mathcal{S}/\mathcal{V}$ which is not less than that in (1.11.6), it is obvious that Λ in (1.6.5) *cannot* be finite even if $\mathcal{F} = \mathcal{M}(\Omega)$. However, since even such a small space as $\mathcal{C}^\infty(\Omega)$ is sequentially dense in $\mathcal{D}'(\Omega)$, with the usual topology on the latter, one can obtain (1.11.6) for *any* infinite index set Λ, whenever for instance $\mathcal{C}^\infty(\Omega) \subset \mathcal{F}$.

Finally, we come to the third quality of spaces of generalized functions. We note that equation (1.11.2) has a *classical solution* $\psi \in \mathcal{C}^m(\Omega)$, if and only if there exists $t \in (\mathcal{C}^m(\Omega))^\Lambda$ such that

$$(1.11.9) \qquad w_t = T(D)t - u(f) \in \mathcal{O}$$

Further, a family of classical functions $s = (\psi_\lambda | \lambda \in \Lambda) \in (\mathcal{M}(\Omega))^\Lambda$ is a *sequential solution* of (1.11.2), if and only if

$$(1.11.10) \qquad \begin{array}{l} \forall \quad t \in \mathcal{S} \cap (\mathcal{C}^m(\Omega))^\Lambda : \\ \qquad t - s \in \mathcal{V} \Longrightarrow w_t = T(D)t - u(f) \in \mathcal{I} \end{array}$$

for certain spaces of generalized functions $E = \mathcal{S}/\mathcal{V}$ and $A = \mathcal{A}/\mathcal{I}$ as in Section 10.

Since $\mathcal{I}$ is an ideal in $\mathcal{A}$, condition (1.11.10) will obviously yield

$$(1.11.11) \qquad \begin{array}{l} \forall \quad t \in \mathcal{S} \cap (\mathcal{C}^m(\Omega))^\Lambda : \\ \quad t - s \in \mathcal{V} \Rightarrow w_t \in \mathcal{I}, \quad \mathcal{A}\cdot w_t \subset \mathcal{I} \end{array}$$

In other words, the *error* in solving the equation (1.11.2), and which is given by

$$(1.11.12) \qquad w_t = T(D)t - u(f)$$

satisfies the *explicit algebraic tests*

$$(1.11.13) \qquad w_t \in \mathcal{I}$$

$$(1.11.14) \qquad \mathcal{A}\cdot w_t \subset \mathcal{I}$$

We note that in the terms of the neutrix calculus, see Section 6 and Appendix 4, condition (1.11.13) means that the *error* w_t is *$\mathcal{I}$-negligible*, while condition (1.11.14) means that each 'projection' $z\cdot w_t$ of the *error* w_t, with $z \in \mathcal{A}$, is also *$\mathcal{I}$-negligible*. Obviously, if $1 \in \mathcal{X}$ then $u(1) \in \mathcal{A}$, according to (1.6.10). Thus (1.11.14) will imply (1.11.13).

We shall call the above *algebraic test on error* in (1.11.13) and (1.11.14) the *exactness* property of the sequential solution s. Obviously, *better exactness* means

(1.11.15) large $\mathcal{A}$ and small $\mathcal{I}$

Remark 6

As seen in (1.11.9), classical solutions have the best exactness property which corresponds to the smallest $\mathcal{I}$, that is $\mathcal{I} = \mathcal{O}$, and thus to the largest $\mathcal{A}$ given by $\mathcal{A} = \mathcal{X}^\Lambda$. That situation can *no longer* occur with nonclassical, that is, generalized solutions. Indeed, in view of the inclusions

$$(1.11.16) \qquad D^p(\mathcal{V} \cap (\mathcal{C}^m(\Omega))^\Lambda) \subset \mathcal{I}, \quad p \in \mathbb{N}^n, \quad |p| \leq m$$

in (1.9.10) which appears in connection with the condition $E \overset{m}{\leq} A$ between the spaces of generalized functions in (1.11.1), it is obvious that *stability* and *exactness* are *conflicting*.

We can conclude that the above mentioned *conflict* between *stability* on the one hand and *generality* and *exactness* on the other, sets up a rather sophisticated *interplay* between these three properties, see Fig 1 below. The way that interplay is handled can depend to a large extent on the particulars of the situations involved in connection with the nonlinear partial differential equations under consideration, see details in Rosinger [2,3].

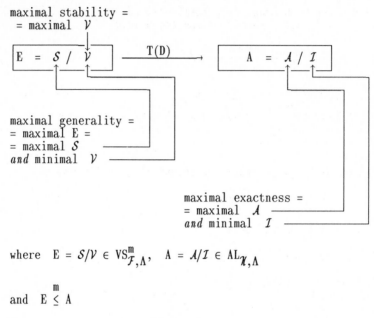

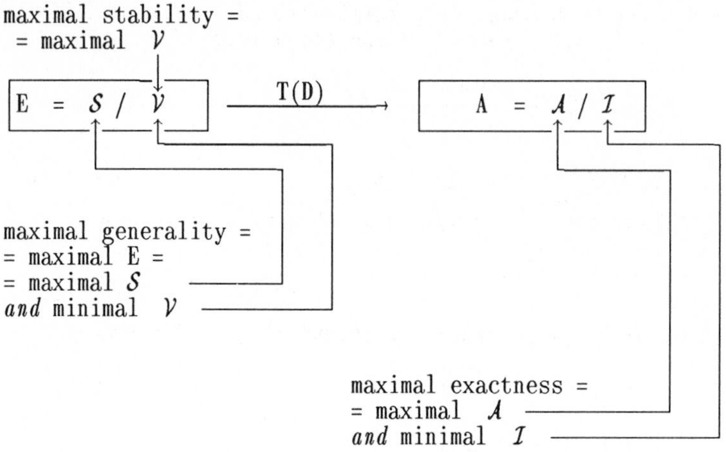

where $E = \mathcal{S}/\mathcal{V} \in VS^{m}_{\mathcal{F},\Lambda}$, $A = \mathcal{A}/\mathcal{I} \in AL_{\mathcal{X},\Lambda}$

and $E \overset{m}{\leq} A$

Figure 1

We should also note the following. Both *stability* and *generality* refer exclusively to the given space $E = \mathcal{S}/\mathcal{V}$ in (1.11.1) and are independent of the linear or nonlinear partial differential operator $T(D)$ and any of its generalized solutions $U \in E = \mathcal{S}/\mathcal{V}$. On the other hand, the conditions (1.11.13) and (1.11.14) defining *exactness*, involve both spaces E and A, as well as the linear or nonlinear partial differential operator $T(D)$ and its sequential solution s.

It is easy to see that, if we replace the framework (1.11.1) with the more particular one in (1.10.12), the conflict between stability, generality and exactness does not become easier, and on the contrary, their interplay has more constraints.

It should be remembered that, no matter how useful particular spaces of generalized functions may be, the *primary interest* is with the linear or nonlinear partial differential equations and their classical or generalized solutions which model physical and other processes. The variety of spaces of generalized functions as well as solution methods are only the *means* constructed to handle the above primary interest.

§12. ALGEBRAIC SOLUTION TO THE NONLINEAR STABILITY PARADOXES

Here we show, based on a *purely algebraic argument*, that if we solve nonlinear partial differential equations within frameworks such as in Section 10, then the mentioned kind of stability paradoxes *cannot* occur any longer.

Suppose given the m-th order polynomial nonlinear partial differential operator $T(D)$ in (1.5.4) and any of its extensions

$$(1.12.1) \qquad E = \mathcal{S}/\mathcal{V} \xrightarrow{T(D)} A = \mathcal{A}/\mathcal{I}$$

where $E = \mathcal{S}/\mathcal{V} \in VS^m_{\mathcal{F},\Lambda}$, $A = \mathcal{A}/\mathcal{I} \in AL_{\mathcal{X},\Lambda}$ and $E \overset{m}{\leq} A$, while $\mathcal{F} \subset \mathcal{M}(\Omega)$ is a vector subspace and $\mathcal{X} \subset \mathcal{M}(\Omega)$ is a subalgebra, such that

$$(1.12.2) \qquad c_i \in \mathcal{X} \text{ for } 1 \leq i \leq h$$

In order to avoid trivial cases, we assume in (1.5.4) that

$$(1.12.3) \qquad k_i \geq 1 \text{ for } 1 \leq i \leq h$$

Now we show that we always have

$$(1.12.4) \qquad U = 0 \in E \Longrightarrow T(D)U = 0 \in A$$

Indeed, let us assume that

$$(1.12.5) \qquad U = s + \mathcal{V} \in E = \mathcal{S}/\mathcal{V}, \quad s \in \mathcal{S}$$

then in view of (1.9.25), (1.9.26), we obtain

$$(1.12.6) \qquad T(D)U = T(D)t + \mathcal{I} \in A = \mathcal{A}/\mathcal{I}$$

where

$$(1.12.7) \qquad t \in \mathcal{S} \cap (\mathcal{C}^m(\Omega))^\Lambda, \quad t - s \in \mathcal{V}$$

But $U = 0 \in E$ and (1.12.5) yield

$$(1.12.8) \qquad s \in \mathcal{V}$$

hence (1.12.7) implies

$$(1.12.9) \qquad t \in \mathcal{V} \cap (\mathcal{C}^m(\Omega))^\Lambda$$

Now we recall that $E \overset{m}{\leq} A$, hence the relations (1.9.10), (1.12.9) yield

(1.12.10) $D^p t \in \mathcal{I}, \quad p \in \mathbb{N}^n, \quad |p| \leq m$

Finally, (1.5.4), (1.12.2) and (1.12.10) will give

(1.12.11) $T(D)t \in \mathcal{I}$

which completes the proof of (1.12.4)

From (1.12.4) it follows in particular that systems such as (1.8.1) or (1.8.76) *cannot* have solutions within any framework (1.12.1).

It is easy to see that similar results will also hold for the general nonlinear partial differential operators in Section 13 next. Within the particular nonlinear theory of generalized functions in Colombeau [1,2], nonlinear stability paradoxes of the kind of those in (1.8.1) or (1.8.76) still can happen, see Chapter 8. However, their unwelcome effects can be satisfactorily handled with the help of a special equivalence relation, called association, defined for a large subset of Colombeau's generalized functions.

§13. GENERAL NONLINEAR PARTIAL DIFFERENTIAL EQUATIONS

The particular *polynomial* form of nonlinearity of the partial differential operators T(D) in (1.5.4) was assumed because it made it easier to set up the natural framework in Sections 5, 6, 8-10.

However, with minor modifications, that framework can accommodate much more *general* nonlinear partial differential operators, which are defined now.

An m-th order *continuous nonlinear* partial differential equation is by definition of the form

(1.13.1) $T(x,U(x),\dots,D^p U(x),\dots) = f(x), \ x \in \Omega$

where $f \in \mathcal{C}^0(\Omega)$ is given, while $p \in \mathbb{N}^n$, $|p| \leq m$ and $T \in \mathcal{C}^0(\Omega \times \mathbb{R}^\ell)$, with

(1.13.2) $\ell = \text{car}\{p \in \mathbb{N}^n \,|\, |p| \leq m\}$

The left hand term in (1.13.1) generates the nonlinear partial differential operator

(1.13.3) $T(D) : \mathcal{C}^m(\Omega) \longrightarrow \mathcal{C}^0(\Omega)$

defined by

(1.13.4) $T(D)U(x) = T(x,U(x),\dots,D^p U(x),\dots), \ U \in \mathcal{C}^m(\Omega), \quad x \in \Omega$

Obviously, the class of all $T(D)$ above contain as a particular case those defined in (1.5.4).

Our aim is to define for $T(D)$ above extensions similar to those in (1.9.23).

Given the spaces of generalized functions

$$E = \mathcal{S}/\mathcal{V} \in VS^m_{\mathcal{F},\Lambda} \quad \text{and} \quad A = \mathcal{A}/\mathcal{I} \in AL_{\mathcal{X},\Lambda}$$

we denote

$$(1.13.5) \qquad E \overset{T(D)}{\leq} A$$

if and only if the following three conditions are satisfied

$$(1.13.6) \qquad T(D)(\mathcal{F} \cap \mathcal{C}^m(\Omega)) \subset \mathcal{X}$$

$$(1.13.7) \qquad T(D)(\mathcal{S} \cap (\mathcal{C}^m(\Omega)^\Lambda) \subset \mathcal{A}$$

and

$$(1.13.8) \qquad \begin{array}{l} \forall \quad t \in \mathcal{S} \cap (\mathcal{C}^m(\Omega))^\Lambda, \quad v \in \mathcal{V} \cap (\mathcal{C}^m(\Omega))^\Lambda : \\ T(D)(t + v) - T(D)t \in \mathcal{I} \end{array}$$

An example for the way this later condition (1.13.8) is satisfied in applications can be seen in Chapter 2.

It is easy to see that in case $E \overset{T(D)}{\leq} A$, one can define the extension

$$(1.13.9) \qquad T(D) : E \longrightarrow A$$

by

$$(1.13.10) \qquad T(D)(s + \mathcal{V}) = T(D)t + \mathcal{I}, \quad s \in \mathcal{S}$$

where

$$(1.13.11) \qquad t \in \mathcal{S} \cap (\mathcal{C}^m(\Omega))^\Lambda, \quad t - s \in \mathcal{V}$$

Once the extension (1.13.9) was defined, one can easily extend the notion of sequential solution defined in Section 10 to the case of $T(D)$ given in (1.13.3).

The connection between the framework above and that in Sections 9, 10 is obvious. Indeed, suppose given the m-th order *polynomial* nonlinear partial differential operator $T(D)$ in (1.5.4) and the extension in (1.9.23)

(1.13.12) $T(D) : E \longrightarrow A$

where $E = \mathcal{S}/\mathcal{V} \in VS^m_{\mathcal{F},\Lambda}$, $A = \mathcal{A}/\mathcal{I} \in AL_{\mathcal{X},\Lambda}$ and $E \overset{m}{\leq} A$.

In that case we obviously have

(1.13.13) $E \overset{T(D)}{\leq} A$

therefore (1.13.12) is an extension in the sense of (1.13.9) as well. Indeed, (1.9.9) and (1.5.4) yield (1.13.6), provided that

(1.13.14) $c_i \in \mathcal{X}, \quad 1 \leq i \leq h$

Further, (1.9.10), (1.5.4) and (1.13.14) yield (1.13.7). Finally, (1.13.8) follows from (1.9.10).

Further details as well as applications can be found in Rosinger [2], where it is also shown the way variable transforms are treated within the general framework presented in Sections 9-12.

§14. SYSTEMS OF NONLINEAR PARTIAL DIFFERENTIAL EQUATIONS

For convenience, we shall only consider systems of *polynomial* nonlinear partial differential equations. The extension of arbitrary continuous nonlinear systems of partial differential equations follows easily by a direct application of the method in Section 13.

Suppose we are given a polynomial nonlinear *system*

(1.14.1) $$\sum_{1 \leq i \leq h_\beta} c_{\beta i}(x) \prod_{i \leq j \leq k_{\beta i}} D^{p_{\beta ij}} U_{\alpha_{\beta ij}}(x) = f_\beta(x), \quad x \in \Omega, \ 1 \leq \beta \leq b$$

where $U = (U_1, \ldots U_a) : \Omega \longrightarrow \mathbb{R}^a$ are unknown functions, while $c_{\beta i}$, $f_\beta \in \mathcal{C}^0(\Omega)$ are given. The *order* of the system in (1.14.1) is by definition

(1.14.2) $$m = \max\{|p_{\beta ij}| \, \big| \, 1 \leq \beta \leq b, \ 1 \leq i \leq h_\beta, \ 1 \leq j \leq k_{\beta i}\}$$

provided that none of the coefficients $c_{\beta i}$ vanishes everywhere on Ω. Let us denote

$$(1.14.3)\qquad T_\beta(D)U(x) = \sum_{1\le i\le h_\beta} c_{\beta i}(x) \prod_{1\le j\le k_{\beta i}} D^{p_{\beta ij}} U_{\alpha_{\beta ij}}(x), \quad x \in \Omega$$

or, in a more compact form

$$(1.14.4)\qquad T(D) = (T_1(D),\dots,T_b(D))$$

which obviously generates a mapping

$$(1.14.5)\qquad T(D) : (\mathcal{C}^m(\Omega))^a \to (\mathcal{C}^0(\Omega))^b$$

It follows that a *classical solution* of (1.14.1) is any family of functions

$$(1.14.6)\qquad \psi = (\psi_1,\dots,\psi_a) \in (\mathcal{C}^m(\Omega))^a$$

for which the mapping (1.14.5) satisfies

$$(1.14.7)\qquad T(D)\psi = f \quad \text{on} \quad \Omega$$

where

$$(1.14.8)\qquad f = (f_1,\dots,f_b)$$

In analogy with (1.10.5), the *sequential solutions* of (1.14.1) will by definition be given by families of functions

$$(1.14.9)\qquad s = (\psi_\lambda \mid \lambda \in \Lambda)$$

with Λ a given *infinite* set, and

$$(1.14.10)\qquad \psi_\lambda = (\psi_{\lambda_1},\dots,\psi_{\lambda a}) \in (\mathcal{C}^m(\Omega))^a, \quad \lambda \in \Lambda$$

In that way

$$(1.14.11)\qquad s \in (\mathcal{C}^m(\Omega))^{\{1,\dots,a\}\times\Lambda}$$

The precise definition can be obtained in analogy to the one in Section 10 as follows.

First, we *rearrange* the functions in s, transforming the family of functions in (1.14.11) into a family of functions

$$(1.14.12)\qquad s = (s_1,\dots,s_a) \in ((\mathcal{C}^m(\Omega))^\Lambda)^a$$

where

(1.14.13) $\quad s_{\alpha\lambda} = \psi_{\lambda\alpha}, \quad 1 \leq \alpha \leq a, \quad \lambda \in \Lambda$

Suppose now given $\mathcal{F}$ and $\mathcal{X}$ respectively a vector subspace and a subalgebra in $\mathcal{M}(\Omega)$, such that (1.9.9) is satisfied, and the coefficients $c_{\beta i}$ as well as the right-hand terms f_β in the system (1.14.1), belong to $\mathcal{X}$. Suppose further given $E = \mathcal{S}/\mathcal{V} \in VS^m_{\mathcal{F},\Lambda}$ and $A = \mathcal{A}/\mathcal{I} \in AL_{\mathcal{X},\Lambda}$ such that $E \overset{m}{\leq} A$. Then, the mapping (1.14.5) can be extended to a mapping

(1.14.14) $\quad T(D) : E^a \longrightarrow A^b$

defined by

(1.14.15) $\quad T(D)(s_1+\mathcal{V},\ldots,s_a+\mathcal{V}) = (T_1(D)t+\mathcal{I},\ldots,T_b(D)t+\mathcal{I}), \quad s_1,\ldots,s_a \in \mathcal{S}$

where

(1.14.16) $\quad t = (t_1,\ldots,t_a)$

(1.14.17) $\quad t_\alpha \in \mathcal{S} \cap (\mathcal{C}^m(\Omega))^\Lambda, \quad s_\alpha - t_\alpha \in \mathcal{V}, \quad 1 \leq \alpha \leq a$

(1.14.18) $\quad T_\beta(D)t = (T_\beta(D)t_1,\ldots,T_\beta(D)t_a), \quad 1 \leq \beta \leq b$

and $T_\beta(D)t_\alpha$ in (1.14.19) is the term by term application of the mapping generated by (1.14.3) to the sequence of functions $t_\alpha \in (\mathcal{C}^m(\Omega))^\Lambda$.

A family of functions

$$s = (s_1,\ldots,s_a) \in \mathcal{S}^a$$

is called a $E \rightarrow A$ *sequential solution* of the system in (1.14.1), if and only if the mapping (1.14.15) maps

$$S = (S_1,\ldots,S_a) \in E^a$$

into

$$f = (f_1,\ldots,f_b) \in A^b$$

where

$$s_\alpha + \mathcal{V} \in E = \mathcal{S}/\mathcal{V}, \quad 1 \leq \alpha \leq a$$

It is easy to see that the above condition is equivalent to

$$
(1.14.19) \qquad \begin{array}{ll} \forall & t = (t_1, \ldots, t_a) \in (\mathcal{S} \cap \mathcal{C}^m(\Omega))^\Lambda)^a : \\ \forall & 1 \le \alpha \le a, \quad 1 \le \beta \le b : \\ & s_\alpha - t_\alpha \in \mathcal{V}, \quad T_\beta(D)t - u(f_\beta) \in \mathcal{I} \end{array}
$$

which is the same as the following condition

$$
(1.14.20) \qquad \begin{array}{l} \forall \quad t \in (\mathcal{S} \cap (\mathcal{C}^m(\Omega))^\Lambda)^a : \\ \left[\begin{array}{ll} \forall & 1 \le \alpha \le a : \\ & s_\alpha - t_\alpha \in \mathcal{V} \end{array} \right] \Rightarrow \left[\begin{array}{ll} \forall & 1 \le \beta \le b : \\ & T_\beta(D)t - u(f_\beta) \in \mathcal{I} \end{array} \right] \end{array}
$$

With the above definition of a sequential solution for a system of polynomial nonlinear partial differential equations, it is obvious that the study of such solutions can actually be reduced to the *simultaneous* study of a *finite number* - in this case given by a - of sequential solutions in the sense of Section 10, which satisfy a *finite number* - in this case given by b - of polynomial nonlinear partial differential equations.

APPENDIX 1

ON HEAVISIDE FUNCTIONS AND THEIR DERIVATIVES

Suppose X is a set with at least two elements, while A is an associative and commutative algebra on $\mathbb{R}$, with unit 1, and with $1 \neq 0 \in A$. Let

$$\mathcal{F} \subset A^X \tag{1.A1.1}$$

be a subalgebra of functions $f : X \longrightarrow A$, such that the constant functions $0, 1 : X \longrightarrow A$ belong to $\mathcal{F}$. A function $H : X \longrightarrow A$ will be called of *Heaviside type*, if an only if

$$H\big|_{X_0} = 0, \quad H\big|_{X_1} = 1 \tag{1.A1.2}$$

where (X_0, X_1) is a partition of X in two nonvoid sets. Obviously, the constant functions 0, 1 and the Heaviside type functions are solutions of the infinite system of equations in $f \in A^X$

$$f^m = f, \quad \forall \; m \in \mathbb{N}, \; m \geq 1 \tag{1.A1.3}$$

We shall now require that

$$\begin{array}{l} \exists \; H \in \mathcal{F} : \\ \quad H \text{ of Heaviside type} \end{array} \tag{1.A1.4}$$

Finally, let

$$D : \mathcal{F} \longrightarrow \mathcal{F} \tag{1.A1.5}$$

be a *derivative*, that is, a linear maping which satisfies the Leibnitz rule of product derivative

$$D(f \cdot g) = (Df) \cdot g + f \cdot (Dg), \quad f, g \in \mathcal{F} \tag{1.A1.6}$$

Then it follows that we always have

$$D0 = D1 = DH = 0 \in \mathcal{F} \tag{1.A1.7}$$

Therefore, the derivative D *fails* to take notice of the nonconstancy and 'discontinuity' of H. Indeed, the fact that $D0 = 0 \in \mathcal{F}$ follows easily from the linearity of $D : \mathcal{F} \longrightarrow \mathcal{F}$. Further, $D1 = 0 \in \mathcal{F}$ follows from (1.A1.6) by taking $f = g = 1 \in \mathcal{F}$ in that relation.

The relations $D0 = D1 = 0 \in \mathcal{F}$ are to be expected, since $0,1 : X \longrightarrow A$ are constant functions. On the other hand, the relation

(1.A1.8) $$DH = 0 \in \mathcal{F}$$

results in a similar way to that used in proving (1.1.10).

The conclusion is that a *differential framework*

(1.A1.9) $$(X,A,\mathcal{F},D)$$

such as in (1.A1.1)-(1.A1.6) is *too coarse* in order to be able to take note of the nonconstancy or 'discontinuity' of Heaviside type functions, and therefore it *cannot* discriminate between the constant and nonconstant solutions of (1.A1.3).

It is interesting to note that the *algebraic roots* of the failure in (1.A1.7) go still deeper. Indeed, let us replace the differential framework in (1.A1.9) with the following more general one

(1.A1.10) $$(\mathcal{F},D)$$

where $\mathcal{F}$ is an associative and commutative algebra on $\mathbb{R}$, with unit 1, and with $1 \neq 0 \in \mathcal{F}$. Thus $\mathcal{F}$ need not be an algebra of functions as in (1.A1.1). Further, we assume that $D : \mathcal{F} \longrightarrow \mathcal{F}$ is linear and satisfies the Leibnitz rule (1.A1.6).

Now, in the proof of (1.1.10), we can note that one only needs the relation

(1.A1.11) $$H^p = H^q = H$$

for certain $p,q \in \mathbb{N}$, $2 \leq p < q$.

Therefore, we call an element $H \in \mathcal{F}$ to be of *extended Heaviside type*, if and only if it satisfies (1.A1.11). It follows that within the framework in (1.A1.10), the relation

(1.A1.12) $$D0 = D1 = DH = 0 \in \mathcal{F}$$

will hold for every extended Heaviside type element $H \in \mathcal{F}$.

Finally, we indicate the nature of the property (1.A1.11) which defines the extended Heaviside type elements in $\mathcal{F}$.

If for given $H \in \mathcal{F}$, we have

(1.A1.13) $$H^2 = H$$

then obviously

(1.A1.14) $$H^m = H, \quad m \in \mathbb{N}, \quad m \geq 1$$

hence (1.A1.11) is satisfied. Further, if for given $H \in \mathcal{F}$ the relation (1.A1.13) does not hold, but we have

(1.A1.15) $$H^p = H$$

for a certain minimal $p \in \mathbb{N}$, $p \geq 3$, then obviously

$$H^{2p-1} = H^{p-1} \cdot H^p = H^{p-1} \cdot H = H^p = H$$

$$H^{3p-2} = H^{p-1} \cdot H^{2p-1} = H^{p-1} \cdot H = H^p = H$$

and in general

(1.A1.16) $$H^{m(p-1)+1} = H, \quad m \in \mathbb{N}, \quad m \geq 1$$

thus (1.A1.12) is again satisfied.

It follows that (1.A1.13) or (1.A1.15) are the only cases when (1.A1.12) can hold.

APPENDIX 2

THE CAUCHY-BOLZANO QUOTIENT ALGEBRA CONSTRUCTION OF THE REAL NUMBERS

A classical example of *quotient algebra* construction is given by the well known Cauchy-Bolzano method, van Rootselaar, for constructing the set of real numbers $\mathbb{R}$ from the set of rational numbers $\mathbb{Q}$.

Let us denote by

$$\mathcal{A}$$

the subalgebra in $\mathbb{Q}^{\mathbb{N}}$ of all the Cauchy sequences $r = (r_0, r_1, \ldots, r_\nu, \ldots)$ of rational numbers and let us denote by

$$\mathcal{I}$$

the ideal in $\mathcal{A}$ of all the sequences $z = (z_0, z_1, \ldots, z_\nu, \ldots)$ of rational numbers which converge to zero.

We shall denote by

$$\mathcal{U}_{\mathbb{Q}}$$

the subalgebra in $\mathcal{A}$ of all the constant sequences. Finally, we denote by $\mathcal{Q}$ the null ideal in $\mathbb{Q}^{\mathbb{N}}$.

Then the following inclusion diagram is valid

$$(1.A2.1) \qquad \begin{array}{ccccc} \mathcal{I} & \longrightarrow & \mathcal{A} & \longrightarrow & \mathbb{Q}^{\mathbb{N}} \\ \uparrow & & \uparrow & & \\ \mathcal{Q} & \longrightarrow & \mathcal{U}_{\mathbb{Q}} & & \end{array}$$

and it satisfies the condition

$$(1.A2.2) \qquad \mathcal{I} \cap \mathcal{U}_{\mathbb{Q}} = \mathcal{Q}$$

Moreover, according to Cauchy-Bolzano, we have that

(1.A2.3) $\quad \mathbb{R}$ and $A = \mathcal{A}/\mathcal{I}$ are isomorphic fields.

As seen in Appendix 4, the condition (1.A2.2) above means that $\mathcal{I}$ is a *neutrix* in $\mathbb{Q}^{\mathbb{N}}$ and the corresponding *neutrix limit* is identical with the usual limit for rational numbers, i.e., the relation holds

(1.A2.4) $$\mathcal{I}\text{ - }\lim_{\nu\to\infty} r_\nu = \lim_{\nu\to\infty} r_\nu$$

whenever $r = (r_0, r_1, \ldots, r_\nu, \ldots) \in \mathbb{Q}^{\mathbb{N}}$ and one of the limits in (1.A2.4) exists.

The ideal $\mathcal{I}$ has several important properties presented now. First we notice that

(1.A2.5) $\mathcal{I}$ is a *maximal* ideal in $\mathcal{A}$

since $\mathrm{A} = \mathcal{A}/\mathcal{I}$ is a field. Furthermore, $\mathcal{I}$ is *subsequence invariant*, that is, a subsequence of a sequence in $\mathcal{I}$ will also belong to $\mathcal{I}$

Let us now denote by

$$\mathcal{B}$$

the subalgebra in $\mathbb{Q}^{\mathbb{N}}$ of all the bounded sequences of rational numbers. Then obviously

(1.A2.6) $\mathcal{A}$ is a subalgebra in $\mathcal{B}$

The special relation between $\mathcal{I}$ and $\mathcal{B}$ is presented in the next two propositions.

Proposition 5

$\mathcal{I}$ is a *maximal subsequence invariant* ideal in $\mathcal{B}$.

Proof

Assume that it is false and $\mathcal{J}$ is a subsequence invariant ideal in $\mathcal{B}$, such that

(1.A2.7) $$\mathcal{I} \subsetneqq \mathcal{J} \subsetneqq \mathcal{B}$$

Let us take then

(1.A2.8) $$z = (z_0, z_1, \ldots, z_\nu, \ldots) \in \mathcal{J}\backslash\mathcal{I}$$

Since $\mathcal{J} \subset \mathcal{B}$, the relation (1.A2.8) gives a point $\xi \in \mathbb{R}\backslash\{0\}$ and a subsequence $z' = (z'_0, z'_1, \ldots, z'_\nu, \ldots)$, such that

(1.A2.9) $$\lim_{\nu\to\infty} z'_\nu = \xi$$

But

(1.A2.10) $$z' = (z'_0, z'_1, \ldots, z'_\nu, \ldots) \in \mathcal{J}$$

since $\mathcal{J}$ is subsequence invariant. Moreover, in view of (1.A2.9) we can obviously assume that

$$|z'| \geq |\xi|/2 > 0, \quad \nu \in \mathbb{N}$$

Therefore, defining $z'' = (z''_0, z''_1, \ldots, z''_\nu, \ldots) \in \mathbb{Q}^{\mathbb{N}}$ by

(1.A2.11) $\qquad z''_\nu = 1/z'_\nu, \quad v \in \mathbb{N}$

we obtain

(1.A2.12) $\qquad z''_\nu = (z''_0, z''_1, \ldots, z''_\nu, \ldots) \in \mathcal{B}$

Now, the relations (1.A2.10)-(1.A2.12) will yield

$$1 = z'' \cdot z'' \in \mathcal{J} \cdot \mathcal{B} \subset \mathcal{J}$$

hence

(1.A2.13) $\qquad \mathcal{J} = \mathcal{B}$

as $\mathcal{J}$ is an ideal $\mathcal{B}$.

Since (1.A2.7) and (1.A2.13) contradict each other, the proof is completed.

□

Proposition 6

$\mathcal{B}$ is *maximal* among all subalgebras in $\mathbb{Q}^{\mathbb{N}}$ in which $\mathcal{I}$ is an ideal.

Proof

Assume that it is false and $\mathcal{C}$ is a subalgebra in $\mathbb{Q}^{\mathbb{N}}$ such that

(1.A2.14) $\qquad \mathcal{B} \subsetneqq \mathcal{C}$

(1.A2.15) $\qquad \mathcal{I}$ is an ideal in $\mathcal{C}$

Let us take then

$$z = (z_0, z_1, \ldots, z_\nu, \ldots) \in \mathcal{C} \backslash \mathcal{B}$$

It follows that there exists a subsequence $z' = (z_{\nu_0}, z_{\nu_1}, \ldots, z_{\nu_\mu}, \ldots)$ in $z = (z_0, z_1, \ldots, z_\nu, \ldots)$, such that

(1.A2.16) $\qquad \lim_{\mu \to \infty} |z_{\nu_\mu}| = \infty$

Obviously, we can assume that

$$(1.A2.17) \qquad \nu_0 < \nu_1 < \ldots < \nu_\mu < \ldots$$

and

$$(1.A2.18) \qquad z_{\nu_\mu} \neq 0, \quad \mu \in \mathbb{N}$$

Then, in view of (1.A2.17) and (1.A2.18), we can define $r = (r_0, r_1, \ldots, r_\nu, \ldots) \in \mathbb{Q}^{\mathbb{N}}$ by

$$(1.A2.19) \qquad r_\nu = 1/z_{\nu_\mu} \quad \text{if} \quad \nu_\mu \leq \nu < \nu_{\mu+1}$$

But, the relations (1.A2.19) and (1.A2.16) yield

$$(1.A2.20) \qquad r = (r_0, r_1, \ldots, r_\nu, \ldots) \in \mathcal{I}$$

therefore, in view of (1.A2.15) we obtain

$$(1.A2.21) \qquad r \cdot z \in \mathcal{I} \cdot \mathcal{C} \subset \mathcal{I}$$

However, in view of (1.A2.19) we obtain

$$(1.A2.22) \qquad r_{\nu_\mu} \cdot z_{\nu_\mu} = 1, \quad \mu \in \mathbb{N}$$

and the relations (1.A2.21), (1.A2.22) obviously contradict each other. Therefore (1.A2.14) cannot hold. □

Corollary 1

$\mathcal{B}$ is the *largest* subalgebra in $\mathbb{Q}^{\mathbb{N}}$ in which $\mathcal{I}$ is an ideal.

Proof

With a slight and obvious extension of the notations in Appendix 7 in the sequel, it follows from Proposition 6 above that

$$\mathcal{B} = \mathcal{A}_{\mathbb{Q},\mathbb{N}}(\mathcal{I})$$ □

APPENDIX 3

HOW 'WILD' SHOULD WE ALLOW THE WORST GENERALIZED FUNCTIONS TO BE?

The quotient algebra structures

$$(1.A3.1) \qquad A = \mathcal{A}/\mathcal{I}$$

which define various algebras of generalized functions, see (1.10.1) and (1.5.24)-(1.5.26), may at first seem to be too general from the point of view of convenient use or interpretation.

In order to better grasp what may in the above situation be indeed the 'minimal necessary level of generality', let us first recall how 'wild' some of the worst nonstandard reals must be in order to allow us the construction of fields $^*\mathbb{R}$ of nonstandard reals in a way that makes them convenient extensions of the classical field $\mathbb{R}$ of real numbers.

One of the ways to obtain nonstandard fields $^*\mathbb{R}$ is the following ultra-power type construction.

First, we *exten[* $\mathbb{R}$ into the *vastly larger* $\mathbb{R}^{\mathbb{N}}$, by the embedding

$$(1.A3.2) \qquad \begin{array}{ccc} \mathbb{R} & \xrightarrow{\ u\ } & \mathbb{R}^{\mathbb{N}} \\ x & \longmapsto & u(x) = (x,x,x,\ldots,x,\ldots) \end{array}$$

and note that $\mathbb{R}^{\mathbb{N}}$ is in a natural way a commutative algebra over $\mathbb{R}$, with respect to the term wise operations on sequences. In this way u in (1.A3.2) is an injective algebra homomorphism and $u(0) = (0,0,0,\ldots,0,\ldots)$ and $u(1) = (1,1,1,\ldots,1,\ldots)$ are the zero and unit elements respectively in $\mathbb{R}^{\mathbb{N}}$.

Unfortunately, with these natural term wise operations, $\mathbb{R}^{\mathbb{N}}$ *fails* to be a field. Indeed, the multiplication in $\mathbb{R}^{\mathbb{N}}$ has many *zero divisors*, for instance

$$(1,0,0,\ldots,0,\ldots)\cdot(0,1,0,\ldots,0,\ldots) = (0,0,0\ldots,0,\ldots)$$

It follows in particular that the algebra $\mathbb{R}^{\mathbb{N}}$ *cannot* be embedded into a field.

Therefore, as a second step, we shall *reduce* the algebra $\mathbb{R}^{\mathbb{N}}$ by factoring it with any suitable ideal $\mathcal{N} \subset \mathbb{R}^{\mathbb{N}}$. The point in this procedure is obvious.

Indeed, as is well known

$$\mathbb{R}^{\mathbb{N}}/\mathcal{N}$$

will be a *field*, if and only if $\mathcal{N}$ is a *maximal ideal* in $\mathbb{R}^{\mathbb{N}}$. Therefore, our problem is to find maximal ideals $\mathcal{N}$ in $\mathbb{R}^{\mathbb{N}}$, such that we end up with *strict inclusions*

(1.A3.3) $$\mathbb{R} \subsetneqq \mathbb{R}^{\mathbb{N}}/\mathcal{N}$$

given by the embedding

(1.A3.4)
$$\begin{array}{ccccc} \mathbb{R} & \xrightarrow{\quad u \quad} & \mathbb{R}^{\mathbb{N}} & \xrightarrow{\text{canonical}} & \mathbb{R}^{\mathbb{N}}/\mathcal{N} \\ x & \longmapsto & u(x) & \longmapsto & u(x) + \mathcal{N} \end{array}$$

In order to help the problem of finding suitable maximal ideals $\mathcal{N}$, we note that all *proper* ideals in $\mathbb{R}^{\mathbb{N}}$ are generated by *filters* on $\mathbb{N}$ in the following way. Given a filter $\mathcal{F}$ on $\mathbb{N}$, let us define

(1.A3.5) $$\mathcal{N}_{\mathcal{F}} = \{a = (a_\nu | \nu \in \mathbb{N}) \in \mathbb{R}^{\mathbb{N}} | \ \zeta(a) \in \mathcal{F}\}$$

where for $a = (a_\nu | \nu \in \mathbb{N}) \in \mathbb{R}^{\mathbb{N}}$ we denoted

(1.A3.6) $$\zeta(a) = \{\nu \in \mathbb{N} | \ a_\nu = 0\}$$

It follows easily that $\mathcal{N}_{\mathcal{F}}$ is a proper ideal in $\mathbb{R}^{\mathbb{N}}$.

Conversely, given a proper ideal $\mathcal{N}$ in $\mathbb{R}^{\mathbb{N}}$, let us define

(1.A3.7) $$\mathcal{F}_{\mathcal{N}} = \left\{F \subset \mathbb{N} \ \middle| \ \begin{array}{c} \exists \ a \in \mathcal{N} : \\ \zeta(a) \subset F \end{array} \right\}$$

It follows easily that $\mathcal{F}_{\mathcal{N}}$ is a filter on $\mathcal{N}$. Furthermore, we have the relation

(1.A3.8) $$\mathcal{N} = \mathcal{N}_{\mathcal{F}_{\mathcal{N}}}$$

Indeed, the inclusion '$\subset$' is immediate. For the converse inclusion '$\supset$', let us take

$$a = (a_\nu | \nu \in \mathbb{N}) \in \mathcal{N}_{\mathcal{F}_{\mathcal{N}}}$$

Then (1.A3.5) yields $\zeta(a) \in \mathcal{F}_{\mathcal{N}}$, hence in view of (1.A3.7) we obtain

$$(1.A3.9) \qquad \zeta(b) \subset \zeta(a)$$

for a certain $b = (b_\nu | \nu \in \mathbb{N}) \in \mathcal{N}$. Let us define now $c = (c_\nu | \nu \in \mathbb{N}) \in \mathbb{R}^{\mathbb{N}}$ by

$$c_\nu = \left| \begin{array}{ll} a_\nu / b_\nu & \text{if} \quad \nu \in \mathbb{N} \setminus \zeta(b) \\ 0 & \text{if} \quad \nu \in \zeta(b) \end{array} \right.$$

which is a correct definition, owing to (1.A3.6). Then in view of (1.A3.9), we have

$$a = b \cdot c$$

thus

$$a = b \cdot c \in \mathcal{N} \cdot \mathbb{R}^{\mathbb{N}} \subset \mathcal{N}$$

and the proof of (1.A3.8) is completed.

The crucial property for us is the following. Given a filter $\mathcal{F}$ on $\mathbb{N}$, then

$$(1.A3.10) \qquad \mathcal{N}_{\mathcal{F}} \text{ maximal ideal } \Longleftrightarrow \mathcal{F} \text{ ultrafilter}$$

as seen by a direct application of (1.A3.5) and (1.A3.7).

The consequence of (1.A3.8), (1.A3.10) is that all maximal ideals in $\mathbb{R}^{\mathbb{N}}$ are of the form

$$(1.A3.11) \qquad \mathcal{N}_{\mathcal{F}}, \text{ with } \mathcal{F} \text{ ultrafilter on } \mathbb{N}$$

Now, on $\mathbb{N}$, and in general on any infinite set, there exist only two kind of ultrafilters. The *fixed* ultrafilters on $\mathbb{N}$ are of the form

$$(1.A3.12) \qquad \mathcal{F}_\alpha = \{F \subset \mathbb{N} | \alpha \in F\}$$

where $\alpha \in \mathbb{N}$ is arbitrary but fixed. The *free* ultrafilters on $\mathbb{N}$ are all the other, non fixed ultrafilters. They can be characterized in the following way. Let us denote

$$\mathcal{FR} = \{F \subset \mathbb{N} | \ \mathbb{N} \setminus F \text{ is a finite subset}\}$$

then $\mathcal{FR}$ is a filter on $\mathbb{N}$, and it is called the Fréchet filter. Now it is easy to see that given an ultrafilter $\mathcal{F}$ on $\mathbb{N}$, we have

$$\text{(1.A3.13)} \qquad \mathcal{F} \text{ free} \iff \mathcal{FR} \subset \mathcal{F} \iff \begin{bmatrix} \forall \ F \in \mathcal{F} : \\ F \text{ infinite} \end{bmatrix}$$

Further, we note that, given any filter $\mathcal{F}$ on $\mathbb{N}$, the mapping, see (1.A3.4)

$$\text{(1.A3.14)} \qquad \begin{array}{ccc} \mathbb{R} & \longrightarrow & \mathbb{R}^{\mathbb{N}}/\mathcal{N}_{\mathcal{F}} \\ x & \longmapsto & u(x) + \mathcal{N}_{\mathcal{F}} \end{array}$$

is indeed an *embedding* of algebras, since for $x \in \mathbb{R}$ we have

$$\text{(1.A3.15)} \qquad u(x) \in \mathcal{N}_{\mathcal{F}} \iff x = 0$$

as follows easily from (1.A3.5) and the fact that $\phi_{\mathbb{N}} \notin \mathcal{F}$.

At this stage a rather surprising thing happens. Namely, for all *fixed* ultrafilters $\mathcal{F}$ on $\mathbb{N}$, the embedding (1.A3.14) gives an *isommorphism*

$$\text{(1.A3.16)} \qquad \mathbb{R} = \mathbb{R}^{\mathbb{N}}/\mathcal{N}_{\mathcal{F}}$$

Indeed, assume that $\mathcal{F} = \mathcal{F}_\alpha$, for a certain fixed $\alpha \in \mathbb{N}$, see (1.A3.12). Then the mapping (1.A3.14) is *onto*. For that, let us take any $a = (a_\nu | \nu \in \mathbb{N}) \in \mathbb{R}^{\mathbb{N}}$, and then denote $x = a_\alpha$. It is immediate then that

$$a - u(x) \in \mathcal{N}_{\mathcal{F}}$$

hence the mapping (1.A3.14) maps $x \in \mathbb{R}$ onto $a + \mathcal{N}_{\mathcal{F}} \in \mathbb{R}^{\mathbb{N}}/\mathcal{N}_{\mathcal{F}}$.

It follows that in order to obtain *proper* extensions (1.A3.3), we have to use *free* ultrafilters $\mathcal{F}$ on $\mathbb{N}$. Fortunately that always works. Indeed, let $\mathcal{F}$ be any free ultrafilter on $\mathbb{N}$ and let us take

$$\text{(1.A3.17)} \qquad a = (0,1,2,3,\ldots,\nu,\ldots) \in \mathbb{R}^{\mathbb{N}}$$

then in view of (1.A3.13), for every $x \in \mathbb{R}$, we obviously have

$$a - u(x) \notin \mathcal{N}_{\mathcal{F}}$$

therefore the mapping (1.A3.14) is not onto, that is, we have a proper embedding

$$\text{(1.A3.18)} \qquad \mathbb{R} \subsetneqq \mathbb{R}^{\mathbb{N}}/\mathcal{N}_{\mathcal{F}}$$

The conclusion is that we encounter the following *dichotomy* :

- either $\mathcal{F}$ is a fixed ultrafilter, and then
$$\mathbb{R} = \mathbb{R}^{\mathbb{N}}/\mathcal{N}_{\mathcal{F}}$$
- or $\mathcal{F}$ is a free ultrafilter, and then
$$\mathbb{R} \subsetneqq \mathbb{R}^{\mathbb{N}}/\mathcal{N}_{\mathcal{F}}$$

In this way we can take

$$(1.A3.19) \qquad {}^{*}\mathbb{R} = \mathbb{R}^{\mathbb{N}}/\mathcal{N}_{\mathcal{F}}$$

for any free ultrafilter $\mathcal{F}$ on $\mathbb{N}$.

At this point we can see how 'wild' some of the worst nonstandard numbers must inevitably be. For that, let us take any

$$(1.A3.20) \qquad a = (a_\nu \mid \nu \in \mathbb{N}) \in \mathbb{R}^{\mathbb{N}}$$

then the equivalence class ${}^{*}a \in \mathbb{R}^{\mathbb{N}}/\mathcal{N}_{\mathcal{F}}$ defined by a is given by the set of all $b = (b_\nu \mid \nu \in \mathbb{N}) \in \mathbb{R}^{\mathbb{N}}$, such that

$$(1.A3.21) \qquad \zeta(b - a) = \{\nu \in \mathbb{N} \mid b_\nu = a_\nu\} \in \mathcal{F}$$

Now we note that in view of (1.A3.13), the sets of indices $\zeta(b - a)$ in (1.A3.21) are *infinite* subsets in $\mathbb{N}$. Therefore, every sequence $b = (b_\nu \mid \nu \in \mathbb{N})$ in the class ${}^{*}a$ has infinitely many identical terms with the sequence $a = (a_\nu \mid \nu \in \mathbb{N})$. In this way, the *completely arbitrary* behaviour of the terms in the sequence a does *not* disappear by the factorization $\mathbb{R}^{\mathbb{N}}/\mathcal{N}_{\mathcal{F}}$ since it is reproduced infinitely many times by the terms of every sequence $b = (b_\nu \mid \nu \in \mathbb{N})$ in the class ${}^{*}a$.

Coming back to generalized functions we note that even in the case of the L. Schwartz distributions on $\mathbb{R}$, we can have such 'wild' elements as for instance

$$(1.A3.22) \qquad \sum_{\nu \in \mathbb{Z}} a_\nu D^{p_\nu} \delta(x-\nu) \in \mathcal{D}'(\mathbb{R})$$

where $a_\nu \in \mathbb{R}$ and $p_\nu \in \mathbb{N}$ can be chosen arbitrarily, δ being the Dirac delta distribution.

In the case of the algebras of generalized functions in (1.5.24)-(1.5.26), the construction proceeds in *two steps* similar to those in (1.A3.2) and (1.A3.3). Namely, first we extend the algebra of smooth functions $\mathcal{X} \subset \mathcal{C}^m(\Omega)$ into the much larger algebra $\mathcal{X}^\Lambda$, with a given, suitably chosen infinite index set Λ. Then, for appropriate subalgebras $\mathcal{A} \subset \mathcal{X}^\Lambda$ and ideals $\mathcal{I}$ in $\mathcal{A}$, we construct the quotient algebra $A = \mathcal{A}/\mathcal{I}$, with the aim of obtaining convenient extensions

(1.A3.23) $\mathcal{C}^m(\Omega) \subset \mathcal{D}'(\Omega) \subset A = \mathcal{A}/\mathcal{I}$

Just as in the case of dealing with nonstandard numbers in $^*\mathbb{R}$ in (1.A3.19), or distributions in $\mathcal{D}'(\Omega)$, in most of the situations when we use generalized functions which are elements of algebras $A = \mathcal{A}/\mathcal{I}$ in (1.A3.23), we shall not have to deal with their most 'wild' instances. However, in order to be able to construct a sufficiently simple and clear general theory, we have to allow the possibility that such 'wild' generalized functions are present.

Such a situation repeats itself often in mathematics. For instance, we define the whole set $\mathbb{Z}$, although we use only moderately large integers. Then we extend $\mathbb{Z}$ into $\mathbb{Q}$, although we use only fractions with moderately large numerators and denominators. We further extend $\mathbb{Q}$ into $\mathbb{R}$, although we shall actually use only a few irrational numbers. But perhaps by far the most significant example is given by Cantor's set theory, where unfathomably large sets are allowed, although those in common mathematical use happen to be of a rather limited size.

One could therefore talk about a kind of mathematical principle according to which definitions of mathematical entities should be allowed to go much farther than the actual use of the respective entities. This principle has its parallel in that mentioned in Appendix 5, according to which representation should be allowed to go much farther than interpretation, see also Section 7.

APPENDIX 4

NEUTRIX CALCULUS AND NEGLIGIBLE SEQUENCES OF FUNCTIONS

Connected with the study of various asymptotic expansions, a 'neutrix calculus' was developed in Van der Corput, aimed to deal in a general and unified way with 'negligible' quantities. The basic idea of his method - presented in the sequel - proved to be of a wider interest, being for instance useful in the theory of distributions, Fisher [1,2].

In the conditions (1.6.6) and (1.6.11) in Section 6, defining the *quotient spaces* respectively *algebras* of generalized functions, we have also made use of the notion of 'negligible' sequences of functions. An other example can be seen in Appendix 2, where $\mathbb{R}$ is defined as a *quotient algebra* of classes of sequences of rational numbers, according to the Cauchy-Bolzano method. In that case the 'negligible' sequences of rational numbers will coincide with the sequences of rational numbers convergent to zero.

And now, the definition of a *neutrix*.

Suppose given an arbitrary non-void set X and an Abelian group G. The object of our study will be the functions

$$f : X \longrightarrow G$$

in other words, the elements of the Cartesian product

$$G^X \tag{1.A4.1}$$

which is in a natural way also an Abelian group.

The problem is to define in a suitable and general manner the notion of 'negligible' function $f \in G^X$.

A given subgroup

$$\mathcal{N} \subset G^X \tag{1.A4.2}$$

will be called a *neutrix*, if and only if

$$\begin{array}{l} \forall \quad f \in \mathcal{N}, \quad \gamma \in G : \\ \quad \left[\begin{array}{ll} \forall & x \in X : \\ & f(x) = \gamma \end{array} \right] \Rightarrow \gamma = 0 \end{array} \tag{1.A4.3}$$

in which case the functions $f \in \mathcal{N}$ will be called $\mathcal{N}$-*negligible*. An interest in the above notion comes from the fact that in case X has a directed partial order $\leq$, one can define a *neutrix limit* for functions in G^X as follows. Suppose given a neutrix $\mathcal{N} \subset G^X$, a function $f \in G^X$ and $\gamma \in G$. Then we define

(1.A4.4) $$\mathcal{N} - \lim_{x \to \infty} f(x) = \gamma$$

if and only if the function $g \in G^X$ defined by

$$g(x) = f(x) - \gamma, \quad x \in X$$

is $\mathcal{N}$-negligible.

In view of (1.A4.3) it is easy to see that the limit (1.A4.4) is unique, whenever it exists.

The condition (1.A4.3) defining a neutrix can be given the following *algebraic* characterization. Let us define the group monomorphism:

$$u : G \longrightarrow G^X$$

by

$$(u(\gamma))(x) = \gamma, \quad \gamma \in G, \quad x \in X$$

and denote

$$\mathcal{U}_G = u(G)$$

Then $\mathcal{U}_G$ is the subgroup of *constant* functions in G^X. Let us denote by

$$\mathcal{Q}$$

the null subgroup in G^X. The following characterization follows easily.

<u>Proposition 7</u>

A subgroup $\mathcal{N} \subset G^X$ is a neutrix if and only if the inclusion diagram

(1.A4.5) $$\begin{array}{ccc} \mathcal{N} & \longrightarrow & G^X \\ \uparrow & & \uparrow \\ \mathcal{Q} & \longrightarrow & \mathcal{U}_G \end{array}$$

satisfies the condition

(1.A4.6) $$\mathcal{N} \cap \mathcal{U}_G = \mathcal{Q}$$

or equivalently, the mapping $\overline{u}$ defined by

$$(1.A4.7) \qquad \begin{array}{ccccc} & & \overline{u} & & \\ G & \xrightarrow[u]{} & G^X & \xrightarrow[\theta]{} & G^X/\mathcal{N} \end{array} \quad (\overline{u} : G \to G^X/\mathcal{N})$$

is a group *monomorphism*, where θ is the canonical quotient epimorphism. □

It is worth noticing that the condition (1.A4.7) is the *opposite* of the condition that the chain of group homomorphisms

$$(1.A4.8) \qquad 0 \longrightarrow G \xrightarrow{u} G^X \xrightarrow{\theta} G^X/\mathcal{N} \longrightarrow 0$$

is exact. Indeed, (1.A4.7) is equivalent to (1.A4.6), while (1.A4.8) is equivalent to

$$(1.A4.9) \qquad \mathcal{U}_G = \mathcal{N}$$

In view of (1.A4.7) and (1.A4.4), in case X is a directed set, we can interpret $G^X/\mathcal{N}$ as a kind of *sequential completion* of G, obtained by using sequences in G with indices in the set X.

If we take now

$$X = \mathbb{N} \quad \text{and} \quad G = \mathcal{M}(\Omega)$$

then the inclusion diagrams (1.6.5) and (1.6.10) are particular cases of (1.A4.5), while the conditions (1.6.6) and (1.6.11) are identical with (1.A4.6) above.

Moreover the *sequential solutions* of polynomial nonlinear partial differential equations defined in Section 10, can be seen as *neutrix limits* in the sense of (1.A4.4) above. Indeed, suppose given the m-th order polynomial nonlinear partial differential equations, see (1.5.1)

$$(1.A4.10) \qquad T(D)u(x) = f(x), \quad x \in \Omega$$

with continuous coefficients and right hand term, and let us consider

$$T(D) : E \longrightarrow A$$

where

$$E = \mathcal{S}/\mathcal{V} \in VS^m_{C^m(\Omega),\mathbb{N}}, \quad A = \mathcal{A}/\mathcal{I} \in AL_{\mathcal{C}^0(\Omega),\mathbb{N}} \quad \text{and} \quad E \overset{m}{\leq} A.$$

Obviously, we can also consider the mapping

$$(1.A4.11) \qquad T(D) : \mathcal{C}^m(\Omega) \longrightarrow \mathcal{C}^0(\Omega)$$

in which case for each given sequence of functions $s = (\psi_\nu \mid \nu \in \mathbb{N}) \in (\mathcal{C}^m(\Omega))^{\mathbb{N}}$ it makes sense to ask whether or not the *neutrix limit* exists

$$(1.A4.12) \qquad \mathcal{I} - \lim_{\nu\to\infty} T(D)\psi_\nu = ?$$

And in case the neutrix limit in (1.A4.12) exists, it will obviously be a function in $\mathcal{C}^0(\Omega)$.

Proposition 8

Suppose given a sequence of functions $s = (\psi_\nu \mid \nu \in \mathbb{N}) \in \mathcal{S}$. Then s is a $E \rightarrow A$ sequential solution of (1.A4.10), if and only if

$$(1.A4.13) \qquad \mathcal{I} - \lim_{\nu\to\infty} T(D)\psi_\nu = f$$

Proof

By definition, the relation (1.A4.13) is equivalent to

$$T(D)s - u(f) \in \mathcal{I}$$

which in view of (1.10.7)-(1.10.11) completes the proof. □

Remark 7

In Van der Corput's 'neutrix calculus' the power of the *neutrix condition* (1.A4.6) is rather explanatory than operative, that is, it can unify known results rather than deliver new ones. The reason for that is the insufficient *algebraic* - that is, only group - structure on G, G^X and $\mathcal{N}$.

However, as soon as G is an *algebra* and $\mathcal{N}$ is an *ideal* in a subalgebra of G^X, the mentioned neutrix condition acquires a surprizing power, as seen in Rosinger [1,2,3], as well as Chapters 2, 3, 6 and 7 in the sequel.

APPENDIX 5

REVIEW OF CERTAIN IMPORTANT REPRESENTATIONS AND INTERPRETATIONS CONNECTED WITH PARTIAL DIFFERENTIAL EQUATIONS.

Since the emergence of the established theory of generalized functions, starting with the Schwartz linear theory of distributions, the study of *generalized solutions* of linear - and later also of certain nonlinear - partial differential equations has become a subdomain of *linear metric topology* on functions spaces, in particular, on Hilbert, Sobolev, Banach or in general, locally convex topological spaces. Here, for the sake of terminological simplicity, we call linear metric topology a topology on a vector space, compatible with it, and which can be defined by a family of semimetrics, that is, a uniform topology, and which may in particular be metrizable, if it can be defined by one single metric.

The apparent advantage of such a course of events seemed obvious, since the respective linear metric topology methods involved in proving the existence of generalized solutions could be expected to lead to *approximation* methods for these solutions, useful in their effective *numerical* computation. Such expectations have more or less been fulfilled: more in the case of fixed point and monotonicity methods for finding generalized solutions, and hardly at all in the case of method proving the existence of generalized solutions based on compactness arguments in function spaces.

Yet, the idea of basing the generalized solution methods on linear metric topology has remained most powerful, to the extent of being nearly exclusive. This situation has most probably been originated and supported by the historical fact that at the dawn of modern mathematics about a century ago, the power of linear metric topological methods on function spaces became quite impressive. Indeed, even in finding classical solutions for nonlinear ordinary differential equations, one of the best methods so far has remained that of Picard's successive approximations, not to mention the elegance and efficiency of the emerging functional analytic methods of those times, used for instance by Fredholm.

As a kind of counterpoint to that, one should however note that no linear metric topology on any function space is used in the proof of the Cauchy-Kovalevskaia theorem, which so far, is one of the most - if not the most - powerful and general local existence, uniqueness and regularity result for solutions of rather arbitrary nonlinear partial differential equations. But is there anything wrong with linear metric topologies on function spaces, when trying to find generalized solutions for partial differential equations?

In the *linear* case, one serious warning came with Lewy's 1957 inexistence result, strengthened by Shapira in 1967. It showed that certain rather simple *linear* variable coefficient partial differential equations cannot have generalized solutions in the space of the Schwartz distributions, or even in more large spaces of hyperfunctions.

Is it that such equations do nevertheless have solutions, but their solutions carry *more information* than distributions or hyperfunctions can handle?

While the above kind of limitations in the case of linear partial differential equations were noted and studied in some detail, a major difficulty in the case of *nonlinear* partial differential equations, presented by *nonlinear stability paradoxes*, has usually been overlooked, except for some recent studies - see Section 8 and the literature cited there - which concentrating on certain special cases of the mentioned difficulty, try remedies within the same linear metric topological approach.

And what is in fact the essence of that linear metric topological approach?

Well, from the point of view of generalized solutions, it comes down to the following. Given a vector space X of classical, real or complex valued functions over a domain $\Omega \subset \mathbb{R}^n$ of independent variables, let us first assume that X is endowed with a metrizable topology defined by a metric

(1.A5.1) $$d : X \times X \longrightarrow [0,\infty)$$

which is compatible with the vector space structure of X. In this case, the space of *generalized functions* $\mathcal{X}$ is the *completion* of X in the given metrizable topology and can be obtained as a *quoteint vector* space

(1.A5.2) $$\mathcal{X} = \mathcal{S}/\mathcal{V}$$

where $\mathcal{S}$ is the set of Cauchy sequences in X, while $\mathcal{V}$ is the set of sequences convergent to zero in X, hence

(1.A5.3) $$\mathcal{V} \subset \mathcal{S} \subset X^{\mathbb{N}}$$

This means that a generalized function $F \in \mathcal{X}$ will have a *quotient representation*

(1.A5.4) $$F = s + \mathcal{V} \in \mathcal{X} = \mathcal{S}/\mathcal{V}, \quad \text{with} \quad s \in \mathcal{S}$$

The *approximation* effect involved in such a situation is obvious: a given generalized function F in (1.A5.4) is the limit of any of the sequences s in (1.A5.4), sequences which in view of (1.A4.3) are of the form

(1.A5.5) $$s = (f_\nu \mid \nu \in \mathbb{N}) \in \mathcal{S}$$

where, for $\nu \in \mathbb{N}$, $f_\nu \in X$ are classical functions, and

(1.A5.6) $$\lim_{\nu\to\infty} d(f,f_\nu) = 0$$

In case the vector space topology on X is not metrizable and it is given by a family

(1.A5.7) $(d_i \mid i \in I)$

of semimetrics, the quotient vector space structure (1.A5.2) of the generalized function space $\mathcal{X}$ still remains valid. Moreover, in many important cases, X will still be sequentially dense in $\mathcal{X}$, thus (1.A5.3)-(1.A5.5) remain valid, while (1.A5.6) will be replaced with

(1.A5.8) $\lim_{\nu\to\infty} d_i(F, f_\nu) = 0$, with $i \in I$

Usually, such an approximation property as in (1.A5.6) or (1.A5.8) can also have a *numerical* meaning in the sense that, although

(1.A5.9) $F(x)$, with $x \in \Omega$, is usually *not* defined

since F is a generalized and *not* a classical function on Ω, nevertheless, the sequences of numbers

(1.A5.10) $(f_\nu(x) \mid \nu \in \mathbb{N})$, with $x \in \Omega$

allow the extraction of some *numerical information* by suitable *limit* processes. A standard example giving the L. Schwartz $\mathcal{D}'(\Omega)$ distributions is obtained as follows. Suppose given a vector space X of classical functions on Ω, such that

(1.A5.11) $\mathcal{C}^\infty(\Omega) \subset X \subset \mathcal{L}^1_{\ell oc}(\Omega)$

and suppose X is endowed with the weak topology induced by $\mathcal{D}'(\Omega)$ through the obvious inclusion $X \subset \mathcal{D}'(\Omega)$ implied by (1.A5.11). Then the sequences of numbers in (1.A5.10) give for each $\psi \in \mathcal{D}(\Omega)$, the number denoted by $F(\psi)$ and defined by

(1.A5.12) $F(\psi) = \lim_{\nu\to\infty} \int_\Omega f_\nu(x)\psi(x)dx$

Moreover, $F(\psi)$ does not depend on s in (1.A5.4), which is equivalent to saying that

(1.A5.13) $\lim_{\nu\to\infty} \int_\Omega g_\nu(x)\psi(x)dx = 0$, with $v = (g_\nu \mid \nu \in \mathbb{N}) \in \mathcal{V}$, $\psi \in \mathcal{D}(\Omega)$

Now, the *numerical* meaning of (1.A5.12) is quite obvious: given $x \in \Omega$ and $\psi \in \mathcal{D}(\Omega)$, then $F(\psi)$ contains an information on the 'value' at and around x of the generalized function F, to the extent that

(1.A5.14) supp ψ is contained in a neighbourhood of x

and

(1.A5.15) $$\int_{\Omega} \psi(x)dx \approx 1$$

in which case $F(\psi)$ can be seen as a kind of *average value* of F in a neighbourhood of x. All that is simply described by

(1.A5.16) $$F \in \mathcal{X} = \mathcal{D}'(\Omega)$$

Let us shortly recall the essence of the above:
A space X of classical functions is extended into a space $\mathcal{X}$ of generalized functions

(1.A5.17) $$X \subset \mathcal{X}$$

The space $\mathcal{X}$ of generalized functions is obtained through a quotient construction

(1.A5.18) $$\mathcal{X} = \mathcal{S}/\mathcal{V}, \quad \mathcal{V} \subset \mathcal{S} \subset X^{\mathbb{N}}$$

The embedding (1.A5.17) is defined by

(1.A5.19) $$X \ni f \longmapsto F = u(f) + \mathcal{V} \in \mathcal{X}$$

where $u(f) = (f,\ldots,f,\ldots)$. Finally, a *quotient representation* of a generalized function

(1.A5.20) $$F = s + \mathcal{V} \in \mathcal{X}, \quad s = (f_\nu | \nu \in \mathbb{N}) \in \mathcal{S}$$

has a certain *approximation* and even *numerical interpretation* in terms of specific *limit processes* involving *numerical values* of the classical functions f_ν, with $\nu \in \mathbb{N}$.

Now, it is *particularly important* to note that there are useful and established *quotient constructions* in Analysis which, although go along lines similar to that in (1.A5.17)-(1.A5.20), are nevertheless less *restrictive and simplistic* in that the construction of the spaces $\mathcal{X}$ is *not* based on numerical, approximation, or in general, topological interpretations. Such interpretations can often be associated later, and usually, only to certain *strict* subsets

(1.A5.21) $$\mathcal{X}_0 \subsetneqq \mathcal{X}$$

Yet, no matter how useful such interpretations of elements of $\mathcal{X}_0$ may be, these interpretations need *not* be employed from the very beginning, when the spaces $\mathcal{X}$ are defined.

One such well known example is given by Nonstandard Analysis, see Stroyan & Luxemburg, as well as the particular nonstandard construction presented in (1.7.2)-(1.7.7), or in Appendix 3.

Prompted by the above, we can conclude as follows.

In certain mathematical models, the *representation* of the entities involved can by itself give one or several *interpretations* leading back to basic relationships between the mathematical model and the system which is modelled. A typical example is the arithmetic of integers, where every not too large number $m \in \mathbb{Z}$ is to a good extent its own interpretation.

A first departure from that situation happens within usual distribution theory, see (1.A5.1)-(1.A5.20), where a generalized function $F \in \mathcal{D}'(\Omega)$ with a quotient *representation* $F = s + \mathcal{V} \in \mathcal{D}'(\Omega) = \mathcal{S}/\mathcal{V}$, $s \in \mathcal{S}$, still has a usual *approximation interpretation* (1.A5.8) in the *function space* X in (1.A5.11), but it can *no* longer have a usual *numerical interpretation* within the real or complex numbers, except for the much more relaxed version given by the weak form (1.A5.12)-(1.A5.15).

In this way we note that representation happens to *go farther* than interpretation, see also (1.A5.21) and (1.7.7).

With Nonstandard Analysis a further departure occurs in so far that a construction such as in (1.7.2) and the resulting representations (1.7.6), are made from the very beginning *without* even an approximation interpretation, see also Appendix 3, although such, as well as a numerical interpretation can later be associated with *part* of the structure, see (1.7.7).

However, the fact that in the above mentioned theories representations are developed farther than interpretations is by no means a drawback, but on the contrary, it is one of the main reasons for the power and applicability of these theories. Indeed, the *rigorous* computations in these theories are and can only be done on the level of *representations*. Therefore, one has an advantage if representations are developed conveniently and far enough.

Coming to Colombeau's version of the nonlinear theory of generalized functions presented shortly in Chapter 8, the issue concerning *approximation* and *numerical interpretations* is as follows. The main instrument is the *relation of association* $\Vdash$ which associates a distribution $T \in \mathcal{D}'$ to every generalized function $F \in \mathcal{G}_0$, where $\mathcal{G}_0$ is a *strict* subset of Colombeau's algebra $\mathcal{G}$ of generalized functions. This relation of association

(1.A5.22) $\quad F \Vdash T$

has a *numerical interpretation* given by

(1.A5.23) $$\int_{\mathbb{R}^n} \psi F \, dx \;\vdash\; \int_{\mathbb{R}^n} \psi T \, dx, \quad \psi \in \mathcal{D}$$

which is similar to (1.A5.12).

Concerning possible *approximation interpretations* associated with Colombeau's generalized functions, one can consult Rosinger [3, Appendices 1 and 2 in Part 2, Chapter 1].

It follows that with respect to the *relationship* between representations and interpretations, Colombeau's nonlinear theory of generalized functions is *one step further* then the usual linear theory of the $\mathcal{D}'$ distributions, since the interpretations used in the former reduce it to the latter, according to (1.A5.22), (1.A5.23). Nevertheless, this can allow for particularly powerful results even in the *numerical analysis* of *nonlinear nonconservative* partial differential equations. Here, we shortly mention one such recent result.

We consider the coupling between the velocity u and the stress σ in a one dimensional homogeneous medium of constant density, modelled by the *nonlinear nonconservative* system

(1.A5.24)
$$\begin{aligned} u_t + u \cdot u_x &= \sigma_x \\ \sigma_t + u \cdot \sigma_x &= k^2 u_x \end{aligned} \qquad t \geq 0, \quad x \in \mathbb{R}$$

where $k > 0$ depends on the medium. As usual, we associate with (1.A5.24) the initial value problem

(1.A5.25)
$$\begin{aligned} u(0,x) &= u_0(x) \\ \sigma(0,x) &= \sigma_0(x) \end{aligned} \quad , \; x \in \mathbb{R}$$

While the first equation in (1.A5.24) being conservative, it could be replaced by the *weak* equation

(1.A5.26)
$$\int_{t \geq 0} \int_{x \in \mathbb{R}} (u\psi_t + \tfrac{1}{2}u^2\psi_x - \sigma\psi_x)dt\, dx = 0, \quad \psi \in \mathcal{D}((0,\infty) \times \mathbb{R})$$

However, the second equation in (1.A5.24) does not offer such a possibility since it has a nonconservative form.

In order to obtain a numerical solution for (1.A5.24), (1.A5.25) a usual two time step, split numerical scheme is used. The first step discretizes the nonlinear convection terms in (1.A5.24) by a usual Lagrangian, Eulerian and antidiffusion correction method. The second step discretizes the wave propagation terms in (1.A5.24). Each step is made through a decentered, Donnor cell scheme. In this way, for given initial values (1.A5.25), we obtain two families of numbers

(1.A5.27) $$(u_i^n \mid n \in \mathbb{N},\ i \in \mathbb{Z}),\ (\sigma_i^n \mid n \in \mathbb{N},\ i \in \mathbb{Z})$$

with the *intended* approximation property

(1.A5.28) $$u(n\Delta t, i\Delta x) \approx u_i^n, \quad \sigma(n\Delta t, i\Delta x) \approx \sigma_i^n, \quad n \in \mathbb{N}, \quad i \in \mathbb{Z}$$

where Δt, $\Delta x > 0$ are the time and space increments respectively.

It is further assumed that

(1.A5.29) $\quad u_0, \sigma_0$ have bounded variation on $\mathbb{R}$

and

(1.A5.30) $\quad |\sigma_0(x)| \leq k u_0(x), \quad x \in \mathbb{R}$

which means that we deal with positive and high velocities, a particularly useful case in certain applications. Then we denote

(1.A5.31) $$M = \sup_{x\in\mathbb{R}} (u_0(x) + |\sigma_0(x)|/k)$$

and assume that

(1.A5.32) $\quad r \leq 1/\max\{M,k\}$

where $r = \Delta t/\Delta x > 0$ is the Courant-Friedrichs-Lewy number of the time and space discretization used.

Under the above conditions (1.A5.29)-(1.A5.32), the numerical method gives numerical solutions (1.A5.27), (1.A5.28) which are stable in the $\mathcal{L}^\infty$ norm, in the total variation in space and in the Tonnelli-Cesari type total variation in time.

Assuming $r > 0$ given, for every $\Delta x > 0$ we can define the piece wise constant functions

(1.A5.33) $\quad u_{\Delta x}, \sigma_{\Delta x} : \{0,\infty) \times \mathbb{R} \longrightarrow \mathbb{R}$

by

(1.A5.34) $$u_{\Delta x}(t,x) = u_i^n, \quad \sigma_{\Delta x}(t,x) = \sigma_i^n, \quad \text{if} \quad t \geq 0, \quad x \in \mathbb{R},$$
$$n \in \mathbb{N}, \quad i \in \mathbb{Z}, \quad |t-n\Delta t| < r\Delta x/2, \ |x-i\Delta x| < \Delta x/2$$

Then, a Helly type compactness argument yields for each $T > 0$ a sequence

(1.A5.35) $$\Delta x_\nu > 0, \quad \nu \in \mathbb{N}, \quad \text{with} \quad \lim_{\nu\to\infty} \Delta x = 0$$

and a pair of functions with bounded variation

(1.A5.36) $\quad u,\sigma : [0,T] \times \mathbb{R} \longrightarrow \mathbb{R}$

such that

(1.A5.37) $$\lim_{\nu\to\infty} u_{\Delta x_\nu} = u, \quad \lim_{\nu\to\infty} \sigma_{\Delta x_\nu} = \sigma$$

in the sense of $\mathcal{L}^1_{\ell oc}([0,T] \times \mathbb{R})$. We note that in view of (1.A5.35)-

(1.A5.37) we have

(1.A5.38) $u, \sigma \in \mathcal{D}'([0,\infty) \times \mathbb{R})$

Now, a well known problem with the above so called *numerical solution* (1.A5.35)-(1.A5.38) of (1.A5.24), (1.A5.25) is that the functions u, σ obtained are *not* smooth enough in order to be *classical* solutions of (1.A5.24), (1.A5.25). Moreover, since the second equation in (1.A5.24) is *nonconservative*, we *cannot* use a weak equation similar to (1.A5.26) in order to *check* whether indeed u and σ are at least *weak* solutions of (1.A5.24), (1.A5.25).

In other words, we simply *cannot* be sure in which ways u and σ may relate to our initial problem (1.A5.24), (1.A5.25), except for the fact that they have been obtained by a *compactness* argument from the stable numerical solutions (1.A5.27), (1.A5.28). And as mentioned earlier in connection with *nonlinear stability paradoxes*, it is in particular with compactness arguments used for obtaining generalized solutions of nonlinear partial differential equations that one has to be specially careful. It is therefore precisely here that Colombeau's nonlinear theory of generalized functions proves to be particularly powerful and useful. Indeed, any family

(1.A5.39) $(u_{\Delta x_\nu}, \sigma_{\Delta x_\nu}), \quad \nu \in \mathbb{N}$

in (1.A5.37) can be associated with two *generalized functions*

(1.A5.40) $U, \Sigma \in \mathcal{G}([0,\infty) \times \mathbb{R})$

having the *quotient representation*

(1.A5.41) $U = f + \mathcal{I}, \ \Sigma = g + \mathcal{I} \in \mathcal{G}([0,\infty) \times \mathbb{R})$

where (f,g) are defined by the family (1.A5.39), such that

(1.A5.42) $U \approx u, \quad \Sigma \approx \sigma$

and

(1.A5.43)
$$\begin{aligned} U_t + U \cdot U_x &\approx \Sigma_x \\ \Sigma_t + U \cdot \Sigma_x &\approx k^2 U_x \end{aligned}$$

where the equivalence relation $\approx$ on $\mathcal{G}([0,\infty) \times \mathbb{R})$ is defined through (1.A5.22), (1.A5.23) by

(1.A5.44) $F \approx G \iff F - G \parallel\!\!-\, 0, \quad F, G \in \mathcal{G}([0,\infty) \times \mathbb{R})$

In this way u and σ do *have* a pointwise *numerical interpretation* through (1.A5.37), although they do *not* satisfy the *equations* (1.A5.24), (1.A5.25) classically or in a weak sense.

On the other hand, U and Σ do *not* have a pointwise *numerical interpretation*, but they *satisfy* the *equations* (1.A5.24) in the modified form (1.A5.43).

The *link* between u, σ and U, Σ is given in (1.A5.42), which as mentioned, has the same *averaging* numerical interpretation with (1.A5.12).

A few important properties should also be mentioned.

The relations (1.A5.42) determine u and σ *uniquely*.

Further, as seen in Chapter 8, the equivalence relation $\approx$ is *not* compatible with the multiplication of generalized functions in $\mathcal{G}$. Nevertheless, in the above case of association in (1.A5.42), we also have the following *stronger* association property

$$(1.A5.45) \qquad P(U,\Sigma) \approx P(u,\sigma)$$

for every two variable constant coefficient polynomial P.

In view of the above, it is obvious that the *information* contained in (1.A5.39) and transmitted to the *quotient representation* (1.A5.40), (1.A5.41), is only *partly* contained in (u,σ) as obtained by the *limit* in (1.A5.37). In other words, the quotient representation (1.A5.40), (1.A5.41) contains *more information* about (1.A5.39) than the limit (u,σ) in (1.A5.37).

Finally, the association property (1.A5.42) has also the following advantage. The functions u, σ cannot be replaced in the equations (1.A5.24) since that would involve multiplications between nonsmooth functions and their distributional derivatives, which is not possible within the linear theory of distributions. The functions u, σ cannot be replaced in weak forms of the equations (1.A5.24), since that system is nonconservative. But the functions u, σ can easily be used in (1.A5.42) which is the simplest possible linear system in these two functions.

Concerning the *interpretations* of the nonlinear theory of generalized functions in Rosinger [1,2,3], it suffices to mention that, although developed somewhat earlier, that theory is a kind of encompassing roof theory for a *large class of possible nonlinear* theories of generalized functions, which among others, contains Colombeau's nonlinear theory as a particular case. Therefore, with respect to the relationship between representations and interpretations discussed above, the theory in Rosinger [1,2,3] is yet one more step further than Colombeau's nonlinear theory, see for details Chapters 2, 3, 6 and 7, as well as Chapter 4, which in the case of a particular but important class of nonlinear hyperbolic partial differential equations goes beyond even the framework in Rosinger [1,2,3].

APPENDIX 6

DETAILS ON NONLINEAR STABILITY PARADOXES, AND ON THE EXISTENCE AND UNIQUENESS OF SOLUTIONS FOR NONLINEAR PARTIAL DIFFERENTIAL EQUATIONS

Let us turn to the general n-dimensional case of the *nonlinear stability paradoxes* illustrated in Section 8.

For that, let us consider the *nonlinear* system

$$(1.A6.1) \qquad \begin{aligned} &U(x) = 0, \quad x \in \mathbb{R}^n \\ &U^2(x) = 1, \quad x \in \mathbb{R}^n \end{aligned}$$

and show that it has solutions given by sequences of $\mathcal{C}^\infty$-smooth functions on $\mathbb{R}^n$

$$(1.A6.2) \qquad v = (\chi_\nu \mid \nu \in \mathbb{N}) \in (\mathcal{C}^\infty(\mathbb{R}^n))^{\mathbb{N}}$$

which are both *weakly* and *strongly* convergent in $\mathcal{D}'(\mathbb{R}^n)$. Indeed, if we take for instance

$$(1.A6.3) \qquad \chi_\nu(x) = \sqrt{2}\cos \nu x_1, \quad x = (x_1, \ldots, x_n) \in \mathbb{R}^n, \quad \nu \in \mathbb{N}$$

we shall obviously have

$$(1.A6.4) \qquad \lim_{\nu\to\infty} \chi_\nu = 0, \quad \lim_{\nu\to\infty} \chi_\nu^2 = 1 \quad \text{in} \quad \mathcal{D}'(\mathbb{R}^n)$$

hence v in (1.A6.2) is a weak and strong solution of (1.A6.1) in $\mathcal{D}'(\mathbb{R}^n)$.

Let us consider now the following nonlinear system which is nontrivially partial differential

$$(1.A6.5) \qquad \begin{aligned} &U(x) = 0, \quad x \in \mathbb{R}^n \\ &U(x)U_{x_1}(x) = \alpha(x_1)\alpha'(x_1), \ x = (x_1, \ldots, x_n) \in \mathbb{R}^n \end{aligned}$$

where as usually, U_{x_1} denotes the partial derivative of U with respect to x_1, while $\alpha \in \mathcal{C}^\infty(\mathbb{R})$ is an arbitrary, given function. We define the sequence of $\mathcal{C}^\infty$-smooth functions

$$(1.A6.6) \qquad w = (\omega_\nu \mid \nu \in \mathbb{N}) \in (\mathcal{C}^\infty(\mathbb{R}^n))^{\mathbb{N}}$$

by

(1.A6.7) $\omega_\nu(x) = \sqrt{2}\alpha(x_1)\cos \nu x_1, \quad x = (x_1,\dots,x_n) \in \mathbb{R}^n, \quad v \in \mathbb{N}$

Then it follows easily that

(1.A6.8) $\lim_{\nu\to\infty} \omega_\nu = 0, \quad \lim_{\nu\to\infty} \omega_\nu(\omega_\nu)_{x_1} = \alpha\alpha' \quad \text{in} \quad \mathcal{D}'(\mathbb{R}^n)$

Therefore w is both a *weak* and *strong* solution of (1.A6.5) in $\mathcal{D}'(\mathbb{R}^n)$.

It is obvious that in a similar way, one can obtain both *weak* and *strong* solutions $v = (\chi_\nu | \nu \in \mathbb{N}) \in (\mathcal{C}^\infty(\mathbb{R}^n))^{\mathbb{N}}$ in $\mathcal{D}'(\mathbb{R}^n)$ for a large variety of *nonlinear partial differential* systems

(1.A6.9)
$$\begin{aligned} &U(x) = 0, \quad x \in \mathbb{R} \\ &\prod_{1\leq j\leq m} D^{q_j}U(x) = \gamma(x), \quad x \in \mathbb{R}^n \end{aligned}$$

where $m \in \mathbb{N}$, $q_j \in \mathbb{N}^n$ and $\gamma \in \mathcal{C}^\infty(\mathbb{R}^n)$, γ not identically zero. Indeed, this follows easily for m *even*, if we take

(1.A6.10) $\chi_\nu(x) = \alpha(x) \cos \nu x_1 + \beta(x) \sin \nu x_1$

for $x = (x_1,\dots,x_n) \in \mathbb{R}^n$, $\nu \in \mathbb{N}$, and with suitably chosen $\alpha,\beta \in \mathcal{C}^\infty(\mathbb{R}^n)$.

It follows that both *weak* and *strong* solutions in $\mathcal{D}'(\mathbb{R}^n)$, given by sequences of functions

(1.A6.11) $v = (\chi_\nu | \nu \in \mathbb{N}) \in (\mathcal{C}^\infty(\mathbb{R}^n))^{\mathbb{N}}$

can be found for a large class of *nonlinear partial differential systems* of the form

(1.A6.12)
$$\begin{aligned} &U(x) = 0, \quad x \in \mathbb{R}^n \\ &\sum_{1\leq i\leq \ell} d_i(x) \prod_{1\leq j\leq m_i} D^{q_{ij}}U(x) = \gamma(x), \quad x \in \mathbb{R}^n \end{aligned}$$

where $\ell, m_i \in \mathbb{N}, q_{ij} \in \mathbb{N}^n, d_i, \gamma \in \mathcal{C}^\infty(\mathbb{R}^n)$, where γ is not identically zero.

Indeed, as follows from (1.A6.9), this can happen if at least one of the m_i is *even*.

Let us show now that (1.A6.12) can have both *weak* and *strong* solutions in $\mathcal{D}'(\mathbb{R}^n)$ even in the case when *all* m_i are *odd*. For that, it suffices to show that (1.A6.9) can have such solutions for m odd. This happens indeed if instead of (1.A6.10), we take

$$(1.A6.13) \qquad \chi_\nu(x) = 1 + \cos \nu x_1 - 2 \cos^2 \nu x_1$$

for $x = (x_1, \ldots, x_n) \in \mathbb{R}^n$, $\nu \in \mathbb{N}$. Then a direct computation gives

$$(1.A6.14) \qquad \lim_{\nu\to\infty} \chi_\nu = 0, \quad \lim_{\nu\to\infty} \chi_\nu^3 = -3/4 \quad \text{in} \quad \mathcal{D}'(\mathbb{R}^n)$$

Thus with (1.A6.13), it follows that

$$(1.A6.15) \qquad v = (\chi_\nu \mid \nu \in \mathbb{N}) \in (\mathcal{C}^\infty(\mathbb{R}^n))^{\mathbb{N}}$$

is a sequence which is both a *weak* and *strong* solution of the nonlinear sytem

$$(1.A6.16) \qquad \begin{aligned} U(x) &= 0, & x \in \mathbb{R}^n \\ U^3(x) &= -3/4, & x \in \mathbb{R}^n \end{aligned}$$

Now the existence of weak and strong solutions for nonlinear systems such as in (1.A6.1), (1.A6.5), (1.A6.9) or (1.A6.12) does have *critical* effects on the *existence and uniqueness* of weak solutions for *nonlinear* partial differential equations, when they are obtained by the usual methods mentioned in Section 8 of this Chapter. Nevertheless, as noted earlier, these critical effects are often *overlooked* in the literature.

In order to illustrate such effects, it suffices to extend the argument in (1.8.17)-(1.8.27) to more general nonlinear partial differential operators (1.5.4), using the solutions of systems (1.A6.12). Indeed, suppose that the equation

$$(1.A6.17) \qquad T(D)U(x) = f(x), \quad x \in \mathbb{R}^n$$

is *claimed* to have a weak solution

$$(1.A6.18) \qquad U$$

given by a sequence

(1.A6.19) $s = (\psi_\nu | \nu \in \mathbb{N}) \in (\mathcal{C}^\infty(\mathbb{R}^n))^{\mathbb{N}}$

for which

(1.A6.20) $U = \lim_{\nu \to \infty} \psi_\nu$ in $\mathcal{D}'(\mathbb{R}^n)$

As usual, this means the existence of the simultaneous limits

(1.A6.21) $s \longrightarrow U,\ T(D)s \longrightarrow f$ in $\mathcal{D}'(\mathbb{R}^n)$

Suppose that the nonlinear partial differential operator $T(D)$ in (1.A6.17) has the form

(1.A6.22) $T(D) = L(D) + Q(D)$

where $L(D)$ is a linear partial differential operator with $\mathcal{C}^\infty$-smooth coefficients, while $Q(D)$ is a nonlinear partial differential operator such as those in the left hand term of the second equation in (1.A6.12).

Then for any sequence

(1.A6.23) $s + v$

with v in (1.A6.11), which satisfies (1.A6.12), we shall have the limit

(1.A6.24) $s + v \longrightarrow U$ in $\mathcal{D}'(\mathbb{R}^n)$

However, we show that in general, we shall *no longer* have the limit

(1.A6.25) $T(D)(s+v) \longrightarrow f$ in $\mathcal{D}'(\mathbb{R}^n)$

In other words, the *so called* weak solution (1.A6.21) does *not extend* to sequences (1.A6.24), where nevertheless, we have $v \longrightarrow 0$ in $\mathcal{D}'(\mathbb{R}^n)$, this being a severe limitation on the *stability* of the weak solution (1.A6.21) and therefore, on the very *meaning* of such a weak solution concept.

Indeed, (1.A6.22) gives

(1.A6.26) $T(D)(s+v) = L(D)s + L(D)v + Q(D)(s+v)$

while obviously

(1.A6.27) $Q(D)(s+v) = Q(D)s + R(D)(s,v) + Q(D)v$

where $R(D)$ is a nonlinear partial differential operator in two arguments. It follows that

(1.A6.28) $$T(D)(s+v) = T(D)s + L(D)v + R(D)(s,v) + Q(D)v$$

But (1.A6.21), (1.A6.11) and (1.A6.12) yield the limits

(1.A6.29) $$T(D)s \to f, \quad L(D)v \to 0, \quad Q(D)v \to \gamma \quad \text{in} \quad \mathcal{D}'(\mathbb{R}^n)$$

If we assume now the existence of the limit

(1.A6.30) $$R(D)(s,v) \to g \quad \text{in} \quad \mathcal{D}'(\mathbb{R}^n)$$

then in view of (1.A6.28), (1.A6.29), we obtain the limit

(1.A6.31) $$T(D)(s+v) \to f + \gamma + g \quad \text{in} \quad \mathcal{D}'(\mathbb{R}^n)$$

which obviously need *not* be the same with (1.A6.25), unless

(1.A6.32) $$\gamma + g = 0 \quad \text{in} \quad \mathcal{D}'(\mathbb{R}^n)$$

It follows that (1.A6.25) will indeed fail, whenever (1.A6.32) or (1.A6.30) fail.

It is now obvious that the *lack of stability* illustrated in (1.A6.24), (1.A6.25) above can lead to the questioning of the usual weak solution methods for nonlinear partial differential equations.

Indeed, concerning the usual constructive proofs for the *existence* of weak solutions, it is obvious that the construction of particular sequences s as in (1.A6.21) can prove to be *irrelevant*, as long as their stability properties in (1.A6.24), (1.A6.25) are not established. Otherwise, we can land in situations when

(1.A6.33) $$s, s + v \to U \quad \text{in} \quad \mathcal{D}'(\mathbb{R}^n)$$

but

(1.A6.34) $$T(D)s \to f, \quad T(D)(s+v) \to g \quad \text{in} \quad \mathcal{D}'(\mathbb{R}^n)$$

and

(1.A6.35) $$f \neq g \quad \text{in} \quad \mathcal{D}'(\mathbb{R}^n)$$

which according to the usual interpretation of weak solutions would mean that $U \in \mathcal{D}'(\mathbb{R}^n)$ is a *simultaneous* weak solution for

(1.A6.36) $$T(D)U = f$$

and

(1.A6.37) $$T(D)U = g$$

It is now also obvious that with difficulties such as in (1.A6.33)-(1.A6.37), the concept of the *uniqueness* of the usual weak solutions becomes even more problematic.

One way to deal with the above situation is the one presented in Sections 9-12 of this Chapter. See also Chapters 2-4 and 6-8.

APPENDIX 7

THE DEFICIENCY OF DISTRIBUTION THEORY FROM THE POINT OF VIEW OF EXACTNESS

We shall show that from the point of view of *exactness*, it is *not* convenient to deal with $E \rightarrow \mathcal{D}'$, or in particular $\mathcal{D}' \rightarrow \mathcal{D}'$ sequential solutions, even when we solve *linear* partial differential equations.

For that purpose we need a few notations.

Suppose given an infinite index set Λ and a subalgebra $\mathcal{X} \subset \mathcal{M}(\Omega)$ as well as a vector subspace $\mathcal{I} \subset \mathcal{X}^\Lambda$. Then we denote by

(1.A7.1) $\qquad \mathcal{A}_{\mathcal{X},\Lambda}(\mathcal{I})$

the largest subalgebra $\mathcal{B} \subset \mathcal{X}^\Lambda$ such that $\mathcal{I} \cdot \mathcal{B} \subset \mathcal{I}$.

Obviously, the following three properties are equivalent:

(1.A7.2) $\qquad \mathcal{I} \subset \mathcal{A}_{\mathcal{X},\Lambda}(\mathcal{I})$

(1.A7.3) $\qquad \mathcal{I}$ subalgebra in $\mathcal{X}^\Lambda$

(1.A7.4) $\qquad \mathcal{I}$ ideal in $\mathcal{A}_{\mathcal{X},\Lambda}(\mathcal{I})$

Let us denote by

(1.A7.5) $\qquad \mathrm{ID}_{\mathcal{X},\Lambda}$

the set of all the vector subspaces $\mathcal{I} \subset \mathcal{X}^\Lambda$ such that, see (1.6.11)

(1.A7.6) $\qquad \mathcal{I} \cap \mathcal{U}_{\mathrm{H},\Lambda} = \mathcal{O}$

and, see (1.6.10)

(1.A7.7) $\qquad \mathcal{I} \oplus \mathcal{U}_{\mathcal{X},\Lambda} \subset \mathcal{A}_{\mathcal{X},\Lambda}(\mathcal{I})$

In view of (1.A7.7), (1.A7.2) and (1.A7.4) it follows that

(1.A7.8) $\qquad \begin{array}{l} \forall \quad \mathcal{I} \in \mathrm{ID}_{\mathcal{X},\Lambda} \\ \qquad \mathcal{I} \text{ ideal in } \mathcal{A}_{\mathcal{X},\Lambda}(\mathcal{I}) \end{array}$

Finally, in view of (1.A7.8), for $\mathcal{I} \in \mathrm{ID}_{\mathcal{X},\Lambda}$ we can define the quotient algebra

(1.A7.9) $\quad A_{\mathcal{X},\Lambda}(\mathcal{I}) = \mathcal{A}_{\mathcal{X},\Lambda}(\mathcal{I})/\mathcal{I}$

Then (1.A7.6), (1.6.10), (1.6.11) yield

(1.A7.10) $\quad \forall\ \mathcal{I} \in ID_{\mathcal{X},\Lambda}\ :\quad A_{\mathcal{X},\Lambda}(\mathcal{I}) \in AL_{\mathcal{X},\Lambda}$

And also *conversely*, it follows easily that

(1.A7.11) $\quad \forall\ A = \mathcal{A}/\mathcal{I} \in AL_{\mathcal{X},\Lambda}\ :\quad \mathcal{I} \in ID_{\mathcal{X},\Lambda}$

In other words, $ID_{\mathcal{X},\Lambda}$ contains exactly all $\mathcal{I} \subset \mathcal{X}^\Lambda$ for which there exists $A = \mathcal{A}/\mathcal{I} \in AL_{\mathcal{X},\Lambda}$. Moreover, it is easy to see that

(1.A7.12) $\quad \forall\ A = \mathcal{A}/\mathcal{I} \in AL_{\mathcal{X},\Lambda}\ :\quad \mathcal{A} \subset \mathcal{A}_{\mathcal{X},\Lambda}(\mathcal{I})$

which means that each quotient algebra $A = \mathcal{A}/\mathcal{I} \in AL_{\mathcal{X},\Lambda}$ is a *subalgebra* in $A_{\mathcal{X},\Lambda}(\mathcal{I})$.

In view of (1.A7.6), (1.A7.7) it is easy to see that we have the following *simple characterization* of $ID_{\mathcal{X},\Lambda}$ as being the set of all $\mathcal{I} \subset \mathcal{X}^\Lambda$ such that

(1.A7.13) $\quad \mathcal{I}$ subalgebra in $\mathcal{X}^\Lambda$

(1.A7.14) $\quad \mathcal{U}_{\mathcal{X},\Lambda} \cdot \mathcal{I} \subset \mathcal{I}$

(1.A7.15) $\quad \mathcal{I} \cap \mathcal{U}_{\mathcal{X},\Lambda} = \mathcal{O}$

Suppose now that we are dealing with a linear partial differential equation with $\mathcal{C}^\infty$-smooth coefficients

(1.A7.16) $\quad L(D)U(x) = f(x),\ x \in \Omega$

where $\Omega \subset \mathbb{R}^n$ is nonvoid, open, while $L(D)$ is given in (1.8.31). Further, suppose given a sequence of functions

(1.A7.17) $\quad s \in (\mathcal{C}^\infty(\Omega))^{\mathbb{N}}$

It is easy to see that in the case of a *linear* partial differential equation (1.A7.16), the definition of $E \to A$ sequential solutions given in (1.10.6)-(1.10.8) can be extended to the case when A is also a quotient vector space and not necessarily a quotient algebra, see also Chapter 4. It is in this sense that we assume that

$$(1.A7.18) \qquad s \text{ is a } E \to \mathcal{D}'(\Omega) \text{ sequential solution of (1.A7.16)}$$

where

$$(1.A7.19) \qquad E = \mathcal{S}/\mathcal{V} \in VS^{m}_{\mathcal{C}^\infty(\Omega),\mathbb{N}}$$

and, see (1.8.43)

$$(1.A7.20) \qquad \mathcal{D}'(\Omega) = \mathcal{S}^\infty(\Omega)/\mathcal{V}^\infty(\Omega)$$

In view of (1.10.10), (1.10.11), the relation (1.A7.18) means the following two conditions

$$(1.A7.21) \qquad s \in \mathcal{S}$$

and

$$(1.A7.22) \qquad \begin{array}{l} \forall \ t \in \mathcal{S} : \\ \quad t-s \in \mathcal{V} \Rightarrow w_t \in \mathcal{V}^\infty(\Omega) \end{array}$$

where

$$(1.A7.23) \qquad w_t = L(D)t - u(f)$$

is the *error* sequence corresponding to $t \in \mathcal{S}$.

Now, the *deficiency* of the approach in (1.A7.18) from the point of view of *exactness* becomes apparent. Indeed, when we try to *strengthen* the condition

$$(1.A7.24) \qquad w_t \in \mathcal{V}^\infty(\Omega)$$

in (1.A7.22) on the *error* sequence w_t by finding conditions of exactness similar to those in (1.11.13), (1.11.14), we have to face the fact that

$$(1.A7.25) \qquad \mathcal{V}^\infty(\Omega) \not\subset \mathcal{A}_{\mathcal{C}^\infty(\Omega),N}(\mathcal{V}^\infty(\Omega))$$

where the notation in (1.A7.1) was used. We note that the relation (1.A7.25) follows easily from, see (1.8.57)

$$(1.A7.26) \qquad (\mathcal{V}^{\infty}(\Omega)\cdot\mathcal{V}^{\infty}(\Omega)) \cap \mathcal{S}^{\infty}(\Omega) \not\subset \mathcal{V}^{\infty}(\Omega)$$

In view of (1.11.15), the relation (1.A7.25) means that we *cannot* obtain a satisfactory *exactness* property for the sequential solution in (1.A7.18).

APPENDIX 8

INEXISTENCE OF LARGEST OFF DIAGONAL VECTOR SUBSPACES OR IDEALS

We give a very simple example of two vector subspaces $\mathcal{V}$ and $\mathcal{V}'$ in $(\mathcal{C}^0(\mathbb{R}))^{\mathbb{N}}$ which are off diagonal, see (1.6.6), but which cannot be contained in an off diagonal vector subspace. Indeed, let $w, w' \in (\mathcal{C}^0(\mathbb{R}))^{\mathbb{N}}$ be defined by

$$(1.A8.1) \qquad w_\nu(x) = 1 + \sin \nu x,\ w'_\nu(x) = \sin \nu x, \quad \text{for} \quad \nu \in \mathbb{N},\ x \in \mathbb{R}$$

and let $\mathcal{V}$ and $\mathcal{V}'$ be the vector subspaces generated by w and w' respectively. Then it follows that

$$(1.A8.2) \qquad \mathcal{V} \cap \mathcal{U}_{\mathcal{C}^0(\mathbb{R}),\mathbb{N}} = \mathcal{V}' \cap \mathcal{U}_{\mathcal{C}^0(\mathbb{R}),\mathbb{N}} = \mathcal{O}$$

since for any $\psi \in \mathcal{C}^0(\mathbb{R})$, the relation $w = u(\psi)$ or $w' = u(\psi)$ implies that ψ vanishes on a dense subset of $\mathbb{R}$.

Now let $\mathcal{V}''$ be a vector subspace in $(\mathcal{C}^0(\mathbb{R}))^{\mathbb{N}}$ which contains $\mathcal{V}$ and $\mathcal{V}'$. Then obviously

$$(1.A8.3) \qquad u(1) = w - w' \in \mathcal{V} + \mathcal{V}' \subset \mathcal{V}''$$

therefore $\mathcal{V}''$ will not satisfy the off diagonality condition (1.6.6).

The above is valid as well for ideals in $(\mathcal{C}^0(\mathbb{R}))^{\mathbb{N}}$. Indeed, let $w, w' \in (\mathcal{C}^0(\mathbb{R}))^{\mathbb{N}}$ be defined by

$$(1.A8.4) \qquad w_\nu(x) = 1 + \sin \nu x,\ w'_\nu(x) = 1 + \cos \nu x, \quad \text{for} \quad \nu \in \mathbb{N},\ x \in \mathbb{R}$$

and let

$$(1.A8.5) \qquad \mathcal{I} = w \cdot (\mathcal{C}^0(\mathbb{R}))^{\mathbb{N}},\ \mathcal{I}' = w' \cdot (\mathcal{C}^0(\mathbb{R}))^{\mathbb{N}}$$

be the ideals in $(\mathcal{C}^0(\mathbb{R}))^{\mathbb{N}}$ generated by w and w' respectively. Then, as above, it is easy to prove that

$$(1.A8.6) \qquad \mathcal{I} \cap \mathcal{U}_{\mathcal{C}^0(\mathbb{R}),\mathbb{N}} = \mathcal{I}' \cap \mathcal{U}_{\mathcal{C}^0(\mathbb{R}),\mathbb{N}} = \mathcal{O}$$

thus the ideals $\mathcal{I}$ and $\mathcal{I}'$ satisfy the off diagonality condition (1.6.11).

Assume now that $\mathcal{I}''$ is an ideal in $(\mathcal{C}^0(\mathbb{R}))^{\mathbb{N}}$ which contains $\mathcal{I}$ and $\mathcal{I}'$. Then obviously

(1.A8.7) $$w'' = w + w' \in \mathcal{I} + \mathcal{I}' \subset \mathcal{I}''$$

But (1.A8.4) yields

(1.A8.8) $$w_\nu''(x) = 2 + \sin \nu x + \cos \nu x > 0, \quad \text{for} \quad \nu \in \mathbb{N},\ x \in \mathbb{R}$$

therefore

(1.A8.9) $$u(1) = w'' \cdot (1/w'') \in \mathcal{I}'' \cdot (\mathcal{C}^0(\mathbb{R}))^{\mathbb{N}} \subset \mathcal{I}''$$

which means that the ideal $\mathcal{I}''$ does not satisfy the off diagonality condition (1.6.11). In fact (1.A8.9) implies that

(1.A8.10) $$\mathcal{I}'' = (\mathcal{C}^0(\mathbb{R}))^{\mathbb{N}}$$

thus, there is no proper ideal in $(\mathcal{C}^0(\mathbb{R}))^{\mathbb{N}}$ which may contain the ideals $\mathcal{I}$ and $\mathcal{I}'$.

CHAPTER 2

GLOBAL VERSION OF THE CAUCHY-KOVALEVSKAIA THEOREM ON ANALYTIC NONLINEAR PARTIAL DIFFERENTIAL EQUATIONS

§1. INTRODUCTION

The aim of this Chapter is to show that *all analytic nonlinear* partial differential equations have generalized solutions on the *whole of their domain of analyticity*.

These generalized solutions are defined in the sense of Chapter 1, Section 10, and are *analytic* on the *whole* of the domain of analyticity of the respective equations, except perhaps for *closed*, *nowhere dense* subsets, which may be chosen to have *zero* Lebesque measure.

This result is *nontrivial* because of at least two reasons. First, there is little understanding of the structure of singularities of analytic functions of several complex variables. Secondly, analytic functions can tend very fast to infinity near to their singularities. These two reasons have so far been sufficient in order to *prevent* the theory of distributions of L. Schwartz from finding *global distributional solutions* for arbitrary analytic nonlinear partial differential equations.

It follows that one may as well look for more general concepts of solutions. Fortunately, the theory in Chapter 1 is sufficient for that purpose.

In this way one obtains for the *first time*, see Rosinger [3], the following result:
The analyticity of solutions of arbitrary analytic nonlinear partial differential equations is a *strongly generic property* of these equations. We recall that a property of a system is called strongly generic, if and only if it holds on an open and dense subset of the domain of definition of that system, see Richtmyer.

The above result may present an interest from several points of view.

Firstly, in case the closed nowhere dense subsets on which the generalized solutions may fail to be analytic have zero Lebesque measure, such solutions can describe a large variety of *shocks*.

Secondly, it is well known that closed, nowhere dense subsets in Euclidean spaces can have arbitrary positive Lebesque measure, Oxtoby. In this case one may expect that the respective generalized solutions may, among others, model *turbulence* or other *chaotical* processes as well. Closed, nowhere dense sets of arbitrary positive Lebesque measure can in particular be various Cantor type sets, encountered as *attractors* in recent studies of asymptotic fluid behaviour, Temam.

Finally, the open problem arises whether the rather large class of generalized solutions constructed in this Chapter may exhaust *all*, *or most* of the solutions corresponding to various solution concepts used so far in the literature, when applied to solving analytic nonlinear partial differential equations. A partial answer is presented in Chapter 7, where the method for dealing with *closed, nowhere dense singularities* developed in this Chapter will be applied to the resolution of singularities of weak solutions of a very large class of polynomial nonlinear partial differential equations.

It may be ideal to find at once a unique and regular solution for a given nonlinear partial differential equation, corresponding to a certain initial and/or boundary value problem. However, as is well known, that kind of ideal situation seldom happens. In fact, it does not happen even in the case of one of the most simple nonlinear partial differential equations, namely, the shock wave equation

$$U_t + U \cdot U_x = 0, \quad t \geq 0, \quad x \in \mathbb{R}$$

with the initial value problem

$$U(0,x) = u(x), \quad x \in \mathbb{R}$$

in which case the Rankine-Hugoniot condition together with the entropy condition may be needed in order to select a unique solution, which typically will fail to be regular, due to the presence of shocks, Smoller.

The above example of the shock wave equation may be useful in understanding the meaning of the results in this Chapter.

Indeed, if we do not ask the Rankine-Hugoniot and entropy conditions, the shock wave equation, for a given initial value problem, may have infinitely many solutions which are well defined outside of the shocks. These solutions can be obtained by *patching up* in the half plane

$$t \geq 0, \quad x \in \mathbb{R}$$

solutions defined by the method of characteristics applied on parts of the space domain corresponding to the initial moment

$$t = 0, \quad x \in \mathbb{R}$$

It only happens that the mentioned Rankine-Hugoniot and entropy conditions can often select out one single solution from an infinity of such patched up solutions.

It follows that, at first, we have to face a multitude of patched up solutions. And then, by using certain additional global principles, such as for instance the Rankine-Hugoniot or entropy condition, be able to select a unique solution.

In the terms of the above example, the aim of this Chapter is to present *one global and universal principle* which can define for arbitrary analytic nonlinear partial differential equations sets of patched up solutions. This principle, namely, to be a *generalized solution* in the sense specified below, does not necessarily lead to the uniqueness of such solutions for given analytic and noncharacteristic Cauchy problems.

It follows therefore that the uniqueness of solutions has to result from the imposition of *further global or local conditions*.

To recapitulate, the aim of this Chapter is to answer the following *question*:

Can one find *sufficiently regular global* solutions for *all analytic nonlinear* partial differential equations ?

This question is nontrivial since classical or distributional global solutions are not available.

As mentioned, the answer to the above question is affirmative. Moreover, for each given analytic nonlinear partial differential equation, one can construct a *large class* of such global solutions. In this way, one is led to the *second question*:

To what extent does this class of global solutions exhaust the various solutions generated by the customary weak or generalized solution concepts used so far in the literature, when applied to analytic nonlinear partial differential equations ?

The answer to this second question remains open.

Several further comments are as follows.

The proof techniques in this Chapter do *not* employ functional analysis. They only use basic constructions in rings of continuous functions on Euclidean spaces, as well as classical calculus and elements of topology in these spaces, added of course to the classical proof of the original, local Cauchy-Kovalevskaia theorem. In particular, the whole construction centers around the so called *nowhere dense ideals*

$$(2.1.1) \qquad \mathcal{I}_{nd}(\mathbb{R}^n) \subset (\mathcal{C}^0(\mathbb{R}^n))^{\mathbb{N}}$$

introduced in Rosinger [1,2,3], which are used for the construction of suitable algebras of generalized functions in the sense defined in (1.6.9)-(1.6.11). See (2.2.4) - (2.2.6) for the detailed definition.

This lack of need for the use of functional analysis should not come as a surprise. Indeed, the classical proof of the Cauchy-Kovalevskaia theorem itself is only using calculus, functions of complex variables and inequality estimates, although that theorem remains one of the most general and powerful nonlinear local existence, uniqueness and regularity results.

It is perhaps the time to become aware of the *blind spot* - if not in fact *relative failure* - of several decades of functional analytic exclusivism in solving nonlinear partial differential equations. Indeed, on the one hand, the classical Cauchy-Kovalevskaia theorem had been proved long before the advent of functional analytic approaches and its proof only uses Abel type estimates of power series, i.e., it is based on elementary properties of the geometric series. On the other hand, the subsequent functional analytic methods, despite their numerous important contributions, have not produced one single comparably general *type independent* and powerful nonlinear local existence, uniqueness and regularity result. Curiously, the existent functional analytic methods are not able to improve on the classical Cauchy-Kovalevskaia result in its given general terms. It appears indeed that, especially in the case of nonlinear partial differential equations, the power of functional analysis is rather limited to the detailed study of various particular, specific classes of equations. Lately, a certain awareness of that limitation seems to emerge in a tentative way. For instance in Evans, methods in measure theory are presented as supplementing the limited power of functional analysis. However, such ideas are still far from identifying the roots of the issues involved in the way that identification becomes possible through the 'algebra first' approach presented in this volume, as well as in Rosinger [1,2,3] and Colombeau [1,2].

The proof of the fact that the global generalized solutions constructed in this Chapter are analytic, except perhaps on closed, nowhere dense subsets, follows quite straightforward from a transfinite induction argument, once the nowhere dense ideals in (2.1.1) are brought into the picture. The respective developments are presented in Sections 2-5, and culminate in Theorem 1 in Section 5. The further refinement in Theorem 2, Section 6, shows that the mentioned closed, nowhere dense singularities can be constructed in such a way as to have zero Lebesque measure. This refinement is based on Proposition 1 in Section 6, contributed by M. Oberguggenberger in a private communication. Curiously, this stronger result in Theorem 2 is constructive and it does not use transfinite induction.

In order to obtain Theorems 1 and 2 in this Chapter, the full generality of the concept of solution introduced in Rosinger [1,2,3] and presented in Chapter 1, Section 10 was used.

The problem remains open whether the mentioned results may as well be obtained within the particular theory of generalized functions introduced in Colombeau [1,2].

§2. THE NOWHERE DENSE IDEALS

The presentation in this Chapter, owing to a simplified notation, is rather *selfcontained*, although it follows, as a particular case, the general construction in Chapter 1.

Given Ω a domain in $\mathbb{R}^n$ and $\ell \in \bar{\mathbb{N}} = \mathbb{N} \cup (\infty)$, we denote by, see Appendix 1, Chapter 5,

(2.2.1) $\qquad \mathcal{S}^{\ell}(\Omega), \ \mathcal{V}^{\ell}(\Omega)$

the set of all the sequences

$$s = (\psi_\nu | \nu \in \mathbb{N}) \in (\mathcal{C}^\ell(\Omega))^{\mathbb{N}}$$

of $\mathcal{C}^\ell$-smooth functions which converge weakly in $\mathcal{D}'(\Omega)$ to a distribution, respectively, to zero. Then, one obtains the vector space isomorphism

(2.2.2) $$\mathcal{S}^\ell(\Omega)/\mathcal{V}^\ell(\Omega) \to \mathcal{D}'(\Omega)$$

defined by

$$s + \mathcal{V}^\ell(\Omega) \longmapsto S, \quad s \in \mathcal{S}^\ell(\Omega)$$

where

$$S(\chi) = \lim_{\nu\to\infty} \int_\Omega \psi_\nu(x)\chi(x)dx, \quad \chi \in \mathcal{D}(\Omega)$$

The basic concept, namely, the *nowhere dense ideal* on Ω is introduced now. We denote by

(2.2.3) $$\mathcal{I}_{nd}(\Omega)$$

the set of all the sequences of continuous functions

$$w = (\omega_\nu | \nu \in \mathbb{N}) \in (\mathcal{C}^0(\Omega))^{\mathbb{N}}$$

which satisfy the condition

(2.2.4) $$\begin{array}{l} \exists \ \Gamma \subset \Omega \ \text{ closed, nowhere dense :} \\ \forall \ x \in \Omega\setminus\Gamma : \\ \exists \ \mu \in \mathbb{N} : \\ \forall \ \nu \in \mathbb{N}, \ \ \nu \geq \mu : \\ \quad \omega_\nu(x) = 0 \end{array}$$

It follows that $\mathcal{I}_{nd}(\Omega)$ is an ideal in $(\mathcal{C}^0(\Omega))^{\mathbb{N}}$, see Proposition 2 in Appendix 1. Moreover, as shown in Appendix 1 based on a Baire category argument in $\mathbb{R}^n$, the condition (2.2.4) in the definition of $\mathcal{I}_{nd}(\Omega)$ can be replaced by the following equivalent one

$$(2.2.5)\qquad \begin{array}{l} \exists\ \Gamma \subset \Omega \text{ closed, nowhere dense :} \\ \forall\ x \in \Omega\backslash\Gamma : \\ \exists\ \mu \in \mathbb{N},\ V \subset \Omega\backslash\Gamma \text{ neighbourhood of } x : \\ \forall\ \nu \in \mathbb{N},\ \nu \geq \mu,\ y \in V : \\ \quad \omega_\nu(y) = 0 \end{array}$$

It is important to note that, for both these properties, the continuity of the functions ω_ν, as well as the condition on Γ being closed and nowhere dense, requested in (2.2.4), are essential, see Appendix 1.

We shall call $\mathcal{I}_{nd}(\Omega)$ the *nowhere dense ideal* on Ω.

An easy consequence of (2.2.5) is the relation

$$(2.2.6)\qquad D^p(\mathcal{I}_{nd}(\Omega) \cap (\mathcal{C}^\ell(\Omega))^{\mathbb{N}}) \subset \mathcal{I}_{nd}(\Omega),\quad p \in \mathbb{N}^n,\ |p| \leq \ell$$

where $\ell \in \bar{\mathbb{N}} = \mathbb{N}$, and D^p is the usual p-th order partial derivative, applied term wise to sequences of smooth functions.

Let us take now a subalgebra $\mathcal{A} \subset (\mathcal{C}^0(\Omega))^{\mathbb{N}}$ which satisfies the following two conditions, see (1.6.10)

$$(2.2.7)\qquad \mathcal{I}_{nd}(\Omega) \subset \mathcal{A}$$

and

$$(2.2.8)\qquad \mathcal{U}(\Omega) \subset \mathcal{A}$$

where $\mathcal{U}(\Omega)$ is the subalgebra of all the sequences with identical terms

$$u(\psi) = (\psi, \psi, \psi, \ldots, \psi, \ldots)$$

corresponding to arbitrary continuous functions $\psi \in \mathcal{C}^0(\Omega)$, see (1.6.4).

Obviously

$$(2.2.9)\qquad A = \mathcal{A}/\mathcal{I}_{nd}(\Omega)$$

is an associative and commutative algebra, with the unit element

$$(2.2.10)\qquad 1_A = u(1) + \mathcal{I}_{nd}(\Omega) \in A$$

Moreover, the mapping

$$(2.2.11)\qquad \mathcal{C}^0(\Omega) \ni \psi \longmapsto u(\psi) + \mathcal{I}_{nd}(\Omega) \in A$$

is an embedding of algebras, since we obviously have satisfied the *neutrix condition*, see (1.6.11)

$$(2.2.12) \qquad \mathcal{I}_{nd}(\Omega) \cap \mathcal{U}(\Omega) = \sigma$$

as shown in Appendix 1.

For $\ell \in \mathbb{N}$ let us define the quotient algebras

$$(2.2.13) \qquad A^{\ell} = (\mathcal{A} \cap (\mathcal{C}^{\ell}(\Omega))^{\mathbb{N}})/(\mathcal{I}_{nd}(\Omega) \cap (\mathcal{C}^{\ell}(\Omega))^{\mathbb{N}})$$

Embeddings of the distributions into algebras of generalized functions, given by

$$(2.2.14) \qquad \mathcal{D}'(\Omega) \subset A^{\infty}$$

are constructed in Rosinger [1-3], as well as in Colombeau [1,2], see also Chapter 6.

Finally we note that one can easily find a subalgebra $\mathcal{A} \subset (\mathcal{C}^0(\Omega))^{\mathbb{N}}$ which satisfies (2.2.7), (2.2.8), by taking for instance $\mathcal{A} = (\mathcal{C}^0(\Omega))^{\mathbb{N}}$.

For the sake of completeness, let us mention the way the partial derivative operators can be extended to the quotient algebras A^{ℓ}.

Let us suppose that for a given $\ell \in \mathbb{N}$, the following additional condition is satisfied by the subalgebra $\mathcal{A}$

$$(2.2.15) \qquad D^p(\mathcal{A} \cap (\mathcal{C}^{\ell}(\Omega))^{\mathbb{N}}) \subset \mathcal{A}, \quad p \in \mathbb{N}^n, \quad |p| \leq \ell$$

This can be obviously secured, if for instance, we take $\mathcal{A} = (\mathcal{C}^0(\Omega))^{\mathbb{N}}$.

Given now $k \in \mathbb{N}$, $p \in \mathbb{N}^n$, $|p| + k \leq \ell$, we can obviously define the mapping

$$(2.2.16) \qquad D^p : A^{\ell} \longrightarrow A^k$$

by

$$(2.2.17) \qquad s + (\mathcal{I}_{nd}(\Omega) \cap (\mathcal{C}^{\ell}(\Omega))^{\mathbb{N}}) \longmapsto D^p s + (\mathcal{I}_{nd}(\Omega) \cap (\mathcal{C}^{k}(\Omega))^{\mathbb{N}})$$

for $s \in \mathcal{A} \cap (\mathcal{C}^{\ell}(\Omega))^{\mathbb{N}}$.

§3. NONLINEAR PARTIAL DIFFERENTIAL OPERATORS ON SPACES OF GENERALIZED FUNCTIONS

Let us specify $m \in \mathbb{N}$, the order of the nonlinear partial differential operators considered in the sequel.

Suppose given an arbitrary continuous function

$$F \in \mathcal{C}^0(\Omega \times \mathbb{R}^{\bar{m}}) \tag{2.3.1}$$

where

$$\bar{m} = \text{car}\{p \in \mathbb{N}^n \mid |p| \leq m\}$$

Then we define the m-th order nonlinear partial differential operator

$$T(D) : \mathcal{C}^m(\Omega) \longrightarrow \mathcal{C}^0(\Omega) \tag{2.3.2}$$

by

$$(T(D)U)(x) = F(x, U(x), \ldots, D^p U(x), \ldots), \quad p \in \mathbb{N}^n, \ |p| \leq m, \ x \in \Omega, \ U \in \mathcal{C}^m(\Omega) \tag{2.3.3}$$

Suppose given $\ell \in \mathbb{N}$, $\ell \geq m$. Let us take

$$\bar{\mathcal{A}} \tag{2.3.4}$$

a subalgebra in $(\mathcal{C}^0(\Omega))^{\mathbb{N}}$ containing $\mathcal{I}_{nd}(\Omega) \cup T(D)(\mathcal{A} \cap (\mathcal{C}^\ell(\Omega))^{\mathbb{N}})$.

Finally, let us define the quotient algebra

$$\bar{A} = \bar{\mathcal{A}}/\mathcal{I}_{nd}(\Omega) \tag{2.3.5}$$

We show now that the mapping (2.3.2) can be extended to a mapping

$$T(D) : A^\ell \longrightarrow \bar{A} \tag{2.3.6}$$

by the definition

$$s + (\mathcal{I}_{nd}(\Omega) \cap (\mathcal{C}^\ell(\Omega))^{\mathbb{N}}) \longmapsto T(D)s + \mathcal{I}_{nd}(\Omega) \tag{2.3.7}$$

for all $s = (\psi_\nu \mid \nu \in \mathbb{N}) \in \mathcal{A} \cap (\mathcal{C}^\ell(\Omega))^{\mathbb{N}}$, where we denoted

$$T(D)s = (T(D)\psi_\nu \mid \nu \in \mathbb{N}) \tag{2.3.8}$$

Indeed, in order to see that (2.3.6) - (2.3.8) give a correct definition, it suffices to note that we have the property

$$(2.3.9) \qquad \begin{aligned} &\forall \quad t \in \mathcal{A} \cap (\mathcal{C}^{\ell}(\Omega))^{\mathbb{N}}, \quad v \in \mathcal{I}_{nd}(\Omega) \cap (\mathcal{C}^{\ell}(\Omega))^{\mathbb{N}} : \\ &\qquad T(D)(t + v) - T(D)t \in \mathcal{I}_{nd}(\Omega) \end{aligned}$$

which follows easily from (2.2.5) and (2.2.6).

The interest in the extensions (2.3.6) is in the fact that the global version of the Cauchy-Kovalevskaia theorem proved in the sequel will give generalized solutions which are elements in A^{ℓ}.

Remark 1

In Sections 2 and 3 above, we have obviously followed the general construction in Sections 6, 9 and 10 in Chapter 1, without however using in full detail the respective notations. The simplified notation in this Chapter may offer the reader a presentation which to a good extent is selfcontained. For the sake of completeness we note that, with the notations in Chapter 1, we have the following. Relation (1.6.4) gives

$$(2.3.10) \qquad \mathcal{U}(\Omega) = \mathcal{U}_{\mathcal{C}^0(\Omega),\mathbb{N}}$$

while (1.6.9)-(1.6.11) together with (2.2.7), (2.2.8), (2.2.12) and (2.2.16) give

$$(2.3.11) \qquad A^{\ell} \in AL_{\mathcal{C}^{\ell}(\Omega),\mathbb{N}}, \quad \ell \in \mathbb{N}$$

Further (1.13.5) together with (2.3.6)-(2.3.9) give

$$(2.3.12) \qquad A^{\ell} \overset{T(D)}{\leq} \bar{A}, \quad \ell \in \mathbb{N}$$

§4. BASIC LEMMA

Given the domain Ω in $\mathbb{R}^n$ and on it, the m-th order *analytic nonlinear* partial differential equation

$$(2.4.1) \qquad D_t^m U(t,y) = G(t,y,\ldots,D_t^p D_y^q U(t,y),\ldots)$$

with $x = (t,y) \in \Omega$, $t \in \mathbb{R}$, $y \in \mathbb{R}^{n-1}$, $m \geq 1$, $0 \leq p < m$, $q \in \mathbb{N}^{n-1}$, $p + |q| \leq m$. Further, given for (2.4.1) the noncharacteristic analytic hypersurface

(2.4.2) $S = \{x = (t,y) \in \Omega | t = t_0\} \neq \phi$

and the analytic Cauchy data

(2.4.3) $D_t^p U(t_0,y) = g_p(y), \quad 0 \leq p < m, \quad (t_0,y) \in S.$

Lemma 1

There exists $\Gamma \subset \Omega$ closed, nowhere dense and an analytic function $\psi : \Omega \backslash \Gamma \to \mathbb{C}$ which is a solution of (2.4.1) on $\Omega \backslash \Gamma$ and satisfies (2.4.3).

Proof

For every $x = (t_0,y) \in S$, the Cauchy-Kovalevskaia theorem, Walter, yields a nonvoid open set

(2.4.4) $\Omega_x \subset \Omega$, with $x \in \Omega_x$

and an anlytic function which is a solution of (2.4.1), (2.4.3) on Ω_x.

It follows by analytic continuation that we obtain an analytic solution of (2.4.1), (2.4.3) on the nonvoid open set

(2.4.5) $\Omega^1 = \bigcup_{x \in S} \Omega_x$

Obviously, we can only have the following two situations.

Case 1. If Ω^1 is dense in Ω, then the proof is completed by taking

(2.4.6) $\Gamma = \Omega \backslash \Omega^1$

Case 2. If Ω^1 is not dense in Ω, then we obtain the partition

(2.4.7) $\Omega = \Omega_1 \cup \Gamma_1 \cup \Omega^1$

where

(2.4.8) Ω_1 = interior of $(\Omega \backslash \Omega^1)$

is nonvoid open, while Γ_1 is closed, nowhere dense.

Now, we can take

(2.4.9) $x_1 = (t_1,y_1) \in \Omega_1$

Then, with $t_1 \in \mathbb{R}$ given by x_1, we define

(2.4.10) $$S_1 = \{x = (t,y) \in \Omega_1 \,|\, t = t_1\} \neq \phi$$

which is a noncharacteristic analytic hypersurface for (2.4.1). On S_1 we can consider any given analytic Cauchy data

(2.4.11) $$D_t^p U(t_1,y) = g_{1p}(y), \quad p \in \mathbb{N}, \quad p < m, \quad (t_1,y) \in S_1 .$$

In this way, we reduced the original problem of proving the Lemma 1 for Ω and S, to the problem of proving it for Ω_1 and S_1.

This reduction obviously sets up an iterative process which can only lead to one of the following two situations.

Alternative 1. After a finite number of iterations, we reach Case 1. More precisely, for some $h \geq 1$, we obtain the finite sequence of partitions

$$\Omega_0 = \Omega_1 \cup \Gamma_1 \cup \Omega^1$$
$$\ldots\ldots$$
$$\Omega_{h-1} = \Omega_h \cup \Gamma_h \cup \Omega^h$$
$$\Omega_h = \Gamma_{h+1} \cup \Omega^{h+1}$$

where $\Omega_0 = \Omega, \Omega_1, \ldots, \Omega_h, \Omega^1, \ldots, \Omega^{n+1}$ are nonvoid open, while $\Gamma_1, \ldots, \Gamma_{h+1}$ are closed, howhere dense.

In that case, we obviously have an analytic solution of (2.4.1), (2.4.3) on the nonvoid open set

$$\Omega' = \Omega^1 \cup \ldots \cup \Omega^{h+1}$$

while

$$\Omega' \text{ is dense in } \Omega$$

therefore, we can take

(2.4.12) $$\Gamma = \Omega \backslash \Omega'$$

and the proof is completed.

Alternative 2. We never reach Case 1, after any finite number of iterations. Then, for an ordinal number $\alpha \geq 1$, we obtain an open set

(2.4.13) $$\Omega_\alpha \subset \Omega$$

in the following way. If $\alpha = \beta + 1$ for a suitable ordinal number β, and the nonvoid open set $\Omega_\beta \subset \Omega$ has already been obtained, then we construct

(2.4.14) $\Omega^\alpha \subset \Omega_\beta$, Ω^α nonvoid, open

according to (2.4.5). Further, we define the nonvoid open set

(2.4.15) $$\Delta_\alpha = \bigcup_{\gamma \leq \alpha} \Omega^\gamma$$

and take

(2.4.16) $$\Omega_\alpha = \text{interior of } (\Omega \backslash \Delta_\alpha)$$

Otherwise, if $\alpha \neq \beta + 1$ for any ordinal number β, instead of (2.4.15), we define the nonvoid open set

(2.4.17) $$\Delta_\alpha = \bigcup_{\beta < \alpha} \Omega^\beta$$

and then again, take Ω_α as in (2.4.16).

This process can be continued until we reach $\Omega_\alpha = \phi$ in (2.4.16). In that case, we obviously have an analytic solution of (2.4.1), (2.4.3) on the nonvoid open set

$$\Omega' = \Delta_\alpha$$

and

$$\Omega' \text{ is dense in } \Omega$$

which means that by taking

(2.4.18) $$\Gamma = \Omega \backslash \Omega'$$

the proof is completed. □

§5. GLOBAL GENERALIZED SOLUTIONS

Given the m-th order analytic nonlinear partial differential equation

$$D_t^m U(t,y) = G(t,y,\dots,D_t^p D_y^q U(t,y), \dots) \tag{2.5.1}$$

with $x = (t,y) \in \Omega$, $t \in \mathbb{R}$, $y \in \mathbb{R}^{n-1}$, $m \geq 1$, $0 \leq p < m$, $q \in \mathbb{N}^{n-1}$, $p + |q| \leq m$, and with the analytic Cauchy data

$$D_t^p U(t_0,y) = g_p(y), \quad 0 \leq p < m, \quad (t_0,y) \in S \tag{2.5.2}$$

on the noncharacteristic analytic hypersurface

$$S = \{x = (t,y) \in \Omega \mid t = t_0\} \neq \phi \tag{2.5.3}$$

Let

$$\psi : \Omega \backslash \Gamma \longrightarrow \mathbb{C} \tag{2.5.4}$$

be an analytic solution of (2.5.1), (2.5.2) as given by the Lemma 1 in Section 4. Then

$$\Gamma \subset \Omega \quad \text{is closed and nowhere dense} \tag{2.5.5}$$

hence, since Γ is closed, it follows that

$$\begin{array}{l} \exists \ \gamma \in \mathcal{C}^\infty(\Omega) \ ; \\ \quad \Gamma = \{x \in \Omega \mid \gamma(x) = 0\} \end{array} \tag{2.5.6}$$

Suppose given a $\mathcal{C}^\infty$-smooth function $\alpha:\mathbb{R} \longrightarrow [0,1]$, such that

$$\begin{array}{ll} \alpha(x) = 0, & \text{for } |x| < a \\ \alpha(x) = 1, & \text{for } |x| > b \end{array} \tag{2.5.7}$$

for certain $0 < a < b < \infty$.

Let us define the sequence of $\mathcal{C}^\infty$-smooth functions

$$s = (\psi_\nu \mid \nu \in \mathbb{N}) \in (\mathcal{C}^\infty(\Omega))^{\mathbb{N}} \tag{2.5.8}$$

where for $\nu \in \mathbb{N}$, we have

$$\psi_\nu(x) = \left| \begin{array}{ll} \alpha((\nu + 1)\gamma(x))\psi(x) & \text{if } x \in \Omega\backslash\Gamma \\ 0 & \text{if } x \in \Gamma \end{array} \right. \tag{2.5.9}$$

Then obviously, this sequence s of $\mathcal{C}^\infty$-smooth functions has the property

$$\begin{aligned}&\forall \quad x \in \Omega\setminus\Gamma : \\ &\exists \quad \mu \in \mathbb{N}, \quad V \subset \Omega\setminus\Gamma \text{ neighbourhood of } x : \\ &\forall \quad \nu \in \mathbb{N}, \quad \nu \geq \mu, \; y \in V : \\ &\quad \psi_\nu(y) = \psi(y)\end{aligned} \tag{2.5.10}$$

Now, similar to (2.3.2), (2.3.3), let us define the nonlinear partial differential operator

$$T(D) : \mathcal{C}^m(\Omega) \longrightarrow \mathcal{C}^0(\Omega) \tag{2.5.11}$$

corresponding to (2.5.1) by

$$(T(D)U)(t,y) = D_t^m U(t,y) - G(t,y,\ldots,D_t^p D_y^q U(t,y),\ldots) \tag{2.5.12}$$

with $x = (t,y) \in \Omega$, $t \in \mathbb{R}$, $y \in \mathbb{R}^{n-1}$, $0 \leq p < m$, $q \in \mathbb{N}^{n-1}$, $p + |q| \leq m$. With the above notation, the equation (2.5.1) can be written in the equivalent form

$$T(D)U(x) = 0, \quad x \in \Omega \tag{2.5.13}$$

Then (2.5.4), (2.5.5) and (2.5.10) obviously imply

$$T(D)s \in \mathcal{I}_{nd}(\Omega) \cap (\mathcal{C}^\infty(\Omega))^{\mathbb{N}} \tag{2.5.14}$$

With the above preparations, we can now construct the algebras of generalized functions within which the analytic nonlinear partial differential equation (2.5.1) has a *global* generalized solution for the noncharacteristic analytic Cauchy data (2.5.2), (2.5.3).

Let us take a subalgebra

$$\mathcal{A} \subset (\mathcal{C}^0(\Omega))^{\mathbb{N}} \tag{2.5.15}$$

which satisfies the conditions

$$s \in \mathcal{A} \tag{2.5.16}$$

and, see (2.2.7), (2.2.8), (2.2.15), (2.2.18)

$$\mathcal{I}_{nd}(\Omega) \cup \mathcal{S}^0(\Omega) \subset \mathcal{A} \tag{2.5.17}$$

The possibility of such a choice is obvious, since $\mathcal{A} = (\mathcal{C}^0(\Omega))^{\mathbb{N}}$ for instance, will satisfy (2.5.16) and (2.5.17).

Now, with the construction in (2.2.16), (2.3.5) and (2.3.6) we obtain the following *global existence* result for the equations (2.5.1), or equivalently (2.5.13).

Theorem 1

For $\ell \in \mathbb{N}$, $\ell \geq m$, the m-th order analytic nonlinear partial differential equation in (2.5.13)

(2.5.18) $$T(D)U(t,y) = 0, \quad x = (t,y) \in \Omega$$

with the noncharacteristic analytic Cauchy data

(2.5.19) $$D_t^p U(t_0,y) = g_p(y), \quad 0 \leq p < m, \quad (t_0,y) \in S$$

and

(2.5.20) $$S = \{x = (t,y) \in \Omega | t = t_0\} \neq \phi$$

has generalized solutions

(2.5.21) $$U \in A^\ell$$

defined on the whole of Ω.

These solutions U are analytic functions

(2.5.22) $$\psi : \Omega \backslash \Gamma \longrightarrow \mathbb{C}$$

when restricted to suitable open, dense subsets $\Omega \backslash \Gamma$, where

(2.5.23) $$\Gamma \subset \Omega \text{ closed, nowhere dense.}$$

Proof

Using the construction in (2.2.16), (2.3.5) and (2.3.6) we can extend (2.5.11) to a mapping

(2.5.24) $$T(D) : A^\ell \longrightarrow \bar{A}$$

where

(2.5.25) $$A^\ell = (\mathcal{A} \cap (\mathcal{C}^\ell(\Omega))^{\mathbb{N}})/(\mathcal{I}_{nd}(\Omega) \cap (\mathcal{C}^\ell(\Omega))^{\mathbb{N}})$$

while

(2.5.26) $$\bar{A} = \bar{\mathcal{A}}/\mathcal{I}_{nd}(\Omega)$$

with $\bar{\mathcal{A}}$ being a subalgebra in $(\mathcal{C}^0(\Omega))^{\mathbb{N}}$ containing $\mathcal{I}_{nd}(\Omega) \cup T(D)(\bar{\mathcal{A}} \cap (\mathcal{C}^{\ell}(\Omega))^{\mathbb{N}})$.

Let us now define

$$(2.5.27) \qquad U = s + (\mathcal{I}_{nd}(\Omega) \cap (\mathcal{C}^{\ell}(\Omega))^{\mathbb{N}}) \in A^{\ell}$$

which is possible in view of (2.5.8) and (2.5.16).

It only remains to show that with the mapping in (2.5.24), we have

$$(2.5.28) \qquad T(D)U = 0 \in \bar{A}$$

But in view of (2.5.27), (2.3.7) and (2.3.8) we have

$$T(D)U = T(D)s + \mathcal{I}_{nd}(\Omega)$$

thus (2.5.28) follows from (2.5.14). □

Remark 2

The result in Lemma 1 in Section 4, and therefore in Theorem 1 above, is an *existence* result.
From the proof of Lemma 1, in particular from the freedom of choice in (2.4.9) and (2.4.11), it is obvious that in general, many solutions (2.5.4), (2.5.5) can be obtained. A first open problem then is whether the class of solutions (2.5.4), (2.5.5) exhausts all the solutions corresponding to various solution concepts which can reasonably be associated with analytic nonlinear partial differential equations. Here one can refer for instance to situations where solution concepts are supposed to accommodate phenomena such as turbulence, strange attractors, etc., see Richtmeyer, Temam and the literature cited there. Within the solution concept used above further open problems concern the *uniqueness* and *regularity* of the global solutions obtained. From the above construction it is obvious that these problems are connected with the appropriate uses of the freedom in the choice of the subalgebras $\mathcal{A}$ and $\bar{\mathcal{A}}$. Indeed, the smaller these subalgebras, within the required conditions, the better the uniqueness and regularity properties of the corresponding global, generalized solutions in (2.5.21). This problem is closely connected with the stability, generality and exactness of generalized solutions for linear and nonlinear partial differential equations, introduced and specifically dealt with in Section 11, Chapter 1. In view of (2.3.12) and (2.5.24)-(2.5.28), the global generalized solution U in (2.5.21) has indeed the meaning defined in Section 13, Chapter 1.

§6. CLOSED NOWHERE DENSE SINGULARITIES WITH ZERO LEBESQUE MEASURE

In this Section, in Theorem 2, we improve on the result in Theorem 1, Section 5, by showing that the closed, nowhere dense singularity Γ in (2.5.22), (2.5.23), outside of which the global generalized solution U in (2.5.21) is analytic, can be chosen in such a way that it has *zero Lebesque measure*, i.e.

$$(2.6.1) \qquad \text{mes } \Gamma = 0$$

The surprising fact about this strengthening of Theorem 1 in Section 5 is that it can be obtained in a *constructive* way, without the use of transfinite induction.

The essential instrument in obtaining this strengthened result in (2.6.1) is presented next in Proposition 1, which was offered by M. Oberguggenberger in a private communication, Oberguggenberger [5]. This result has as well an obvious interest in itself, since it is, according to our best knowledge, the *first result* which gives the rather sharp kind of information in (2.6.1) on the size of the subsets on which the classical Cauchy-Kovalevskaia theorem may fail to yield analytic solutions.

Proposition 1 (Oberguggenberger [5])

Given the *analytic* nonlinear partial differential equation (2.5.1), there exists $\Gamma \subset \Omega$ with

$$(2.6.2) \qquad \Gamma \text{ closed, nowhere dense in } \Omega$$

$$(2.6.3) \qquad \text{mes } \Gamma = 0$$

and $W : \Omega \backslash \Gamma \longrightarrow \mathbb{C}$ an *analytic* solution of (2.5.1) on $\Omega \backslash \Gamma$.

Proof

Assume given $(t,y) \in \Omega$. If we choose some initial values on an analytic hypersurface passing through (t,y), then the Cauchy-Kovalevskaia theorem yields

$$(t,y) \in I = \prod_{1 \leq i \leq n} (\alpha_i, \beta_i) \subset \Omega$$

and an analytic solution $W : I \longrightarrow \mathbb{C}$ of the equation (2.5.1) on I.

Assume given $K \subset \Omega$ open, such that its closure $\overline{K}$ is compact and $\overline{K} \subset \Omega$. Then applying the above to points $(t,y) \in \overline{K}$, we obtain

$$I_j = \prod_{1 \leq i \leq n} (\alpha_{ji}, \beta_{ji}) \subset \Omega, \quad 1 \leq j \leq J$$

and analytic solutions $W_j : I_j \longrightarrow \mathbb{C}$ of the equation (2.5.1) on each I_j, with $1 \leq j \leq J$, such that

$$K \subset \bigcup_{1\leq j\leq J} I_j$$

Assume given $1 \leq i \leq n$ and let

$$(2.6.4) \qquad \gamma_{0,i} < \dots < \gamma_{j(i),i}$$

be the set of pairwise different elements in $\alpha_{1i},\dots,\alpha_{Ji},\beta_{1i},\dots,\beta_{Ji}$ taken in increasing order. Let us denote by

$$\mathcal{I}$$

the set of all

$$I = \prod_{1\leq i\leq n} (\gamma_{j_i-1,i}\ \gamma_{j_i,i})$$

where $1 \leq j_i \leq j(i)$, with $1 \leq i \leq n$. Further, let us denote by

$$Q_1 ,\dots, Q_H$$

all the $I \in \mathcal{I}$ such that

$$I \subset \bigcup_{1\leq j\leq J} I_j$$

Then obviously

$$(2.6.5) \qquad K \subset \bigcup_{1\leq j\leq J} I_j \subset \overline{\bigcup_{1\leq h\leq H} Q_h} \subset \Omega$$

But in view of (2.6.4), $Q_1,\dots,Q_H$ are pairwise disjoint and

$$\forall\ 1 \leq h \leq H :$$
$$\exists\ 1 \leq j_h \leq J :$$
$$Q_h \subset I_{j_h}$$

and then

$$(2.6.6) \qquad W_{j_h}\big|_{Q_h} : Q_h \longrightarrow \mathbb{C}$$

is an analytic solution of equation (2.5.1) on Q_h.

Let us now denote for $1 \leq h \leq H$

$$R_h = Q_h \cap K$$

which are obviously open and pairwise disjoint. We shall show that

(2.6.7) $$\overline{\bigcup_{1 \leq h \leq H} R_h} = \overline{K}$$

Indeed, first we note that the inclusion '$\subset$' is immediate. Now we prove the inclusion '$\supset$'. Obviously

$$\bigcup_{1 \leq h \leq H} R_h = \bigcup_{1 \leq h \leq H} (Q_h \cap K) = (\bigcup_{1 \leq h \leq H} Q_h) \cap K$$

If we denote

$$A = \bigcup_{1 \leq h \leq H} Q_h, \quad B = K$$

and apply Lemma 2 below to (2.6.5), we obtain

$$\overline{(\bigcup_{1 \leq h \leq H} Q_h) \cap K} \supset \overline{K}$$

and the proof of (2.6.7) is completed.

Since $R_1, \ldots, R_H$ are pairwise disjoint, we obtain from (2.6.6) an analytic solution

(2.6.8) $$W : \bigcup_{1 \leq h \leq H} R_h \longrightarrow \mathbb{C}$$

of the equation (2.5.1) on an open, dense subset K' of K, where

(2.6.9) $$K' = \bigcup_{1 \leq h \leq H} R_h$$

From (2.6.4) it is obvious that

(2.6.10) $$\text{mes } (K \backslash K') = 0$$

It only remains now to extend (2.6.8)-(2.6.10) from K to Ω.

Let us take a sequence

$$K_0, K_1, \ldots K_\nu, \ldots$$

of nonvoid open subsets in Ω, such that

$$\overline{K}_\nu \subset K_{\nu+1}, \ \overline{K}_\nu \ \text{compact}, \quad \nu \in \mathbb{N}$$

and

(2.6.11) $$\bigcup_{\nu \in \mathbb{N}} K_\nu = \Omega$$

We apply (2.6.8)-(2.6.10) successively to

$$K = K_0 \ \text{ and } K = K_\nu \backslash \overline{K}_{\nu-1}, \ \nu \in \mathbb{N}, \quad \nu \geq 1$$

and obtain

$$W_\nu : K'_\nu \longrightarrow \mathbb{C}, \quad v \in \mathbb{N}$$

analytic solutions of (2.5.1) on K'_ν, with

$$K'_\nu = \bigcup_{1 \leq h \leq H_\nu} R_{\nu h}, \quad \nu \in \mathbb{N}$$

Since K'_ν, with $\nu \in \mathbb{N}$ are obviously pairwise disjoint, we only have to show that

$$\Omega \subset \overline{\bigcup_{\nu \in \mathbb{N}H} K'_\nu}$$

For that we note the following obvious relations

$$\overline{\bigcup_{\nu \in \mathbb{N}} K'_\nu} = \overline{\bigcup_{\substack{\nu \in \mathbb{N} \\ 1 \leq h \leq H_\nu}} R_{\nu h}} \supset \bigcup_{\nu \in \mathbb{N}} \overline{\bigcup_{1 \leq h \leq H_\nu} R_{\nu h}} = K_0 \cup \bigcup_{\substack{\nu \in \mathbb{N} \\ \nu \geq 1}} \overline{K_\nu \backslash \overline{K}_{\nu-1}} \supset \overline{\Omega}$$

We also note that in view of (2.6.10), we have

$$\text{mes}(K_\nu \backslash K'_\nu) = 0, \quad \nu \in \mathbb{N}$$

which together with (2.6.11) yield

$$\text{mes}(\Omega \backslash \bigcup_{\nu \in \mathbb{N}} K'_\nu) = 0$$ □

Lemma 2

Let A, B be subsets in a topological space. If B is open and $\overline{B} \subset \overline{A}$, then also

$$\overline{B} \subset \overline{A \cap B}$$

Proof

Take $x \in \overline{B} \setminus \overline{A \cap B}$. Then $x \in \overline{B}$ yields

$$\forall \quad Q \quad \text{open} : x \in Q \Longrightarrow B \cap Q \neq \phi$$

while $x \notin \overline{A \cap B}$ implies

$$\exists \quad Q_0 \quad \text{open} : x \in Q_0 \quad \text{and} \quad A \cap B \cap Q_0 = \phi$$

But in view of the above

$$Q_1 = B \cap Q_0 \neq \phi$$

and obviously Q_1 is open. If we take now $y \in Q_1 \subset B$, we obtain $y \notin \overline{A}$, since $A \cap Q_1 = \phi$. Hence $B \setminus \overline{A} \neq \phi$, which is absurd □

An now we come to the *strengthened* form of the globalized version of the classical Cauchy-Kovalevskaia theorem.

Theorem 2

Given the m-th order analytic nonlinear partial differential equation

$$(2.6.12) \qquad D_t^m U(t,y) = G(t,y,\ldots,D_t^p D_y^q U(t,y),\ldots)$$

with $x = (t,y) \in \Omega$, $t \in \mathbb{R}$, $y \in \mathbb{R}^{n-1}$, $m \geq 1$, $0 \leq p < m$, $q \in \mathbb{N}^{n-1}$, $p + |q| \leq m$, and with the noncharacteristic analytic Cauchy data

$$(2.6.13) \qquad D_t^p U(t_0,y) = g_p(y), \quad 0 \leq p < m, \quad (t_0,y) \in S$$

where

$$(2.6.14) \qquad S = \{x = (t,y) \in \Omega \mid t = t_0\} \neq \phi$$

Then for $\ell \in \mathbb{N}$, $\ell \geq m$, there exist generalized solutions

$$(2.6.15) \qquad U \in A^{\ell}$$

defined on the whole of Ω.

These solutions U are analytic functions

(2.6.16) $\psi \in \Omega\backslash\Gamma \longrightarrow \mathbb{C}$

when restricted to suitable open, dense subsets $\Omega\backslash\Gamma$, where

(2.6.17) $\Gamma \subset \Omega$ closed, nowhere dense

and

(2.6.18) $\text{mes}(\Gamma) = 0$

Proof

It follows easily from Proposition 1 above and the proof of Theorem 1 in Section 5.

§7. STRANGE PHENOMENA IN PARTIAL DIFFERENTIAL EQUATIONS

§7.1 Too Many Equations and Solutions?!

The results in Theorems 1 and 2 in this Chapter may at first seem strange owing to the apparent *excessive wealth* of the patched up type generalized solutions which they provide.

Let us further enquire into that phenomenon of 'excessive wealth' of solutions of differential equations, and cite two recent results, one concerning *ordinary*, and the other *partial* differential equations.

A rather related fact which comes to attention when observing the various linear and nonlinear ordinary or partial differential equations which appear in the customary mathematical models is their *low order*, as well as relatively *simple nonlinearity*. For instance, the Navier-Stokes equations or those of general relativity are of *second* order and have *quadratic* nonlinearities, etc. In view of that, one may as well ask whether indeed we need *general nonlinear* theories capable of handing partial differential equations of *arbitrary* order and nonlinearity. The question becomes only aggravated when we learn about several recent results which seem to indicate that partial differential equations of high order and nonlinearity could hardly be models of physical theories, since they present *strange properties* concerning their *sets of solutions*. In the next two subsections we shall present two such recent results. They seem to indicate the existence of a kind of *upper bound* on the complexity of partial differential equations which may be suitable for modelling physical phenomena.

§7.2 Universal Ordinary Differential Equations

As if to make things worse, the mentioned kind of upper bound seems to exist even in the particular world of nonlinear ordinary differential equations of polynomial type. We mention in this respect the following recent result, Rubel.

There exists a nontrivial *fourth order* differential equation

$$P(u',u'',u''',u'^{v}) = 0,\ x \in \mathbb{R}, \tag{2.7.1}$$

where P is a polynomial in four variables with integer coefficients, such that for every $f \in C^0(\mathbb{R})$ and $w \in C^0(\mathbb{R})$, with $w(x) > 0$, for $x \in \mathbb{R}$, there exists a *solution* $u \in C^\infty(\mathbb{R})$ of (2.7.1) with the property

$$|f(x) - u(x)| \leq w(x),\quad x \in \mathbb{R} \tag{2.7.2}$$

In other words, (2.7.1) has *so many* C^∞-smooth solutions u that these solution can approximate every continuous function f on $\mathbb{R}$ in the rather strong sense of (2.7.2).

An additional *disturbing* feature of the above result is in the simplicity and directness of its proof, whose main steps we reproduce now. Assume given f and w as above. Obviously we can find a pricewise affine continuous function $g \in C^0(\mathbb{R})$, such that $|f(x) - g(x)| \leq w(x)/2$, with $x \in \mathbb{R}$. Therefore, it suffices to find a solution $u \in C^\infty(\mathbb{R})$ of (2.7.1), such that $|g(x) - u(x)| \leq w(x)/2$, with $x \in \mathbb{R}$. This can be done easily as follows. Let us define $\psi, \chi \in C^\infty(\mathbb{R})$ by

$$\psi(x) = \begin{cases} e^{-1/(1-x^2)} & \text{if } x \in (-1,1) \\ 0 & \text{if } x \in \mathbb{R}\setminus(-1,1) \end{cases}$$

$$\xi(x) = \int_{-\infty}^{x} \psi(y)dy,\ x \in \mathbb{R}$$

Given $\alpha, \beta, A, B \in \mathbb{R}$, let us define $\chi \in C^\infty(\mathbb{R})$ by

$$\chi(x) = A\xi(\alpha x + \beta) + B,\quad x \in \mathbb{R} \tag{2.7.3}$$

Then, by an ingenious sequence of differentiaions and eliminations, aimed at α, β, A and B, one obtains (2.7.1) as the equation which admits (2.7.3) as solution, for all the values of α, β, A and B.

Assume now given a finite interval $I = [a,b] \subset \mathbb{R}$ on which g is affine. Obviously, one can choose $\alpha, \beta \in \mathbb{R}$, such that χ has the constant value A for $a \leq x \leq a+\epsilon$ and has the constant value B for $b-\epsilon \leq x \leq b$, where $\epsilon > 0$ is suitably chosen. Of course, we can choose $A = g(a)$ and

$B = g(b)$ and then g and χ are both monotonously increasing or decreasing on $I = [a,b]$. Now, by choosing I sufficiently small, the continuity of g and χ will yield

$$(2.7.4) \qquad |g(x) - \chi(x)| \leq w(x)/2, \quad x \in I$$

Finally, the solution $u \in \mathcal{C}^\infty(\mathbb{R})$ of (2.7.1) is obtained by the concatenation of all χ. Then (2.7.2) follows from (2.7.4).

§7.3 Universal Partial Differential Equations

After the result in subsection 7.3 above, the following similar property of nonlinear partial differential equations, Buck, may no longer seem surprising. This time however the proof is rather hard, as it uses Kolmogorov's solution of Hilbert's Thirteenth Problem. In view of this, we shall omit the proof and only present the result which reads as follows.

For every $n \in \mathbb{N}$, $n \geq 2$, there exists a nontrivial polynomial nonlinear partial differential equation

$$(2.7.5) \qquad P(U,\ldots,D^pU,\ldots) = 0, \quad x \in [0,1]^n$$

where P is a polynomial with real coefficients and degree at most $d(n)$, with the order of (2.7.5) at most $m(n)$, i.e., $p \in N^n$, $|p| \leq m(n)$ in (2.7.5), and such that for every $f \in \mathcal{C}^0([0,1]^n)$ and $\epsilon > 0$, there exists a *solution* $U \in \mathcal{C}^\infty([0,1]^n)$ of (2.7.5), with the property

$$(2.7.6) \qquad |f(x) - U(x)| \leq \epsilon, \quad x \in [0,1]^n$$

In this way, similar to (2.7.1) and (2.7.2), the algebraic partial differential equation in (2.7.5) has *so many* $\mathcal{C}^\infty$-smooth solutions U that they can *uniformly approximate* every continuous function f on $[0,1]^n$

Unlike in subsection 7.2 above, a somewhat comforting thing with the result in (2.7.5) and (2.7.6) is that - as known so far - the values of the order $m(n)$ and degree $d(n)$ are quite large even for small values of n, such as for instance $n = 2$, 3 or 4.

§7.4 Final Remark

The results in the above subsection 7.2 and 7.3 show the existence of polynomial ordinary or partial differential equations which are *universal* in the sense that their $\mathcal{C}^\infty$-smooth solutions are *sufficiently many* in order to approximate uniformly continuous function. A *difficulty* arising from that situation is the following. A given ordinary or partial differential equation which appears in the modelling of a physical phenomenon will define by its solutions the possible states of this phenomenon. Therefore the set of these solutions should have a similar size with that of the set of states. Now, the set of possible states of a given physical phenomenon is usually much more *restricted* than the set of all continuous functions, for instance. In view of this, it may appear that equations such as (2.7.1) or (2.7.5) may fail to model any usual physical phenomena.

APPENDIX 1

ON THE STRUCTURE OF THE NOWHERE DENSE IDEALS

Proposition 2

$\mathcal{I}_{nd}(\Omega)$ is an ideal in $(\mathcal{C}^0(\Omega))^{\mathbb{N}}$.

Proof

In view of (2.2.4) it is obvious that

$$\mathcal{I}_{nd}(\Omega) \cdot (\mathcal{C}^0(\Omega))^{\mathbb{N}} \subset \mathcal{I}_{nd}(\Omega)$$

It only remains to show that

$$\mathcal{I}_{nd}(\Omega) + \mathcal{I}_{nd}(\Omega) \subset \mathcal{I}_{nd}(\Omega) \tag{2.A1.1}$$

Let $w, w' \in \mathcal{I}_{nd}(\Omega)$, then (2.2.4) yields

$$\Gamma, \Gamma' \subset \Omega \text{ closed, nowhere dense} \tag{2.A1.2}$$

such that

$$\begin{array}{l} \forall\ x \in \Omega \backslash \Gamma,\ \ x' \in \Omega \backslash \Gamma' : \\ \exists\ \mu, \mu' \in \mathbb{N}\ ; \\ \forall\ \nu, \nu' \in \mathbb{N},\ \nu \geq \mu,\ \nu' \geq \mu' : \\ \quad w_\nu(x) = w'_{\nu'}(x') = 0 \end{array} \tag{2.A1.3}$$

Let us denote

$$w'' = w + w', \quad \Gamma = \Gamma \cup \Gamma'$$

then obviously $\Gamma'' \subset \Omega$ is again closed and nowhere dense. Moreover, in view of (2.A1.3), w'' and Γ'' will also satisfy (2.2.4). Therefore, (2.A1.1) holds. □

Proposition 3

The nowhere dense ideal satisfies the *neutrix condition*

$$\mathcal{I}_{nd}(\Omega) \cap \mathcal{U}(\Omega) = \mathcal{O} \tag{2.A1.4}$$

Proof

Take $\psi \in \mathcal{C}^0(\Omega)$ such that

$$u(\psi) \in \mathcal{I}_{nd}(\Omega)$$

Then (2.2.4) yields

$$(2.A1.5) \qquad \Gamma \subset \Omega \text{ closed, nowhere dense}$$

such that

$$(2.A1.6) \qquad \begin{array}{l} \forall \ x \in \Omega \backslash \Gamma : \\ \quad \psi(x) = 0 \end{array}$$

But the continuity of ψ on Ω and the relations (2.A1.5), (2.A1.6) will obviously yield

$$\psi(x) = 0, \quad x \in \Omega \qquad \square$$

An important property of the nowhere dense ideal $\mathcal{I}_{nd}(\Omega)$ is presented now.

Proposition 4

Every sequence $w \in \mathcal{I}_{nd}(\Omega)$ of continuous functions on Ω satisfies the following *stronger* version of (2.2.4)

$$(2.A1.7) \qquad \begin{array}{l} \exists \ \Gamma \subset \Omega \text{ closed, nowhere dense :} \\ \forall \ x \in \Omega \backslash \Gamma : \\ \exists \ \mu \in \mathbb{N}, \ V \subset \Omega \backslash \Gamma \text{ neighbourhood of } x : \\ \forall \ \nu \in \mathbb{N}, \ \nu \geq \mu, \ y \in V : \\ \quad w_\nu(y) = 0 \end{array}$$

Proof

It follows easily from Lemma 3 below. Indeed, assume $w \in \mathcal{I}_{nd}(\Omega)$ and $\Gamma \subset \Omega$ as given by (2.2.4). Then $\Omega' = \Omega \backslash \Gamma$ is open and dense in Ω. Further

$$\begin{array}{l} \forall \ x \in \Omega' : \\ \exists \ \mu \in \mathbb{N} : \\ \forall \ \nu \in \mathbb{N}, \ \nu \geq \mu \ ; \\ \quad w_\nu(x) = 0 \end{array}$$

Let $I \subset \Omega'$ be nonvoid, closed. Then Lemma 3 yields a nonvoid, relatively open $\Delta_I \subset I$ such that w_ν vanishes on Δ_I, for $\nu \in \mathbb{N}$ sufficiently large.

Let Ω'' be the union of all the nonvoid, open subsets $\Delta \subset \Omega'$ on which w_ν vanishes for $\nu \in \mathbb{N}$ sufficiently large.

But Ω'' is dense in Ω' since in the argument above, one can choose I as ranging over all the closed balls in Ω'. Hence Ω'' is dense in Ω, since Ω' is dense in Ω.

Now $\Gamma' = \Omega \backslash \Omega''$ will satisfy (2.A1.7) □

Lemma 3

Suppose E is a complete metric space, E' is a topological space and we are given the continuous functions $f : E \longrightarrow E'$, and $f_\nu : E \longrightarrow E'$, with $\nu \in \mathbb{N}$, such that

$$\begin{aligned} &\forall \ x \in E : \\ &\exists \ \nu \in \mathbb{N} : \\ &\forall \ \nu \in \mathbb{N}, \ \nu \geq \mu : \\ &\quad f_\nu(x) = f(x) \end{aligned}$$

Then for every nonvoid, closed subset $I \subset E$, there exists a nonvoid, relatively open subset $\Delta \subset I$ and $\nu \in \mathbb{N}$, such that

$$\begin{aligned} &\forall \ x \in \Delta, \ \nu \in \mathbb{N}, \ \nu \geq \mu : \\ &\quad f_\nu(x) = f(x) \end{aligned}$$

Proof

It follows easily from a Baire category argument.

Indeed, for $I \subset E$ nonvoid, closed and for $\nu \in \mathbb{N}$, let us denote

$$I_\mu = \left\{ x \in I \ \middle| \ \begin{array}{l} \forall \ \nu \in \mathbb{N}, \ \nu \geq \mu : \\ \quad f_\nu(x) = f(x) \end{array} \right\}$$

Then the hypothesis obviously yields

$$I = \bigcup_{\mu \in \mathbb{N}} I_\mu$$

But for $\mu \in \mathbb{N}$, I_μ is closed in I, since f and all f_ν are continuous. And I being closed in E, it is in itself a complete metric space. Thus the Baire category argument implies that for at least one $\mu \in \mathbb{N}$, the interior of I_μ relative to I is not void □

The utility of Proposition 4 is in the next result.

Corollary 1

For $\ell \in \mathbb{N}$, we have the relations

$$D^p(\mathcal{I}_{nd}(\Omega) \cap (\mathcal{C}^\ell(\Omega))^{\mathbb{N}}) \subset \mathcal{I}_{nd}(\Omega), \; p \in \mathbb{N}, \; |p| \leq \ell \qquad (2.A1.8)$$

Proof

It follows at once from (2.A1.7) □

CHAPTER 3

ALGEBRAIC CHARACTERIZATION FOR THE SOLVABILITY OF NONLINEAR PARTIAL DIFFERENTIAL EQUATIONS

§1. INTRODUCTION

In this Chapter we consider arbitrary polynomial nonlinear partial differential equations with continuous coefficients

$$(3.1.1) \qquad \sum_{1 \leq i \leq h} c_i(x) \prod_{1 \leq j \leq k_i} D^{p_{ij}} U(x) = f(x), \quad x \in \Omega$$

where $\Omega \subset \mathbb{R}^n$ is nonvoid open, $c_i, f \in \mathcal{C}^0(\Omega)$ are given, $p_{ij} \in \mathbb{N}^n$, while $U : \Omega \longrightarrow \mathbb{R}$ is the unknown function.

The aim of the Chapter is to establish a necessary and sufficient condition for the existence of generalized solutions for partial differential equations in (3.1.1). This characterization happens to be of a simple and purely algebraic nature and it is given by a special case of the *neutrix condition*.

Obviously, the nonlinear partial differential equations in (3.1.1) are partly more general than the arbitrary analytic nonlinear partial differential equations. To that extent, the existence of generalized solutions, more precisely, their characterization, proved in this Chapter is an extension of the global version of the Cauchy-Kovalevskaia theorem presented earlier in Chapter 2.

The reason we consider the particular, *polynomial* form of nonlinear partial differential equations in (3.1.1) is that they allow for a rather simple and direct algebraic treatment leading to the mentioned neutrix characterization for the existence of generalized solutions. At the cost of certain technical complications , a similar result may be obtained for arbitrary continuous nonlinear partial differential equations

$$(3.1.2) \qquad F(x, U(x), \ldots, D^p U(x), \ldots) = 0, \quad x \in \Omega$$

where F is any real valued function, continuous in all its arguments. Such a result will obviously give a full scale extension of the mentioned global version of the Cauchy-Kovalevskaia theorem.

The question remains open whether existence results such as those in this Chapter can be obtained within the particular framework in Colombeau [1,2].

§2. THE NOTION OF GENERALIZED SOLUTION

For convenience, and in order to make the presentation of this Chapter more selfcontained, we recall the relevant general entities defined in Chapter 1, and do so by using again a simplified notation.

Let us denote by $T(D)$ the polynomial nonlinear partial differential operator in the left hand term of (3,1.1) and define its *order* by

$$m = \max\{|p_{ij}| \;\big|\; 1 \leq i \leq h,\ 1 \leq j \leq k_i\} \tag{3.2.1}$$

The spaces of generalized functions used in the sequel will be quotient vector spaces

$$E = \mathcal{S}/\mathcal{V} \tag{3.2.2}$$

or quotient algebras

$$A = \mathcal{A}/\mathcal{I} \tag{3.2.3}$$

where we have

$$\mathcal{V} \subset \mathcal{S} \subset (\mathcal{C}^m(\Omega))^{\mathbb{N}} \tag{3.2.4}$$

with $\mathcal{V}$ and $\mathcal{S}$ vector subspaces, and respectively

$$\mathcal{I} \subset \mathcal{A} \subset (\mathcal{C}^0(\Omega))^{\mathbb{N}} \tag{3.2.5}$$

where $\mathcal{A}$ is a subalgebra, while $\mathcal{I}$ is an ideal in $\mathcal{A}$.

Further, we shall require that the inclusion diagrams hold

$$\begin{array}{ccc} \mathcal{V} & \longrightarrow & \mathcal{S} \\ \uparrow & & \uparrow \\ \mathcal{O} & \longrightarrow & \mathcal{U}^m(\Omega) \end{array} \tag{3.2.6}$$

and

$$\begin{array}{ccc} \mathcal{I} & \longrightarrow & \mathcal{A} \\ \uparrow & & \uparrow \\ \mathcal{O} & \longrightarrow & \mathcal{U}^0(\Omega) \end{array} \tag{3.2.7}$$

where $\mathcal{O}$ denotes the null vector subspace, while $\mathcal{U}^m(\Omega)$ is the *diagonal* in $(\mathcal{C}^m(\Omega))^{\mathbb{N}}$, that is

(3.2.8) $\mathcal{U}^m(\Omega) = \{u(\psi) \mid \psi \in \mathcal{C}^m(\Omega)\}$

with $u(\psi) = (\psi,,\ldots,\psi,\ldots) \in (\mathcal{C}^m(\Omega))^{\mathbb{N}}$, and similarly for $\mathcal{U}^0(\Omega)$. Note that $\mathcal{U}^m(\Omega) \subset \mathcal{U}^0(\Omega)$ and both are subalgebras.

Finally, each of the inclusion diagrams (3.2.6) and (3.2.7) is supposed to satisfy the *neutrix condition*, that is

(3.2.9) $\mathcal{V} \cap \mathcal{U}^m(\Omega) = \mathcal{O}$

respectively

(3.2.10) $\mathcal{I} \cap \mathcal{U}^0(\Omega) = \mathcal{O}$

It is easy to see that (3.2.9) and (3.2.10) respectively are the necessary and sufficient conditions for the existence of the vector space and algebra embeddings

(3.2.11) $$\begin{array}{ccc} \mathcal{C}^m(\Omega) & \longrightarrow & E = \mathcal{S}/\mathcal{V} \\ \psi & \longmapsto & u(\psi) + \mathcal{V} \end{array}$$

and

(3.2.12) $$\begin{array}{ccc} \mathcal{C}^0(\Omega) & \longrightarrow & A = \mathcal{A}/\mathcal{I} \\ \psi & \longmapsto & u(\psi) + \mathcal{I} \end{array}$$

We recall that from a geometric point of view, the neutrix condition (3.2.9) means that the vector subspace $\mathcal{V}$ is *off diagonal* in the cartesian product $(\mathcal{C}^m(\Omega))^{\mathbb{N}}$. Similarly, the neutrix conditon (3.2.10) means that the ideal $\mathcal{I}$ is *off diagonal* in the cartesian product $(\mathcal{C}^0(\Omega))^{\mathbb{N}}$.

Assuming now that

(3.2.13) $D^p\mathcal{V} \subset \mathcal{I}, \quad D^p\mathcal{S} \subset \mathcal{A}, \quad \forall \; p \in \mathbb{N}^n, \quad |p| \le m,$

we can obviously extend the classical partial derivative operators

(3.2.14) $D^p : \mathcal{C}^m(\Omega) \to \mathcal{C}^0(\Omega), \quad \forall \, p \in \mathbb{N}^n, \; |p| \le m$

to

(3.2.15) $D^p : E \to A, \quad \forall \, p \in \mathbb{N}^n, \quad |p| \le m$

by defining them according to

(3.2.16) $$D^p(s + \mathcal{V}) = D^p s + \mathcal{I}, \quad \forall\, s \in \mathcal{S}, \quad p \in \mathbb{N}^n, \quad |p| \leq m$$

where $D^p s = (D^p s_0, \ldots, D^p s_\nu, \ldots)$ for every $s = (s_0, \ldots, s_\nu, \ldots) \in (\mathcal{C}^m(\Omega))^{\mathbb{N}}$.

It follows that, as an extension of the classical partial differential operator

(3.2.17) $$T(D) : \mathcal{C}^m(\Omega) \longrightarrow \mathcal{C}^0(\Omega)$$

we can define the mapping

(3.2.18) $$T(D) : E \longrightarrow A$$

by

(3.2.19) $$T(D)(s + \mathcal{V}) = T(D)s + \mathcal{I}, \quad \forall \quad s \in \mathcal{S}$$

where $T(D)s = (T(D)s_0, \ldots, T(D)s_\nu, \ldots)$ for every $s = (s_0, \ldots, s_\nu, \ldots) \in (\mathcal{C}^m(\Omega))^{\mathbb{N}}$. Indeed, it is easy to see that the definition in (3.2.18), (3.2.19) is correct. For that, we note the following. Let $s, v \in (\mathcal{C}^m(\Omega))^{\mathbb{N}}$, then in view of (3.1.1), we obtain

(3.2.20) $$T(D)(s + v) = T(D)s + \sum_\alpha s_\alpha \cdot D^{p_\alpha} v$$

where s_α are products of c_i, $D^{p_{ij}} s$ and possibly $D^{p_{ij}} v$, while p_α are some of the p_{ij}. In this way, if $s \in \mathcal{S}$ and $v \in \mathcal{V}$, then the inclusions (3.2.13) imply that

(3.2.21) $$\sum_\alpha s_\alpha \cdot D^{p_\alpha} v \in \mathcal{I}$$

therefore

(3.2.22) $$T(D)(s + v) - T(D)s \in \mathcal{I}$$

hence (3.2.19) is a valid definition.

Now, as an extension of the notion of classical solution, we can define the generalized solutions for the nonlinear partial differential equation in (3.1.1) as being given by all generalized functions

(3.2.23) $$U = s + \mathcal{V} \in E = \mathcal{S}/\mathcal{V}$$

such that, in the sense of (3.2.18), we have

(3.2.24) $\quad T(D)U = f$

where we note that, in view of (3.2.12), we have $f \in A$.

As shown in Rosinger [1,2,3] and later in Chapter 6, this notion of generalized solution is an extension of the notion of distributional or weak solution. In particular, one can construct the spaces of generalized functions in (3.2.2) and (3.2.3) in such a way that they contain the distributions, that is

(3.2.25) $\quad \mathcal{D}'(\Omega) \subset E, \ \mathcal{D}'(\Omega) \subset A$

Moreover, in (3.2.18), and therefore in (3.2.23), E itself can be an algebra. In fact, one can have $E = A$, in which case they are differential algebras and contain the distributions.

However, for the purposes of this Chapter, it is convenient to allow for the generality of the framework in (3.2.18) and (3.2.23).

We note that within this framework, the only connection needed between E and A is that in (3.2.13), which for convenience we shall denote by

(3.2.26) $\quad E \overset{m}{\leq} A$

In rest, E can be arbitrary within the conditions (3.2.2), (3.2.4), (3.2.6) and (3.2.9). We shall denote by

(3.2.27) $\quad VS^m(\Omega)$

the set of all such quotient vector spaces E of generalized functions. Similarly, A can be arbitrary, provided that it satisfies the conditions (3.2.3), (3.2.5), (3.2.7) and (3.2.10). And we denote by

(3.2.28) $\quad AL(\Omega)$

the set of all these quotient algebras A of generalized functions.

§3. THE PROBLEM OF SOLVABILITY OF NONLINEAR PARTIAL DIFFERENTIAL EQUATIONS

Given the polynomial nonlinear partial differential equation with continuous coefficients in (3.1.1)

(3.3.1) $\quad T(D)U(x) = f(x), \quad x \in \Omega$

the *problem of its solvability* will be formulated as follows.

For sequences of smooth functions

(3.3.2) $\quad s \in (\mathcal{C}^m(\Omega))^{\mathbb{N}}$

we shall find the *necessary and sufficient condition* for the existence of spaces of generalized functions $E = \mathcal{S}/\mathcal{V} \in VS^m(\Omega)$ and $A = \mathcal{A}/\mathcal{I} \in AL(\Omega)$ with

(3.3.3) $\quad E \overset{m}{\leq} A$

and such that the generalized function

(3.3.4) $\quad U = s + \mathcal{V} \in E = \mathcal{S}/\mathcal{V}$

defined by the sequence s in (3.3.2), satisfies the nonlinear partial differential equation

(3.3.5) $\quad T(D)U = f$

in the sense of (3.2.18).

For convenience, we shall denote the nonlinear partial differential equation in (3.3.5) by $\mathcal{E}$.

§4. NEUTRIX CHARACTERIZATION FOR THE SOLVABILITY OF NONLINEAR PARTIAL DIFFERENTIAL EQUATIONS

Let us now make more explicit the above solvability problem in (3.3.2)-(3.3.5). Condition (3.3.3) is nothing but (3.2.13), while condition (3.3.4) is equivalent to

(3.4.1) $\quad s \in \mathcal{S}$

Finally, in view of (3.2.19), condition (3.3.5) is equivalent to

(3.4.2) $\quad w_s = T(D)s - u(f) \in \mathcal{I}$

It follows that the solvability problem in (3.3.2)-(3.3.5) is equivalent to finding $E = \mathcal{S}/\mathcal{V} \in VS^m(\Omega)$ and $A = \mathcal{A}/\mathcal{I} \in AL(\Omega)$ such that (3.2.13), (3.4.1) and (3.4.2) are satisfied.

In view of the fact that condition (3.4.2) only involves the equation $\mathcal{E}$, the sequence s and the ideal $\mathcal{I}$, without relating to $E = \mathcal{S}/\mathcal{V}$ or $\mathcal{A}$, we shall deal with it first. For that purpose, given a sequence of continuous functions $w \in (\mathcal{C}^0(\Omega))^{\mathbb{N}}$, let us define the quotient algebra

(3.4.3) $\quad A_w = \mathcal{A}_w/\mathcal{I}_w$

where $\mathcal{A}_w$ is the subalgebra in $(\mathcal{C}^0(\Omega))^{\mathbb{N}}$ generated by $\{w\} \cup \mathcal{U}^0(\Omega)$, while $\mathcal{I}_w$ is the ideal in $\mathcal{A}_w$ generated by w, thus

$$(3.4.4) \qquad \mathcal{I}_w = w \cdot \mathcal{A}_w$$

It follows easily that

$$(3.4.5) \qquad A_w = \mathcal{A}_w/\mathcal{I}_w \in AL(\Omega) \Leftrightarrow \mathcal{I}_w \cap \mathcal{U}^0(\Omega) = \mathcal{O}$$

The interest in the quotient algebra A_w comes from the following characterization.

Proposition 1

If $w \in (\mathcal{C}^0(\Omega))^{\mathbb{N}}$ then the three conditions below are equivalent

$$(3.4.6) \qquad \exists\ A = \mathcal{A}/\mathcal{I} \in AL(\Omega) \ : \ w \in \mathcal{I}$$

$$(3.4.7) \qquad A_w = \mathcal{A}_w/\mathcal{I}_w \in AL(\Omega)$$

$$(3.4.8) \qquad \mathcal{I}_w \cap \mathcal{U}^0(\Omega) = \mathcal{O}$$

Further, if (3.4.6) holds, then we have the inclusion diagram

$$(3.4.9) \qquad \begin{array}{ccccc} \mathcal{I} & \longrightarrow & \mathcal{A} & \longrightarrow & (\mathcal{C}^0(\Omega))^{\mathbb{N}} \\ \uparrow & & \uparrow & & \\ \mathcal{I}_w & \longrightarrow & \mathcal{A}_w & & \\ \uparrow & & \uparrow & & \\ \mathcal{O} & \longrightarrow & \mathcal{U}^0(\Omega) & & \end{array}$$

Proof

In view of (3.4.5) the conditions (3.4.7) and (3.4.8) are equivalent.

Let us assume (3.4.6). Then (3.2.7) yields $\{w\} \cup \mathcal{U}^0(\Omega) \subset \mathcal{I} \cup \mathcal{A} \subset \mathcal{A}$ therefore $\mathcal{A}_w \subset \mathcal{A}$. Further, in view of (3.4.4) we obviously have

$$w \in \mathcal{I} \Rightarrow w \cdot \mathcal{A} \subset \mathcal{I} \Rightarrow w \cdot \mathcal{A}_w \subset \mathcal{I} \Rightarrow \mathcal{I}_w \subset \mathcal{I}$$

Thus in view of (2.10) we obtain

$$\mathcal{I}_w \cap \mathcal{U}^0(\Omega) \subset \mathcal{I} \cap \mathcal{U}^0(\Omega) = \mathcal{O}$$

and then (3.4.5) ends the proof of (3.4.8). Meanwhile we note that (3.4.9) has been proved as well.

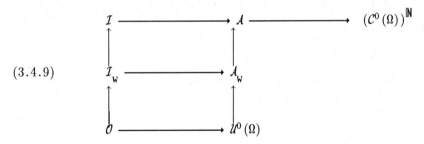

Conversely, if (3.4.8) holds, then we can take $A = A_w$ in (3.4.6). □

Returning to condition (3.4.2), it is now obvious that it can be written in the equivalent *neutrix* or *off diagonal* form

$$(3.4.10) \qquad \mathcal{I}_{w_s} \cap \mathcal{U}^0(\Omega) = \mathcal{O}$$

and in view of (3.4.9), $\mathcal{I}_{w_s}$ is the *smallest* ideal for which (3.4.10) has to hold.

In order to further explicitate the condition (3.4.10) it is useful to impose the following rather natural and mild restriction on the plynomial nonlinear partial differential operators T(D) in $\mathcal{E}$.

We call T(D) *nontrivial* on Ω, if and only if, when restricted to any $\mathcal{C}^m(\Omega')$, with $\Omega' \subset \Omega$ nonvoid and open, the range of the mapping T(D) in (3.2.17) is an infinite subset of $\mathcal{C}^0(\Omega')$.

It is equally convenient to impose a similarly natural and mild restriction on the sequences s in (3.3.2), which through (3.3.4) are supposed to give the generalized solutions for $\mathcal{E}$. For that, we shall replace (3.3.2) by the condition

$$(3.4.11) \qquad s \in (\mathcal{C}^m(\Omega))^{\mathbb{N}}_{\mathcal{E}}$$

Here $(\mathcal{C}^m(\Omega))^{\mathbb{N}}_{\mathcal{E}}$ is the subset of all the sequences $s \in (\mathcal{C}^m(\Omega))^{\mathbb{N}}$ which do *not* satisfy the condition

$$(3.4.12) \qquad \begin{array}{l} \exists \ \Omega' \subset \Omega \text{ nonvoid, open:} \\ \exists \ t \text{ subsequce in } s,\ g \in \mathcal{C}^0(\Omega'),\ g \neq f : \\ \forall \ \nu \in \mathbb{N} : \\ \quad T(D)t_\nu = g, \text{ on } \Omega' \end{array}$$

Proposition 2

If T(D) is nontrivial on Ω, then

$$(3.4.13) \qquad (\mathcal{C}^m(\Omega))^{\mathbb{N}}_{\mathcal{E}} \neq \phi$$

Proof

Assume that

$$(\mathcal{C}^m(\Omega))^{\mathbb{N}}_{\mathcal{E}} = \phi$$

then (3.4.12) implies that

$$\begin{aligned}
&\forall \quad s \in (\mathcal{C}^m(\Omega))^{\mathbb{N}} : \\
&\exists \quad \Omega' \subset \Omega \quad \text{nonvoid, open} : \\
(3.4.14) \qquad &\exists \quad t \quad \text{subsequence in} \quad s,\ g \in \mathcal{C}^0(\Omega'),\ g \neq f : \\
&\exists \quad \nu \in \mathbb{N} : \\
&\qquad T(D)t_\nu = g, \quad \text{on} \quad \Omega'
\end{aligned}$$

But (3.4.14) obviously implies that the range of the mapping $T(D)$ in (3.2.17), when restricted to $\mathcal{C}^m(\Omega')$, is a finite subset of $\mathcal{C}^0(\Omega')$ □

At this stage we are led to introduce, see Rosinger [2, p. 39]

(3.4.15) $\mathcal{R}(\Omega)$

as the set of all the sequences $w \in (\mathcal{C}^0(\Omega))^{\mathbb{N}}$ for which the following condition does *not* hold

$$\begin{aligned}
&\exists \quad \Omega' \subset \Omega \quad \text{nonvoid, open}: \\
(3.4.16) \qquad &\exists \quad z \quad \text{subsequence in} \quad w,\ h \in \mathcal{C}^0(\Omega'),\ h \neq 0 : \\
&\forall \quad \nu \in \mathbb{N} : \\
&\qquad z_\nu = h, \quad \text{on} \quad \Omega'
\end{aligned}$$

Indeed, we obviously have the property

$$\begin{aligned}
&\forall \quad s \in (\mathcal{C}^m(\Omega))^{\mathbb{N}} : \\
(3.4.17) \qquad &\quad s \in (\mathcal{C}^m(\Omega))^{\mathbb{N}}_{\mathcal{E}} \Leftrightarrow w_s \in \mathcal{R}(\Omega)
\end{aligned}$$

However, the important property of the set of sequences $\mathcal{R}(\Omega)$ related to condition (3.4.10) is the following.

Proposition 3

If $w \in \mathcal{R}(\Omega)$ then

(3.4.18) $\mathcal{I}_w \cap \mathcal{U}^0(\Omega) = \mathcal{O}$

<u>Proof</u>

It is easy to see that an element of the intersection in (3.4.18) has the form

(3.4.19) $u(\psi) = w \cdot u(\psi_0) + \ldots + w^{\ell+1} \cdot u(\psi_\ell)$

where $\ell \in \mathbb{N}$ and $\psi, \psi_0, \ldots, \psi_\ell \in \mathcal{C}^0(\Omega)$. Therefore in order to prove (3.4.18), it suffices to show that $\psi = 0$ on Ω. Assume that this is not the case and $\Omega' \subset \Omega$ is nonvoid, open, such that

(3.4.20) $\psi(x) \neq 0, \quad \forall\, x \in \Omega'$

Denoting by w_ν, with $\nu \in \mathbb{N}$, the continuous functions on Ω that are the terms in the sequence w, the relation (3.4.19) written term by term, yields

(3.4.21) $(w_\nu(x))^{\ell+1} \cdot \psi_\ell(x) + \ldots + w_\nu(x) \cdot \psi_0(x) + (-\psi(x)) = 0, \ \forall\, \nu \in \mathbb{N},\ x \in \Omega$

Therefore (3.4.20) will imply that the infinite matrix

$$\begin{bmatrix} (w_0(x))^{\ell+1} & \cdots\cdots\cdots & w_0(x) & 1 \\ \vdots & & \vdots & \vdots \\ (w_\nu(x))^{\ell+1} & \cdots\cdots\cdots & w_\nu(x) & 1 \\ \vdots & & \vdots & \vdots \end{bmatrix}$$

has rank at most $\ell + 1$, for any given $x \in \Omega'$.

Now, a well-known property of Vandermonde determinants implies that the infinite sequence of *numbers*

$$w_0(x)\ ,\ldots,\ w_\nu(x)\ ,\ldots\ldots$$

contains at most $\ell + 1$ different terms, for any given $x \in \Omega'$. Therefore Lemma 1 below will grant the existence of a closed, nowhere dense subset $\Gamma' \subset \Omega'$, such that each $x \in \Omega' \setminus \Gamma'$ possesses an open neighbourhood $\Omega'' \subset \Omega' \setminus \Gamma'$, with the property that the infinite sequence of *functions*

$$w_0, \ldots, w_\nu, \ldots\ldots$$

when restricted to Ω'', contains only a *finite number of different functions*. In other words, there exists a subsequence w'' in w and $\psi'' \in \mathcal{C}^0(\Omega'')$ such that

(3.4.22) $w'' = u(\psi'')$ on Ω''

Now, $w \in \mathcal{R}(\Omega)$ will imply that $\psi'' = 0$ on Ω''. And then (3.4.22) together with (3.4.21) will contradict (3.4.20). □

Lemma 1

Suppose the sequence $w = (w_0, \ldots, w_\nu, \ldots)$ of continuous functions on Ω is such that for any given $x \in \Omega$, the sequence of *numbers*

$$w_0(x), \ldots, w_\nu(x), \ldots\ldots$$

contains only a finite number of different terms. Then there exists a closed, nowhere dense subset $\Gamma \subset \Omega$, such that the sequence of *functions*

$$w_0, \ldots, w_\nu, \ldots\ldots$$

restricted to a suitable neighbourhood of any given $x \in \Omega \backslash \Gamma$, contains only a *finite number of different terms*.

Proof

Denote by Γ the set of all points $x \in \Omega$ such that the sequence of functions

$$w_0, \ldots, w_\nu, \ldots\ldots$$

when restricted to any neighbourhood of x, contains infinitely many different terms. It is easy to see that Γ is closed. Therefore it only remains to prove that Γ has no interior. It suffices to show that

(3.4.23) $\Gamma \neq \Omega$

Indeed, denote $\Omega' = \text{int}\,\Gamma$ and assume $\Omega' \neq \phi$. Then Γ' corresponding as above to Ω', will satisfy $\Gamma' = \Omega'$, hence contradicting (3.4.23).

In order to obtain (3.4.23), the Baire category argument will be used in *two successive steps*.

First, for $\mu \in \mathbb{N}$, define the closed set

$$\Delta_\mu = \left\{x \in \Omega \;\middle|\; \begin{array}{l} \forall \;\; \nu \in \mathbb{N}, \;\; \nu \geq \mu + 1 : \\ \exists \;\; \lambda \in \mathbb{N}, \;\; \lambda \leq \mu : \\ \quad w_\nu(x) = w_\lambda(x) \end{array}\right\}$$

then obviously

$$\Omega = \bigcup_{\mu \in \mathbb{N}} \Delta_\mu$$

therefore the Baire category argument implies that

$$\text{int } \Delta_\mu \neq \phi$$

for a certain $\mu \in \mathbb{N}$. Denote $\Omega' = \text{int } \Delta_\mu$. We shall prove that

(3.4.24) $\qquad \Omega' \cap (\Omega \backslash \Gamma) \neq \phi.$

Denote for $\rho \in \mathbb{N}$

$$\Delta'_\rho = \left\{x \in \Omega' \;\middle|\; \begin{array}{l} \forall \;\; \lambda, \; \nu \in \mathbb{N}, \; \lambda < \nu \leq \mu : \\ \quad w_\lambda(x) \neq w_\nu(x) \Rightarrow |w_\lambda(x) - w_\nu(x)| \geq 1/(\rho+1) \end{array}\right\}$$

then we have

(3.4.25) $\qquad \Omega' = \bigcup_{\rho \in \mathbb{N}} \Delta'_\rho$

Indeed, denote for $x \in \Omega'$

$$M_x = \{(\lambda,\nu) \in \mathbb{N} \times \mathbb{N} \mid \lambda < \nu \leq \mu, \; w_\lambda(x) \neq w_\nu(x)\}$$

and take $\rho \in \mathbb{N}$, such that

$$1/(\rho+1) \leq \min\{|w_\lambda(x) - w_\nu(x)| \;\big|\; (\lambda,\nu) \in M_x\}$$

then obviously $x \in \Delta'_\rho$.

Now we show that

(3.4.26) $\qquad \Delta'_\rho$ closed, $\forall \; \rho \in \mathbb{N}$

Indeed, denoting

$$M = \{(\lambda,\nu) \in \mathbb{N} \times \mathbb{N} \mid \lambda < \nu \leq \mu\}$$

we have

$$\Delta'_\rho = \bigcup_{K\subset M}\Big(\bigcup_{(\lambda,\nu)\in K}\{x\in\Omega' \,\Big|\, |w_\lambda(x) - w_\nu(x)| \geq 1/(\rho+1)\}\cap$$

$$\bigcup_{(\lambda,\nu)\in M\setminus K}\{x\in\Omega' | w_\lambda(x) = w_\nu(x)\}\Big).$$

But (3.4.25) and (3.4.26) together with the Baire category argument imply that

$$\text{int }\Delta'_\rho \neq \phi$$

for a certain $\rho\in\mathbb{N}$. Denote then $\Omega'' = \text{int }\Delta'_\rho$. The proof of (3.4.24) will obviously be complete if we show that

$$(3.4.27)\qquad \Omega''\subset\Omega\setminus\Gamma$$

Assume therefore $x\in\Omega''$ and $V\subset\Omega''$ an open, connected neighbourhood of x. We shall prove that the sequence of functions

$$w_0,\ldots, w_\nu,\ldots\ldots$$

when restricted to V, contains at most $\mu+1$ different terms. Indeed, if $\nu\in\mathbb{N}$, $\nu\geq\mu+1$, then $w_\nu(x) = w_\lambda(x)$, for a certain $\lambda\in\mathbb{N}$, $\lambda\leq\mu$, since

$$x\in V\subset\Omega''\subset\Delta'_\rho\subset\Omega'\subset\Delta_\mu$$

But then

$$(3.4.28)\qquad w_\nu = w_\lambda \quad\text{on}\quad V$$

Assume indeed that (3.4.28) is false. Then $w_\nu(y)\neq w_\lambda(y)$, for a certain $y\in V$. Denote

$$V' = \{x'\in V \mid w_\nu(x') = w_\lambda(x')\}$$
$$V'' = \{x''\in V \mid w_\nu(x'')\neq w_\lambda(x'')\}$$

then $x\in V'$, $y\in V''$, $V = V'\cup V''$, $V'\cap V'' = \phi$ and V' is obviously closed. But V'' is also closed, since

$$(3.4.29)\qquad V'' = \{x''\in V \,\Big|\, |w_\nu(x'') - w_\lambda(x'')| \geq 1/(\rho+1)\}$$

the inclusion $\supset$ being obvious, while the converse results as follows. Take $x''\in V''$, then there exists $\sigma\in\mathbb{N}$, $\sigma\leq\mu$, such that $w_\nu(x'') = w_\sigma(x'')$, since $\nu\geq\mu+1$ and

$$x''\in V''\subset V\subset\Omega''\subset\Delta'_\rho\subset\Omega'\subset\Delta_\mu$$

Hence $w_\sigma(x'') \neq w_\lambda(x'')$, therefore $\sigma, \lambda \leq \mu$ and

$$x'' \in V'' \subset V \subset \Omega'' \subset \Delta'_\rho$$

will imply that

$$|w_\nu(x'') - w_\lambda(x'')| = |w_\sigma(w'') - w_\lambda(x'')| \geq 1/(\rho+1)$$

and this completes the proof of (3.4.29).

As the decomposition $V = V' \cup V''$ that has been obtained contradicts the connectedness of V, it follows that (3.4.28) holds.

Now, (3.4.28) implies (3.4.27), which completes the proof of (3.4.24). Thus finally (3.4.23) has been proved. □

Remark 1

The extent to which Lemma 1 above is *sharp* can be seen in the examples presented in Appendix 1.

For the sake of completeness, we note the following obvious property of the set $\mathcal{R}(\Omega)$ of sequences of continuous functions on Ω

(3.4.30)
$$\forall \; w \in (\mathcal{C}^0(\Omega))^{\mathbb{N}} : \quad w \in \mathcal{R}(\Omega) \Leftrightarrow \left[\begin{array}{l} \forall \; \Omega' \subset \Omega \text{ nonvoid open, } z \text{ subsequence in } w : \\ \quad z\big|_{\Omega'} \in \mathcal{R}(\Omega') \end{array} \right]$$

where, for $z = (z_0, \ldots, z_\nu, \ldots) \in (\mathcal{C}^0(\Omega))^{\mathbb{N}}$, we denote

$$z\big|_{\Omega'} = (z_0\big|_{\Omega'}, \ldots, z_\nu\big|_{\Omega'}, \ldots)$$

with

$$\mathcal{C}^0(\Omega) \ni k \longmapsto k\big|_{\Omega'} \in \mathcal{C}^0(\Omega')$$

being the usual restriction of the function k to Ω'. Further, for any subset $\mathcal{I} \subset (\mathcal{C}^0(\Omega))^{\mathbb{N}}$ let us denote

$$\mathcal{I}\big|_{\Omega'} = \{z\big|_{\Omega'} \mid z \in \mathcal{I}\}$$

Based on (3.4.30), we obtain

Corollary 1

If $w \in (\mathcal{C}^0(\Omega))^{\mathbb{N}}$ then

$$(3.4.31) \quad w \in \mathcal{R}(\Omega) \quad \Leftrightarrow \left[\begin{array}{l} \forall \ \Omega' \subset \Omega \text{ nonvoid open, z subsequence in w :} \\ \mathcal{I}_z \Big|_{\Omega'} \cap \mathcal{U}^0(\Omega') = \mathcal{O} \end{array} \right]$$

Proof

The implication '$\Rightarrow$' follows easily from (3.4.18) and (3.4.30).

Conversely, assume that $w \notin \mathcal{R}(\Omega)$. Then in view of (3.4.16) we obtain a subsequence z in w, such that for a certain nonvoid and open $\Omega' \subset \Omega$ and $h \in \mathcal{C}^0(\Omega')$, $h \neq 0$, we have

$$u(h) \in \mathcal{I}_z \Big|_{\Omega'} \cap \mathcal{U}^0(\Omega')$$

which contradicts the hypotesis. □

Connected with the neutrix condition (3.4.10) we can now obtain the following result.

Corollary 2

If $s \in (\mathcal{C}^m(\Omega))^{\mathbb{N}}_{\mathcal{E}}$ then, for every $\Omega' \subset \Omega$ nonvoid and open, we have

$$(3.4.32) \qquad \mathcal{I}_{w_s} \Big|_{\Omega'} \cap \mathcal{U}^0(\Omega') = \mathcal{O}$$

Proof

It follows from (3.4.17), (3.4.18) and (3.4.31) □

We note that in view of (3.4.17), we have the following correspondent of property (3.4.30)

$$(3.4.33) \quad \begin{array}{l} \forall \ s \in (\mathcal{C}^m(\Omega))^{\mathbb{N}} : \\ s \in (\mathcal{C}^m(\Omega))^{\mathbb{N}}_{\mathcal{E}} \Leftrightarrow \left[\begin{array}{l} \forall \ \Omega' \subset \Omega \text{ nonvoid open, t subsequence in s:} \\ t \Big|_{\Omega'} \in (\mathcal{C}^m(\Omega'))^{\mathbb{N}}_{\mathcal{E}} \end{array} \right] \end{array}$$

which leads to

Corollary 3

If $s \in (\mathcal{C}^m(\Omega))^{\mathbb{N}}$ then

(3.4.34) $$s \in (\mathcal{C}^m(\Omega))^{\mathbb{N}}_{\mathcal{E}} \Leftrightarrow \left[\begin{array}{l} \forall \ \Omega' \subset \Omega \text{ nonvoid open, t subsequence in s :} \\ \mathcal{I}_{w_t}\Big|_{\Omega'} \cap \mathcal{U}^0(\Omega') = \mathcal{O} \end{array} \right]$$

Proof

The implication '$\Rightarrow$' follows from (3.4.32) and (3.4.33).

Conversely, (3.4.31) implies that

$$w_s \in \mathcal{R}(\Omega)$$

thus (3.4.17) completes the proof □

We can summarize the above as follows.

Proposition 4

If $T(D)$ is nontrivial on Ω, then $(\mathcal{C}^m(\Omega))^{\mathbb{N}}_{\mathcal{E}} \neq \phi$ and

(3.4.35) $$\begin{array}{l} \forall \ s \in (\mathcal{C}^m(\Omega))^{\mathbb{N}}_{\mathcal{E}} : \\ A_{w_s} = \mathcal{A}_{w_s}/\mathcal{I}_{w_s} \in AL(\Omega) \end{array}$$

Proof

It follows from Proposition 2 and Corollary 2 □

At this point we can turn to the conditions (3.3.3) and (3.4.1).

Obviously, it is easy to find $E = \mathcal{S}/\mathcal{V} \in VS^m(\Omega)$ such that (3.4.1) is satisfied.

Therefore, we concentrate on satisfying condition (3.3.3), that is (3.2.13).

For that, it is useful to note that the existence of a vector space E in (3.2.2) and of an algebra A in (3.2.3) satisfying the conditions (3.2.13), (3.4.1) and (3.4.2) is equivalent to the existence of an algebra A in (3.2.3), satisfying the following two conditions

(3.4.36) $$A = \mathcal{A}/\mathcal{I} \in AL(\Omega)$$

$$(3.4.37) \qquad w_s \in \mathcal{I}, \quad \{D^p s \mid p \in \mathbb{N}^n,\ |p| \le m\} \subset \mathcal{A}$$

In this case we can define $E = \mathcal{S}/\mathcal{V}$ by

$$(3.4.38) \qquad \mathcal{S} = \left\{ t \in (\mathcal{C}^m(\Omega))^{\mathbb{N}} \;\middle|\; \begin{array}{l} \forall\ p \in \mathbb{N}^n,\ |p| \le m : \\ D^p t \in \mathcal{A} \end{array} \right\}$$

and

$$(3.4.39) \qquad \mathcal{V} = \left\{ v \in (\mathcal{C}^m(\Omega))^{\mathbb{N}} \;\middle|\; \begin{array}{l} \forall\ p \in \mathbb{N}^n,\ |p| \le m : \\ D^p v \in \mathcal{I} \end{array} \right\}$$

Indeed, it is easy to see that

$$(3.4.40) \qquad E = \mathcal{S}/\mathcal{V} \in \mathrm{VS}^m(\Omega) \Leftrightarrow \mathcal{V} \cap \mathcal{U}^m(\Omega) = \mathcal{O}$$

Therefore, we prove now that the neutrix condition in the right hand term of (3.4.40) does indeed hold. For that, assume the relation $v = u(\psi)$, for a certain $v \in \mathcal{V}$ and $\psi \in \mathcal{C}^m(\Omega)$. Then, for $p = (0,\dots,0) \in \mathbb{N}^n$, the relation (3.4.39) yields $v \in \mathcal{I}$. Thus

$$u(\psi) = v \in \mathcal{I} \cap \mathcal{U}^0(\Omega)$$

since $\psi \in \mathcal{C}^m(\Omega) \subset \mathcal{C}^0(\Omega)$. Now, in view of (3.4.36) and hence (3.2.10), we obtain $v \in \mathcal{O}$.

Finally, we are near to the solution of the problem (3.3.2)-(3.3.5). Indeed, in view of Proposition 1, the conditions (3.4.36) and (3.4.37) are equivalent to the existence of the inclusion diagram

$$(3.4.41) \qquad \begin{array}{ccccc} \mathcal{I} & \longrightarrow & \mathcal{A} & \longrightarrow & (\mathcal{C}^0(\Omega))^{\mathbb{N}} \\ \uparrow & & \uparrow & & \\ \mathcal{I}_{w_s} & \longrightarrow & \mathcal{A}_{w_s} & & \\ \uparrow & & \uparrow & & \\ \mathcal{O} & \longrightarrow & \mathcal{U}^0(\Omega) & & \end{array}$$

satisfying the additional two conditions

$$(3.4.42) \qquad \{D^p s \mid p \in \mathbb{N}^n,\ |p| \le m\} \subset \mathcal{A}$$

and

(3.4.43) $\mathcal{I} \cap \mathcal{U}^0(\Omega) = \mathcal{O}$

We show now that in case (3.4.41)-(3.4.43) hold, one can find the *smallest* subalgebra $\mathcal{A}$ and ideal $\mathcal{I}$ in $\mathcal{A}$ which satisfy these conditions. Indeed, let us denote by

(3.4.44) $\mathcal{A}_s$

the subalgebra in $(\mathcal{C}^0(\Omega))^{\mathbb{N}}$ generated by

(3.4.45) $\{D^p s \mid p \in \mathbb{N}^n, |p| \leq m\} \cup \mathcal{U}^0(\Omega)$

Then (3.4.42) and (3.2.7) imply that

(3.4.46) $\mathcal{A}_s \subset \mathcal{A}$

Let us now denote by

(3.4.47) $\mathcal{I}_s$

the ideal in $\mathcal{A}_s$ generated by $\mathcal{I}_{w_s}$. Then obvisously $\mathcal{I}_s$ is the vector subspace generated by

(3.4.48) $\mathcal{I}_{w_s} \cdot \mathcal{A}_s$

therefore (3.4.41) results in the inclusion

(3.4.49) $\mathcal{I}_s \subset \mathcal{I}$

In this way, whenever (3.4.41) holds, it can always be augmented to become the following inclusion diagram

$$
(3.4.50)\qquad
\begin{array}{ccccc}
\mathcal{I} & \longrightarrow & \mathcal{A} & \longrightarrow & (\mathcal{C}^0(\Omega))^{\mathbb{N}} \\
\uparrow & & \uparrow & & \\
\mathcal{I}_s & \longrightarrow & \mathcal{A}_s & & \\
\uparrow & & \uparrow & & \\
\mathcal{I}_{w_s} & \longrightarrow & \mathcal{A}_{w_s} & & \\
\uparrow & & \uparrow & & \\
\mathcal{O} & \longrightarrow & \mathcal{U}^0(\Omega) & &
\end{array}
$$

which will automatically satisfy (3.4.42), as a consequence of (3.4.45).

In conclusion, the *existence* of inclusion diagrams (3.4.41) which satisfy the conditions (3.4.42) and (3.4.43) is equivalent to the *neutrix condition*

(3.4.51) $\qquad \mathcal{I}_s \cap \mathcal{U}^0(\Omega) = \mathcal{O}$

Now we can summarize the results in this section and obtain the following *neutrix characterization* for the *existence of generalized solutions.*

<u>Theorem 1</u>

If $T(D)$ is nontrivial on Ω, then

(3.4.52) $\qquad (\mathcal{C}^m(\Omega))^{\mathbb{N}}_{\mathcal{E}} \neq \phi$

Given any sequence of functions

(3.4.53) $\qquad s \in (\mathcal{C}^m(\Omega))^{\mathbb{N}}_{\mathcal{E}}$

there exist $E = \mathcal{S}/\mathcal{V} \in VS^m(\Omega)$ and $A = \mathcal{A}/\mathcal{I} \in AL(\Omega)$ such that

(3.4.54) $\qquad E \overset{m}{\leq} A$

(3.4.55) $\qquad U = s + \mathcal{V} \in E = \mathcal{S}/\mathcal{V}$

and

(3.4.56) $\qquad T(D)U = f \quad$ (see (3.2.18))

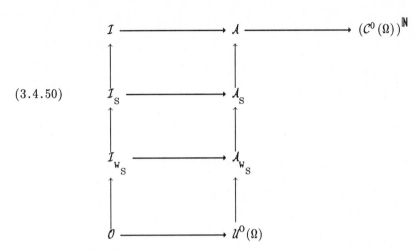

if and only if the *neutrix condition* is satisfied (see (3.4.47))

(3.4.57) $\mathcal{I}_s \cap \mathcal{U}^0(\Omega) = \mathcal{O}$ □

Remark 2

Let us recapitulate the main steps leading to the above *algebraic* characterization in (3.4.57) for the existence of generalized solutions.

Given on Ω a nontrivial polynomial nonlinear PDE $\mathcal{E}$ of order m and with continuous coefficients

(3.4.58) $T(D)U(x) = f(x), \quad x \in \Omega$

and a sequence of functions

(3.4.59) $s \in (\mathcal{C}^m(\Omega))^{\mathbb{N}}_{\mathcal{E}}$

one constructs $\mathcal{A}_s$ in (3.4.44) and $\mathcal{I}_s$ in (3.4.47). Then, the *neutrix condition*

(3.4.60) $\mathcal{I}_s \cap \mathcal{U}^0(\Omega) = \mathcal{O}$

is necessary and sufficient for having

(3.4.61) $A_s = \mathcal{A}_s/\mathcal{I}_s \in AL(\Omega)$

in which case one can take

(3.4.62) $A = A_s$

Further, one can take

(3.4.63) $E = E_s$

where

(3.4.64) $E_s = \mathcal{S}_s/\mathcal{V}_s \in VS^m(\Omega)$

with

(3.4.65) $\mathcal{S}_s = \left\{ t \in (\mathcal{C}^m(\Omega))^{\mathbb{N}} \;\middle|\; \begin{array}{l} \forall \; p \in \mathbb{N}^n, \; |p| \leq m : \\ D^p t \in \mathcal{A}_s \end{array} \right\}$

and

$$(3.4.66) \qquad \mathcal{V}_s = \left\{ v \in (\mathcal{C}^m(\Omega))^{\mathbb{N}} \;\middle|\; \begin{array}{l} \forall \;\; p \in \mathbb{N}^n, \; |p| \leq m : \\ D^p v \in \mathcal{I}_s \end{array} \right\}$$

In this way one obtains

$$(3.4.67) \qquad E \overset{m}{\leq} A$$

and for the generalized function

$$(3.4.68) \qquad U = s + \mathcal{S} \in E = \mathcal{S}/\mathcal{V}$$

the equation $\mathcal{E}$ namely

$$(3.4.69) \qquad T(D)U = f$$

is satisfied.

The above construction, which summarizes the proof of Theorem 1, gives in fact a particular pair of spaces of generalized functions $E = \mathcal{S}/\mathcal{V}$ and $A = \mathcal{A}/\mathcal{I}$. The existence of other pairs which satisfy the conditions (3.4.54)-(3.4.56) can be obtained from a detailed study of the *stability*, *generality* and *exactness* of generalized solutions, see Rosinger [2, pp. 13-16, 163-172], Rosinger [3, pp. 224-229], and also Oberguggenberger [6].

Remark 3

Concerning the problems of uniqueness, regularity - or more generally, coherence, see Colombeau [1,2], Rosinger [2,3], Oberguggenberger [6] and Chapter 4 next - of the generalized solutions whose existence is characterized in Theorem 1, the situation at present is roughly as follows. Within the more particular framework of Colombeau [1,2], rather strong uniqueness, regularity and coherence results have been obtained for large classes of linear and nonlinear PDEs. On the other hand, in Oberguggenberger [6], see the next Chapter, the framework used is still more general than in this Chapter, in the sense that the mappings (3.2.18) are replaced by mappings

$$(3.4.70) \qquad T(D) : E_1 \longrightarrow E_2$$

where $E_2 = \mathcal{S}/\mathcal{V} \in VS^0(\Omega)$ is a quotient vector space which *need not* be a quotient algebra.

Within this more general setting (3.4.70), and for rather general semi-linear hyperbolic systems in two independent variables, existence and uniqueness is proved for the Cauchy problem with rough initial data. Rather surprisingly, precisely because of the more general nature of the framework in (3.4.70), particularly *strong coherence* results are proved for the unique generalized solutions of the mentioned type of Cauchy problems, results which are shown to be impossible within the framework of Colombeau

[1,2], this being one of the outstanding features of Oberguggenberger [6] as seen next in Chapter 4.

However, the choice of the setting in (3.2.18), (3.4.70) or that in Colombeau [1,2] may be influenced by considerations other than the strongest possible coherence of generalized solutions. Indeed, for the global version of the Cauchy-Kovalevskaia theorem in Chapter 2, the setting which proves to be useful is the following particular form of (3.2.18)

$$(3.4.71) \quad T(D() : A_1 \longrightarrow A_2$$

where *both* A_1 and A_2 are quotient *algebras* of generalized functions.

Remark 4

The results in Chapter 2 which, within the setting of (3.4.71), prove the global existence of generalized solutions on the whole of the domain of analyticity of arbitrary analytic nonlinear partial differential equations, can be seen as a particular case of the result in Theorem 1 above.

Indeed, the essence of the proofs in Chapter 2 comes down to the fact that one can construct sequences of functions s in (3.3.2)-(3.3.5) such that

$$(3.4.72) \qquad w_s \in \mathcal{I}_{nd}(\Omega)$$

where we recall that $\mathcal{I}_{nd}(\Omega)$ is the set of all the sequences $w \in (\mathcal{C}^0(\Omega))^{\mathbb{N}}$ which satisfy the condtion, see (2.2.4)

$$(3.4.73) \qquad \begin{array}{l} \exists \;\; \Gamma \subset \Omega \;\text{ closed, nowhere dense :} \\ \forall \;\; x \in \Omega \setminus \Gamma \;: \\ \exists \;\; \mu \in \mathbb{N},\; V \subset \Omega\setminus\Gamma \;\text{ neighbourhood of }\; x \;: \\ \forall \;\; \nu \in \mathbb{N},\; \nu \geq \mu,\; y \in V \;: \\ \quad w_\nu(y) = 0 \end{array}$$

The important fact is that $\mathcal{I}_{nd}(\Omega)$ is an *ideal* in $(\mathcal{C}^0(\Omega))^{\mathbb{N}}$, called the *nowhere dense ideal*, which satisfies the neutrix conditon (3.2.10), see (2.2.12). This makes it possible to easily secure the neutrix condition (3.4.57) in Theorem 1, and then set up the framework in (3.4.71).

§5. THE NEUTRIX CONDITION AS A DENSELY VANISHING CONDITION ON IDEALS

The neutrix conditon (3.2.10) first comes into the picture as the rather trivial necessary and sufficient condition for the algebra embedding in (3.2.12), see also (1.6.14). Then, in (3.4.57), it nevertheless proves to give the characterization for the existence of generalized solutions of a large class of nonlinear partial differential equations.

This power of the neutrix conditon (3.2.10) when applied to ideals $\mathcal{I}$ in subalgebras $\mathcal{A}$ of $(\mathcal{C}^0(\Omega))^{\mathbb{N}}$ should not come as a surprise. Indeed, as seen in Rosinger [2,pp. 75-88] and Rosinger [3, pp. 306-315], as well as in Chapter 6 in this volume, the neutrix condition happens also to characterize the existence of a very large class of chains of differential algebras of generalized functions, chains which contain the L. Schwartz distributions, and incorporate as particular cases various distribution multiplications encountered in the literature, such as for instance that in Colombeau [1,2], for details on such multiplications see the references in Oberguggenberger [1].

The form of the neutrix conditon (3.2.10) has the significant advantage of being a particularly simple algebraic condition on the ideal $\mathcal{I}$, with the clear geometric meaning that $\mathcal{I}$ is off diagonal in $(\mathcal{C}^0(\Omega))^{\mathbb{N}}$. However, except for that, the neutrix conditon (3.2.10) does not give an explicit insight into the structure of the respective ideals $\mathcal{I}$. It is therefore of special interest to obtain *alternative* characterizations for the ideals $\mathcal{I}$ which satisfy the neutrix condition, characterizations which can give deeper insight into their structure. Indeed, as seen in (3.4.2) and (3.4.57), the structure of the ideals $\mathcal{I}$ which appear in quotient algebras $A = \mathcal{A}/\mathcal{I} \in \mathrm{AL}(\Omega)$ can give a direct and explicit understanding of the conditions of solvability of nonlinear partial differential equations.

In this Section, such an alternative characterization is presented, according to which, in a certain sense specified later, the sequences of functions $w \in \mathcal{I}$ have to *vanish asymptotically on dense subsets* of Ω.

Let us now turn to the details. Our problem is the following. Given an inclusion diagram

(3.5.1)
$$\begin{array}{ccccc} \mathcal{I} & \longrightarrow & \mathcal{A} & \longrightarrow & (\mathcal{C}^0(\Omega))^{\mathbb{N}} \\ \uparrow & & \uparrow & & \\ \mathcal{O} & \longrightarrow & \mathcal{U}^0(\Omega) & & \end{array}$$

where $\mathcal{A}$ is a subalgebra and $\mathcal{I}$ is an ideal in $\mathcal{A}$ which satisfies the neutrix condition

(3.5.2) $$\mathcal{I} \cap \mathcal{U}^0(\Omega) = \mathcal{O}$$

we want to find equivalent conditions on $\mathcal{I}$, which give a better explicit understanding of the structure of $\mathcal{I}$.

It should be noted from the beginning that this problem of structural characterizations of $\mathcal{I}$ seems to be highly nontrivial. Indeed, it can often happen that when constructing generalized solutions for linear or nonlinear partial differential equations, we shall encounter in (3.5.1) the situation where

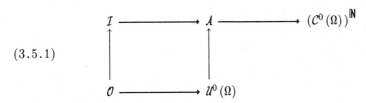

(3.5.3) $$\mathcal{A} \subsetneqq (\mathcal{C}^0(\Omega))^{\mathbb{N}}$$

and

(3.5.4) $$\mathcal{I} \text{ is } \textit{not} \text{ an ideal in } (\mathcal{C}^0(\Omega))^{\mathbb{N}}$$

Since obviously

(3.5.5) $$(\mathcal{C}^0(\Omega))^{\mathbb{N}} = \mathcal{C}(\mathbb{N} \times \Omega)$$

the situation in (3.5.3), (3.5.4) means that we are dealing with the structure of ideals $\mathcal{I}$ in *strict* subalgebras $\mathcal{A}$ of a ring of continuous functions, namely $\mathcal{C}(\mathbb{N} \times \Omega)$, a problem well known for its difficulty, see Gillman & Jerison, Walker.

Finally, we should note that from the point of view of the structure of $\mathcal{I}$, the problem in (3.5.1), (3.5.2) has the following equivalent and simpler formulation, in which the subalgebra $\mathcal{A}$ does no longer appear : find the structure of *subalgebras*

(3.5.6) $$\mathcal{I} \subset (\mathcal{C}^0(\Omega))^{\mathbb{N}}$$

which satisfy the conditions

(3.5.7) $$\mathcal{I} \cdot \mathcal{U}^0(\Omega) \subset \mathcal{I}$$

and

(3.5.8) $$\mathcal{I} \cap \mathcal{U}^0(\Omega) = \mathcal{O}$$

Obviously, the neutrix condition (3.5.8) will imply that $\mathcal{I}$ is a *strict* subalgebra in $(\mathcal{C}^0(\Omega))^{\mathbb{N}}$.

And now, we can turn to the 'dense vanishing' characterization of subalgebras $\mathcal{I}$ in (3.5.6)-(3.5.8).

The conjecture about it has emerged earlier, following the study of the essence of various *sufficient* conditions for obtaining (3.5.6)-(3.5.8), see for details Rosinger [1, pp. 39-59, 132-138], Rosinger [2, pp. 173-198]. Indeed, for $w \in (\mathcal{C}^0(\Omega))^{\mathbb{N}}$, let us define its *vanishing* set by

(3.5.9) $$\Omega_w = \{x \in \Omega \mid \inf_{\nu \in \mathbb{N}} |w_\nu(x)| = 0\}$$

Then, the following *sufficient* condition for (3.5.8) results easily.

Proposition 5

Given any subset $\mathcal{I} \subset (\mathcal{C}^0(\Omega))^{\mathbb{N}}$, if

$$\text{(3.5.10)} \qquad \begin{array}{l} \forall \ w \in \mathcal{I} : \\ \quad \Omega_w \ \text{is dense in} \ \Omega \end{array}$$

then

$$\text{(3.5.11)} \qquad \mathcal{I} \cap \mathcal{U}^0(\Omega) \subset \mathcal{O}$$

Proof

Assume that for certain $w \in \mathcal{I}$ and $\psi \in \mathcal{C}^0(\Omega)$ we have

$$w = u(\psi)$$

then

$$\psi(x) = w_\nu(x), \quad \forall \ \nu \in \mathbb{N}, \quad x \in \Omega$$

hence (3.5.9) implies that

$$\text{(3.5.12)} \qquad |\psi(x)| = \inf_{\nu \in \mathbb{N}} |w_\nu(x)| = 0, \quad \forall \ x \in \Omega_w$$

Now (3.5.12) and (3.5.10) as well as the continuity of ψ on Ω will yield

$$\psi(x) = 0, \quad \forall \ x \in \Omega$$

□

It is easy to see that $\mathcal{I}_{nd}(\Omega)$ defined in (3.4.73), does satisfy the 'densely vanishing' conditon (3.5.10).

As the main result of this Section, in Theorem 2 it will be shown that the 'dense vanishing' condition (3.5.10) is *equivalent to the neutrix condition*, within a rather large and natural class of inclusion diagrams (3.5.1). The basic property used for that purpose is presented now.

Proposition 6

If $w \in (\mathcal{C}^0(\Omega))^{\mathbb{N}}$ and Ω_w is not dense in Ω, then

$$\text{(3.5.13)} \qquad \begin{array}{l} \exists \ \Omega' \subset \Omega \ \text{nonvoid, open,} \ \ c > 0 : \\ \forall \ \nu \in \mathbb{N}, \ \ x \in \Omega' : \\ \quad |w_\nu(x)| \geq c \end{array}$$

Proof

As Ω_w is not dense in Ω, it follows that

$$(3.5.14) \qquad \begin{array}{l} \exists \ x \in \Omega, \ \ \epsilon > 0 : \\ \quad B(x,\epsilon) \cap \Omega_w = \phi \end{array}$$

Define now $\delta : \Omega \longrightarrow [0,\infty)$ by

$$(3.5.15) \qquad \delta(y) = \inf_{\nu \in \mathbb{N}} |w_\nu(y)|, \quad \forall \ y \in \Omega$$

Obviously δ is upper semicontinuous. Therefore, the set $D \subset \Omega$ of discontinuities of δ is of first Baire category in Ω. In this way it follows that

$$B(x,\epsilon) \backslash D \neq \phi$$

thus we can find $y \in B(x,\epsilon)$ such that δ is continuous at y. But then, in view of (3.5.14), we have

$$\delta(y) > 0$$

hence (3.5.13) follows by taking (3.5.15) into account. □

At this stage, we shall restrict the class of inclusion diagrams (3.5.1), (3.5.2) by noting that many of the spaces of generalized functions have a *sheaf* structure, see Rosinger [3, pp. 131-133]. For instance, in the case of the L. Schwartz distributions, we have a natural mapping

$$(3.5.16) \qquad \Omega \supset \Omega' \text{ open } \longmapsto \mathcal{D}'(\Omega')$$

which turns $\mathcal{D}'(\Omega)$ into a sheaf of sections over Ω, see Appendix 2 and Seebach et. al.

In view of the above, a subalgebra $\mathcal{I}$ in $(\mathcal{C}^0(\Omega))^{\mathbb{N}}$, see (3.5.6), will be called *local*, if and only if

$$(3.5.17) \qquad \begin{array}{l} \forall \ \ \Omega' \subset \Omega \text{ nonvoid, open } : \\ \quad \mathcal{I}\Big|_{\Omega'} \cap \mathcal{U}^0(\Omega') = \mathcal{O} \end{array}$$

Here we recall that we have

$$\mathcal{I}\Big|_{\Omega'} = \{w\Big|_{\Omega'} \mid w \in \mathcal{I}\}$$

where, for $w = (w_0,\dots,w_\nu,\dots) \in (\mathcal{C}^0(\Omega))^{\mathbb{N}}$, we denoted

$$w\Big|_{\Omega'} = (w_0\Big|_{\Omega'},\dots,w_\nu\Big|_{\Omega'},\dots)$$

with

$$\mathcal{C}^0(\Omega) \ni h \longmapsto h\Big|_{\Omega'} \in \mathcal{C}^0(\Omega')$$

being the usual restriction of the function h to Ω'.

It follows easily that $\mathcal{I}_{nd}(\Omega)$ is local.

Finally, a subalgebra $\mathcal{A}$ in $(\mathcal{C}^0(\Omega))^{\mathbb{N}}$ is called *full*, if and only if

$$\text{(3.5.18)}\qquad \begin{array}{l} \forall \quad \Omega' \subset \Omega \text{ nonvoid, open, } t \in \mathcal{A} : \\ \left[\begin{array}{l} \exists \quad c > 0 : \\ \forall \quad \nu \in \mathbb{N},\quad x \in \Omega' : \\ \quad |t_\nu(x)| \geq c \end{array}\right] \Rightarrow \frac{1}{t\Big|_{\Omega'}} \in \mathcal{A}\Big|_{\Omega'} \end{array}$$

Obviously $\mathcal{A} = (\mathcal{C}^0(\Omega))^{\mathbb{N}}$ is full. Therefore $\mathcal{I}_{nd}(\Omega)$ is a local ideal in the full algebra $(\mathcal{C}^0(\Omega))^{\mathbb{N}}$.

We also note that conditions (3.5.17) and (3.5.18), obviously recall (3.4.12), (3.4.16) and the definition of a *nontrivial* T(D) in Section 4, in all of which an arbitrary nonvoid, open $\Omega' \subset \Omega$ is present.

And now, the main result in this Section.

Theorem 2

Given $\mathcal{A}$ a *full* subalgebra in $(\mathcal{C}^0(\Omega))^{\mathbb{N}}$ and $\mathcal{I}$ a *local* ideal in $\mathcal{A}$.

Then, the *neutrix condition*

$$\text{(3.5.19)}\qquad \mathcal{I} \cap \mathcal{U}^0(\Omega) = \mathcal{O}$$

and the 'densely vanishing' conditon

$$\text{(3.5.20)}\qquad \begin{array}{l} \forall \quad w \in \mathcal{I} : \\ \quad \Omega_w \text{ is dense in } \Omega \end{array}$$

are equivalent.

Proof

The implication (3.5.20) $\Rightarrow$ (3.5.19) follows from Proposition 5.

Conversely, assume that Ω_w is not dense in Ω, for some $w \in \mathcal{I}$. Then we have (3.5.13). But obviously

$$w\big|_{\Omega'} \in \mathcal{I}\big|_{\Omega'} \subset \mathcal{A}\big|_{\Omega'}$$

therefore (3.5.18) implies that

$$\frac{1}{w\big|_{\Omega'}} \in \mathcal{A}\big|_{\Omega'}$$

It follows that

$$u(1) = (w\big|_{\Omega'}) \cdot (\frac{1}{w\big|_{\Omega'}}) \in \mathcal{I}\big|_{\Omega'} \cdot \mathcal{A}\big|_{\Omega'} \subset \mathcal{I}\big|_{\Omega'} \tag{3.5.21}$$

since $\mathcal{I}\big|_{\Omega'}$ is an ideal in $\mathcal{A}\big|_{\Omega'}$.

But (3.5.21) contradicts (3.5.17). □

As mentioned in Appendix 4 to Chapter 1, the 'neutrix calculus', in particular, the *neutrix conditon* (3.2.10), has first been introduced in Van der Corput, in connection with an abstract model for the study of large classes of asymptotic expansions, and the setting is in essence the following. Given an Abelian group G and an arbitrary infinite set X, a subgroup

$$\mathcal{N} \subset G^X$$

is called a *neutrix*, if and only if

$$\forall \; f \in \mathcal{N}, \; \gamma \in G :$$

$$\begin{bmatrix} \forall \; x \in X : \\ f(x) = \gamma \end{bmatrix} \Rightarrow \gamma = 0 \tag{3.5.22}$$

in which case the functions $f \in \mathcal{N}$ will be called *$\mathcal{N}$-negligible*.

However, the power of the neutrix condition (3.2.10) comes into play in a significant manner within the more particular framework of (3.2.7), when it is applied to ideals $\mathcal{I}$, see Rosinger [2, pp. 75-88], Rosinger [3, pp. 306-315], as well as Chapter 6 below.

§6. DENSE VANISHING IN THE CASE OF SMOOTH IDEALS

In a rather surprising manner, it happens that the 'densely vanishing' property (3.5.20) can be significantly strenghtened in the case of sub-algebras in $(\mathcal{C}^\infty(\Omega))^{\mathbb{N}}$. Indeed, as in Theorem 2, suppose given a full subalgebra $\mathcal{A}$ in $(\mathcal{C}^0(\Omega))^{\mathbb{N}}$ and a local ideal $\mathcal{I}$ in $\mathcal{A}$. Let us define

$$(3.6.1) \qquad \mathcal{I}^\infty = \left\{ w \in \mathcal{I} \cap (\mathcal{C}^\infty(\Omega))^{\mathbb{N}} \;\middle|\; \begin{array}{l} \forall\ p \in \mathbb{N}^n : \\ D^p w \in \mathcal{I} \end{array} \right\}$$

which in view of the Leibnitz rule of product derivative, will be an ideal in

$$\mathcal{A}^\infty = \left\{ t \in \mathcal{A} \cap (\mathcal{C}^\infty(\Omega))^{\mathbb{N}} \;\middle|\; \begin{array}{l} \forall\ p \in \mathbb{N}^n : \\ D^p t \in \mathcal{A} \end{array} \right\}$$

It follows easily that, as an example, we have

$$(3.6.2) \qquad (\mathcal{I}_{nd}(\Omega))^\infty = \mathcal{I}_{nd}(\Omega) \cap (\mathcal{C}^\infty(\Omega))^{\mathbb{N}}, \quad ((\mathcal{C}^0(\Omega))^{\mathbb{N}})^\infty = (\mathcal{C}^\infty(\Omega))^{\mathbb{N}}$$

As seen in Rosinger [1-3], as well as in Chapters 2 and 7 in this volume, the ideal $\mathcal{I}_{nd}(\Omega)$ plays a crucial role in the construction of generalized solutions for wide classes of nonlinear partial differential equations.

Now, for $w \in \mathcal{I}^\infty$ let us denote

$$(3.6.3) \qquad \Omega(w) = \left\{ x \in \Omega \;\middle|\; \begin{array}{l} \forall\ p \in \mathbb{N}^n : \\ \inf\limits_{\nu \in \mathbb{N}} |D^p w_\nu(x)| = 0 \end{array} \right\} = \bigcap_{p \in \mathbb{N}^n} \Omega_{D^p w}$$

<u>Theorem 3</u>

The ideal $\mathcal{I}^\infty$ satisfies the following 'densely vanishing' condition

$$(3.6.4) \qquad \begin{array}{l} \forall\ w \in \mathcal{I}^\infty : \\ \quad \Omega(w) \text{ is dense in } \Omega \end{array}$$

<u>Proof</u>

Assume (3.6.4) is false and take $w \in \mathcal{I}^\infty$, $x \in \Omega$ and $\epsilon > 0$ such that

$$(3.6.5) \qquad B(x,\epsilon) \cap \Omega(w) = \phi$$

For $p \in \mathbb{N}^n$ we define $\delta_p : \Omega \longrightarrow [0,\infty)$ by

$$(3.6.6) \qquad \delta_p(y) = \inf_{\nu \in \mathbb{N}} |D^p w_\nu(y)|, \quad \forall \ y \in \Omega$$

Then δ_p is upper semicontinuous. Therefore the set $D_p \subset \Omega$ of discontinuities of δ_p is of first Baire category in Ω. In this way

$$D = \bigcup_{p \in \mathbb{N}^n} D_p \quad \text{is of first Baire category in } \Omega$$

It follows that we can take

$$y \in B(x,\epsilon) \backslash D$$

in which case (3.6.5) yields

$$\exists \ p \in \mathbb{N}^n :$$
$$y \notin \Omega_{D^p w}$$

and in view of (3.6.6), we obtain

$$\delta_p(y) > 0$$

thus (3.5.13) will hold for $t = D^p w$.

But $w \in \mathcal{I}^\infty$ and (3.6.1) imply that

$$t = D^p w \in \mathcal{I} \subset \mathcal{A}$$

and we can apply (3.5.18), obtaining the relation

$$\frac{1}{t|_{\Omega'}} \in \mathcal{A}|_{\Omega'}$$

Then, as in (3.5.21), it follows that

$$u(1) \in \mathcal{I}|_{\Omega'}$$

and (3.5.17) is contradicted. □

The 'densely vanishing' conditon (3.6.4) can further be strengthened. A subalgebra $\mathcal{I}$ in $(\mathcal{C}^0(\Omega))^{\mathbb{N}}$ is called *circled*, if and only if

$$(3.6.7) \qquad \begin{array}{l} \forall \; w \in \mathcal{I} : \\ \quad |w| \in \mathcal{I} \end{array}$$

Obviously $\mathcal{I}_{nd}(\Omega)$ is circled.

Given a vector subspace $\mathcal{W} \subset \mathcal{I}^{\infty}$ let us denote

$$(3.6.8) \qquad \Omega(\mathcal{W}) = \bigcap_{w \in \mathcal{W}} \Omega(w)$$

Theorem 4

Under the conditions in Theorem 3, suppose that $\mathcal{I}$ is circled.

Then the ideal $\mathcal{I}^{\infty}$ satisfies the following 'densely vanishing' condition

$$(3.6.9) \qquad \begin{array}{l} \forall \; \mathcal{W} \subset \mathcal{I}^{\infty} \text{ countably infinite dimensional vector subspace :} \\ \quad \Omega(\mathcal{W}) \text{ is dense in } \Omega \end{array}$$

Proof

Assume that (3.6.9) is false for a certain countably infinite dimensional vector subspace $\mathcal{W} \subset \mathcal{I}^{\infty}$ generated by a Hamel basis

$$(3.6.10) \qquad w^0, \ldots, w^m, \ldots \in \mathcal{W}$$

Let us take $x \in \Omega$ and $\epsilon > 0$ such that

$$(3.6.11) \qquad B(x,\epsilon) \cap \Omega(\mathcal{W}) = \phi$$

For $m \in \mathbb{N}$ and $p \in \mathbb{N}^n$ we dedfine

$$(3.6.12) \qquad w^{m,p} = |D^p w^0| + \ldots + |D^p w^m| \in \mathcal{I}$$

and $\delta_{m,p} : \Omega \longrightarrow [0,\infty)$ by

$$(3.6.13) \qquad \delta_{m,p}(y) = \inf_{\nu \in \mathbb{N}} |(w^{m,p})_{\nu}(y)|, \quad \forall \; y \in \Omega$$

In this way $\delta_{m,p}$ are upper semicontinuous. Therefore, denoting by

$$D_{m,p} \subset \Omega$$

the set of discontinuities of $\delta_{m,p}$, it follows that

$$D_{m,p} \quad \text{is of first Baire category in} \quad \Omega$$

In this way

$$D = \bigcup_{m \in \mathbb{N}} \bigcup_{p \in \mathbb{N}^n} D_{m,p} \quad \text{is of first Baire category in} \quad \Omega$$

and we can take

$$(3.6.14) \qquad y \in B(x,\epsilon)\backslash D$$

Then (3.6.11) implies that

$$(3.6.15) \qquad \begin{array}{l} \exists \quad w \in \mathcal{W},\ p \in \mathbb{N}^n : \\ \quad y \notin \Omega_{D^p w} \end{array}$$

But according to (3.6.10), we obtain

$$(3.6.16) \qquad w = \lambda_0 w^0 + + \ldots + \lambda_m w^m$$

for suitable $\lambda_0,\ldots,\lambda_m \in \mathbb{R}$. Therefore, in view of (3.6.12), (3.6.16) and Lemma 2 below, we have

$$\Omega_{w^{m,p}} \subset \Omega_{D^p w}$$

hence (3.6.15) implies that

$$y \notin \Omega_{w^{m,p}}$$

and then (3.6.13) will give

$$\delta_{m,p}(y) > 0$$

It follows that (3.5.13) holds for $t = w^{m,p}$. But (3.6.12) implies that

$$t = w^{m,p} \in \mathcal{I} \subset \mathcal{A}$$

Thus in view of (3.5.18) we obtain

$$\frac{1}{t|_{\Omega'}} \in \mathcal{A}|_{\Omega'}$$

and similar to (3.5.21), the relation results

$$u(1) \in \mathcal{I}\Big|_{\Omega'}$$

which contradicts (3.5.17) □

Lemma 2

If $w \in (\mathcal{C}^0(\Omega))^{\mathbb{N}}$ then

(3.6.17) $$\Omega_{|w|} = \Omega_w$$

More generally, if $w^0,\ldots,w^m \in (\mathcal{C}^0(\Omega))^{\mathbb{N}}$ and $\lambda_0,\ldots,\lambda_m \in \mathbb{R}$ then

(3.6.18) $$\Omega_{|w^0|+\ldots+|w^m|} \subset \Omega_{\lambda_0 w^0+\ldots+\lambda_m w^m}$$

Proof

In view of (3.5.9), the relation (3.6.17) is obvious.

Take now $\nu \in \mathbb{N}$ and $x \in \Omega$, then

$$|\lambda_0(w^0)_\nu(x) +\ldots+ \lambda_m(w^m)_\nu(x)| \leq$$

(3.6.19) $$\leq |\lambda_0|\cdot|(w^0)_\nu(x)| +\ldots+ |\lambda_m|\cdot|(w^m)_\nu(x)| \leq$$

$$\leq |\lambda|\cdot(|(w^0)_\nu(x)| +\ldots+ |(w^m)_\nu(x)|)$$

for every $\lambda \in \mathbb{R}$, such that

$$\max\{|\lambda_0|,\ldots,|\lambda_m|\} \leq |\lambda|$$

But (3.5.9), (3.6.17) and (3.6.19) obviously imply (3.6.18) □

§7. THE CASE OF NORMAL IDEALS

The increasingly stronger 'densely vanishing' conditions (3.5.20), (3.6.4) and (3.6.9) seem to point to a deeper property involved, whose full explicitation is still an open problem. This is illustrated for instance by the fact that the above 'densely vanishing' conditions (3.5.20), (3.6.4) and (3.6.9) can be obtained under the following alternative assumptions, when $\mathcal{I}$ is a subalgebra in $(\mathcal{C}^0(\Omega))^{\mathbb{N}}$ which satisfies the neutrix condition

(3.7.1) $$\mathcal{I} \cap \mathcal{U}^0(\Omega) = \mathcal{O}$$

and in addition, it is also *normal*, Köthe, that is, it has the property

(3.7.2) $$\begin{array}{l} \forall \quad w \in (\mathcal{C}^0(\Omega))^{\mathbb{N}} : \\ \left[\begin{array}{l} \exists \quad z \in \mathcal{I} : \\ \forall \quad \nu \in \mathbb{N},\ x \in \Omega : \\ \qquad |w_\nu(x)| \leq |z_\nu(x)| \end{array} \right] \Rightarrow w \in \mathcal{I} \end{array}$$

Indeed, the proofs of Theorems 2-4 will go through with the following modification. When obtaining (3.5.13) in the respective proofs, we no longer use (3.5.18). Instead we note that we can use the property

(3.7.3) $$\begin{array}{l} \exists \quad \alpha \in \mathcal{C}^0(\Omega) \ ; \\ \quad *) \quad \alpha \neq 0 \\ \quad **) \quad \forall \quad \nu \in \mathbb{N}, \quad x \in \Omega : \\ \qquad\qquad 0 \leq \alpha(x) \leq |w_\nu(x)| \end{array}$$

which in view of (3.7.2) will imply

$$u(\alpha) \in \mathcal{I} \cap \mathcal{U}^0(\Omega)$$

Then owing to *) in (3.7.3), the neutrix condition (3.7.1) is contradicted.

We note as an example that $\mathcal{I}_{nd}(\Omega)$ is obviously normal.

§8. CONCLUSIONS

The results on 'dense vanishing' obtained in Sections 5 and 7 can further be strengthened and systematized in the following way, indicated by M. Oberguggenberger in a private communication.

Given a subalgebra

(3.8.1) $\mathcal{I} \subset (\mathcal{C}^0(\Omega))^{\mathbb{N}}$

let us consider the following *three properties* encountered in Sections 5 and 7:

(DS) $\forall\ w \in \mathcal{I} : \Omega_w$ dense in Ω

(LC) $\mathcal{I}$ local (see (3.5.7))

(NX) $\mathcal{I} \cap \mathcal{U}^0(\Omega) = \mathcal{O}$

Then the following implications hold.

Theorem 5

We always have

(3.8.2) (DS) $\Rightarrow$ (LC) $\Rightarrow$ (NX)

If $\mathcal{I}$ is *normal*, see (3.7.2), then

(3.8.3) (DS) $\Longleftrightarrow$ (LC) $\Longleftrightarrow$ (NX)

Further, if $\mathcal{I}$ is an *ideal* in a *full* subalgebra $\mathcal{A} \subset (\mathcal{C}^0(\Omega))^{\mathbb{N}}$, see (3.5.18), then

(3.8.4) (DS) $\Longleftrightarrow$ (LC) $\Rightarrow$ (NX)

Finally, if $\mathcal{I}$ is an *ideal* in a subalgebra $\mathcal{A} \subset (\mathcal{C}^0(\Omega))^{\mathbb{N}}$, and $\mathcal{A} \supset \mathcal{U}^0(\Omega)$, then

(3.8.5) (DS) $\Rightarrow$ (LC) $\Longleftrightarrow$ (NX)

Proof

Let us prove (3.8.2). This implication (LC) $\Rightarrow$ (NX) follows from the definition (3.5.17). For the implication (DS) $\Rightarrow$ (LC), let us take $\Omega' \subset \Omega$ nonvoid, open. Given $w \in (\mathcal{C}^0(\Omega))^{\mathbb{N}}$, it is obvious from (3.5.9) that

(3.8.6) $$\Omega'_{w_{|\Omega'}} = \Omega_w \cap \Omega'$$

Now, if (DS) holds for $\mathcal{I}$, than (3.8.6) yields

$$\forall \ w \in \mathcal{I} : \ \Omega'_{w_{|\Omega'}} \text{ dense in } \Omega'$$

thus Propisition 5 implies that

$$\mathcal{I}_{|\Omega'} \cap \mathcal{U}^0(\Omega') = \mathcal{O}$$

therefore $\mathcal{I}$ is indeed local.

We prove now (3.8.3). We recall (3.8.2) and in addition, we prove the implication (NX) $\Rightarrow$ (DS).

For that, let us assume the failure of (DS). Then it follows that (3.5.13) holds. But in view of (3.7.3) and (3.7.2), we obtain $u(\alpha) \in \mathcal{I} \cap \mathcal{U}^0(\Omega)$. And then *) in (3.7.3) will contradict (NX).

For the proof of (3.8.4) it suffices to show the implication (LC) $\Rightarrow$ (DS), which follows obviously from Theorem 2.

Finally, in order to prove (3.8.5), we only have to show the implication (NX) $\Rightarrow$ (LC). For that, let us assume the failure of (LC). Then we can take $\Omega' \subset \Omega$ nonvoid, open, $w \in \mathcal{I}$ and $\psi \in \mathcal{C}^0(\Omega')$, such that

$$\psi(x) \neq 0, \quad x \in \Omega'$$

and

$$w_\nu\big|_{\Omega'} = \psi, \quad \nu \in \mathbb{N}$$

Let us take $\alpha \in \mathcal{C}^0(\Omega)$ such that $\operatorname{supp} \alpha \subset \Omega'$ and for certain $x_0 \subset \Omega$, we have

(3.8.7) $$\alpha(x_0)\psi(x_0) \neq 0$$

It follows that

$$u(\alpha)\cdot w \in \mathcal{U}(\Omega)\cdot\mathcal{I} \subset \mathcal{A}\cdot\mathcal{I} \subset \mathcal{I}$$

while also

$$u(\alpha)\cdot w = u(\alpha\cdot\psi) \in \mathcal{U}^0(\Omega)$$

hence

$$u(\alpha)\cdot w \in \mathcal{I} \cap \mathcal{U}^0(\Omega)$$

thus in view of (3.8.7), the assumption (NX) is contradicted. □

A convenient way to summarize the results in Theorem 5 above is as follows.

Corollary 4

If one of the following two conditions holds

(3.8.8) $\mathcal{I}$ is a normal subalgebra in $(\mathcal{C}^0(\Omega))^{\mathbb{N}}$

or

(3.8.9) $\mathcal{I}$ is an ideal in a full subalgebra $\mathcal{A} \subset (\mathcal{C}^0(\Omega))^{\mathbb{N}}$, with $\mathcal{A} \subset \mathcal{U}^0(\Omega)$,

then for $\mathcal{I}$ we have the *equivalences*

(3.8.10) (DS) $\Longleftrightarrow$ (LC) $\Longleftrightarrow$ (NS) □

APPENDIX 1

ON THE SHARPNESS OF LEMMA 1 IN SECTION 4

In view of the fact that, on the one hand, Lemma 1 in Section 4 plays a fundamental role in the algebraic characterization of the solvability of nonlinear partial differential equations given in Theorem 1 in Section 4, while on the other hand, it appears to be unknown in the earlier literature, it is useful to try to analyze its sharpness. For that purpose, we shall present two examples, with Ω open subsets in $\mathbb{R}$.

Example 1

There exist sequences $w = (w_0, \ldots, w_\nu, \ldots) \in (\mathcal{C}^0(\Omega))^{\mathbb{N}}$ of continuous functions on Ω, such that the following three conditions are satisfied:

$$\text{(3.A1.1)} \qquad \begin{array}{l} \forall \; x \in \Omega : \\ \quad \operatorname{car}(w_\nu(x) \mid \nu \in \mathbb{N}) \leq 2 \end{array}$$

while for a certain $v \in \Omega$, we have

$$\text{(3.A1.2)} \qquad \begin{array}{l} \forall \; V \subset \Omega, \; V \text{ neighbourhood of } v : \\ \quad \operatorname{car}\{w_\nu|_{V \setminus \{v\}} \mid \nu \in \mathbb{N}\} = \infty \end{array}$$

as well as

$$\text{(3.A1.3)} \qquad \begin{array}{l} \forall \; x \in \Omega, \; x \neq v : \\ \exists \; V \subset \Omega, \; V \text{ neighbourhood of } x : \\ \quad \operatorname{car}\{w_\nu|_V \mid \nu \in \mathbb{N}\} = 1 \end{array}$$

Indeed, take $\Omega = \mathbb{R}$, $v = 0 \in \Omega$ and $\alpha \in \mathcal{C}^0(\mathbb{R})$, with $\operatorname{supp} \alpha \subset [0,1]$. Define then $w \in (\mathcal{C}^0(\Omega))^{\mathbb{N}}$ by

$$w_\nu(x) = \alpha((\nu + 1)((\nu + 2)x - 1)), \quad \nu \in \mathbb{N}, \quad x \in \Omega$$

It follows easily that (3.A1.1), (3.A1.2) and (3.A1.3) are satisfied.

The above example shows the sharpness of Lemma 1, in the case of Ω connected. Indeed, in view of (3.A1.2), the relation (3.A1.3) only holds for $x \neq \nu = 0 \in \Omega$. In other words, for w in Example 1, we must have

$$\text{(3.A1.4)} \qquad \Gamma \neq \phi$$

with the notation in Lemma 1, since the relation (3.A1.2) obviously implies $v = 0 \in \Gamma$.

On the other hand, relation (3.A1.1) shows that w in Example 1 satisfies the assumption of Lemma 1 in a way which is most inconvenient for (3.A1.2) and (3.A1.4) to happen, yet these two latter relations still hold.

For Ω not connected, we have:

Example 2

There exist sequences $w = (w_0, \ldots, w_\nu, \ldots) \in (\mathcal{C}^0(\Omega))^{\mathbb{N}}$ of continuous functions on Ω such that the following three conditions are satisfied:

(3.A1.5)
$$\forall \; x \in \Omega : \quad \text{car}\{w_\nu(x) \mid \nu \in \mathbb{N}\} \leq 2$$

and with the notation in Lemma 1

(3.A1.6) $\Gamma = \phi$

while in the same time

(3.A1.7)
$$\forall \; \epsilon > 0 : \quad \text{car}\{w_\nu\big|_{\Omega \cap (0,\epsilon)} \mid \nu \in \mathbb{N}\} = \infty$$

Indeed, let us take

$$\Omega = \bigcup_{\nu \in \mathbb{N}} (1/(2\nu+2),\; 1/(2\nu+1))$$

and define $w \in (\mathcal{C}^0(\Omega))^{\mathbb{N}}$ by

$$w_\nu(x) = \begin{vmatrix} 1 & \text{if} & x \in (1/(2\nu+2),\; 1/(2\nu+1)) \\ 0 & \text{if} & x \in \Omega \setminus (1/(2\nu+2),\; 1/(2\nu+1)) \end{vmatrix}$$

for $\nu \in \mathbb{N}$. Then (3.A1.5)-(3.A1.7) follow easily.

The interest in Example 2 comes from the fact that, although (3.A1.5) and (3.A1.6) are most inconvenient for (3.A1.7) to happen, that latter relation does nevertheless hold.

APPENDIX 2

SHEAVES OF SECTIONS

The *localization* property of the Schwartz distributions $\mathcal{D}'(\mathbb{R}^n)$, see Appendix 1 to Chapter 5, gives them a structure of *sheaf* of *sections* over $\mathbb{R}^n$, as follows from the definition below. It is easy to see that various classical spaces of functions, such as $\mathcal{C}^p$, with $p \in \mathbb{N}$, as well as the analytic functions have a similar structure.

It should be recalled that, as mentioned in Rosinger [3], one encounters a *localization principle* on the very level of the usual reduction of the integro-differential balance equations of physics to the corresponding partial differential equations. And the use of such a localization principle seems to be unavoidable if the continuous formulation of physical laws is used, see Abbott for the history of discrete and continuous formulations of Newtonian laws.

To the extent that *local* and *global* phenomena are interrelated in continuously formulated physical laws, the presence of a sheaf structure on various spaces of functions and generalized functions can be particularly useful. Indeed, as pointed out for instance in Seebach et. al, sheaf theory is an effective tool in areas where problems have to be approached based on local structure and information.

For convenience, here we recall the definition of a *sheaf of sections*. For details, as well as for the definition of the associated notion of sheaf of germs, one can consult Seebach et. al., which presents a convenient introduction aimed at a larger, mathematically trained readership.

Suppose give a topological space X and a set $\mathcal{S}$ of spaces S. Suppose given a mapping

$$(3.A2.1) \qquad X \supset U \text{ open } \stackrel{\sigma}{\longmapsto} \sigma(U) = S \in \mathcal{S}$$

We call $S = \sigma(U)$ a *section* over U.

Finally, suppose that for each pair of open sets $U,V \subset X$, with $U \subset V$, we have *restriction* mappings

$$(3.A2.2) \qquad \rho_{U,V} : \sigma(V) \rightarrow \sigma(U)$$

Then $(\sigma,\rho_{U,V})$ is called a *sheaf of sections* over X, if and only if the following *four* conditions are satisfied:

For every open $U \subset X$ we have

$$(3.A2.3) \qquad \rho_{U,U} = \mathrm{id}_{\sigma(U)} : \sigma(U) \rightarrow \sigma(U)$$

For every open $U,V,W \subset X$ such that $U \subset V \subset W$, we have

(3.A2.4) $$\rho_{U,V} \circ \rho_{V,W} = \rho_{U,W}$$

For every family of open $U_i \in X$, with $i \in I$, and $s,t \in \sigma(\bigcup_{i\in I} U_i)$, we have

(3.A2.5) $$\left[\begin{array}{l} \forall \;\; i \in I : \\ \\ \rho_{U_i,U}\, s = \rho_{U_i,U}\, t \end{array}\right] \Rightarrow s = t$$

where $U = \bigcup_{i\in I} U_i$.

And finally, for every family of open $U_i \subset X$, and $s_i \in \sigma(U_i)$, with $i \in I$, we have the property: if

(3.A2.6) $$\begin{array}{l} \forall \;\; i,j \in I : \\ U_i \cap U_j \neq \sigma \Rightarrow \rho_{U_i\cap U_j, U_i}\, s_i = \rho_{U_i\cap U_j, U_j}\, s_j \end{array}$$

then

(3.A2.7) $$\begin{array}{l} \exists \;\; s \in \sigma(\bigcup_{i\in I} U_i) : \\ \forall \;\; i \in I : \\ \rho_{U_i,U}\, s = s_i \end{array}$$

Now, in order to show that the Schwartz distributions have a natural *sheaf of sections* structure, we shall take with the above notations

(3.A2.8) $$X = \mathbb{R}^n$$

(3.A2.9) $$\sigma(\Omega) = \mathcal{D}'(\Omega), \quad \text{for open } \Omega \subset \mathbb{R}^n$$

Finally, for open $\Omega \subset \Delta \subset \mathbb{R}^n$, we define

(3.A2.10) $$\rho_{\Omega,\Delta} : \mathcal{D}'(\Delta) \longrightarrow \mathcal{D}'(\Omega)$$

(3.A2.11) $$\rho_{\Omega,\Delta} F = F\big|_{\Omega}, \quad F \in \mathcal{D}'(\Delta)$$

where $F|_\Omega$ is the restriction of the distribution $F \in \mathcal{D}'(\Delta)$ to the open subset $\Omega \subset \Delta$.

It is easy to check that (3.A2.8)-(3.A2.11) satisfy (3.A2.3)-(3.A2.7).

CHAPTER 4

GENERALIZED SOLUTIONS OF SEMILINEAR WAVE EQUATIONS WITH ROUGH INITIAL VALUES

§1. INTRODUCTION

As mentioned in Remark 3, Section 4, Chapter 3 - see (3.4.70) - it can be useful to *further extend* the concept of generalized solution introduced in Section 10, Chapter 1, concept which proved to be so effective in the results presented in Chapters 2 and 3.

The aim of this Chapter is to indicate one possible such extension, introduced recently in Oberguggenberger [6]. This extension, made in the spirit of (3.4.70), turns out to be particularly effective in solving for rough initial values semilinear hyperbolic systems of the form

(4.1.1) $$U_t(t,x) + A(t,x)U_x(t,x) = F(t,x,U(t,x)), \quad t \geq 0, \ x \in \mathbb{R}$$

Here

$$U : [0,\infty) \times \mathbb{R} \longrightarrow \mathbb{R}^n$$

is the unknown function, while the given are the diagonal, $n \times n$ matrix of functions

$$A : \mathbb{R}^2 \longrightarrow \mathbb{R}^{n^2}$$

and the right hand term

$$F : \mathbb{R}^2 \times \mathbb{R}^n \longrightarrow \mathbb{R}^n$$

with both A and F being $\mathcal{C}^\infty$-smooth.

The semilinear hyperbolic system (4.1.1) is supposed to be solved with the initial value problem

(4.1.2) $$U = u \quad \text{at} \quad t = 0$$

The interest in the problem (4.1.1), (4.1.2) comes from the fact that the initial value

$$u = (u_1, \ldots, u_n)$$

in (4.1.2) can be chosen in a *quite rough* manner, namely $u_1, \ldots, u_n$ can be rather arbitrary *generalized functions* on the domain $\mathbb{R}$ of the space variable x. Then, owing to the *nonlinearity* of the system (4.1.1), one is faced with the highly nontrivial problem of establishing the precise way of the *propagation of singularities* in the solution $U(t,x)$, with $t \geq 0$ and $x \in \mathbb{R}$, singularities caused at $t = 0$ by the rough initial value u.

Recently, a particular case of rough initial values leading to the so called *delta waves* has been studied in Oberguggenberger [6] and Rauch & Reed.

The extension of the concept of generalized solution introduced in Oberguggenberger [6] and presented in this Chapter proves to have two rather striking qualities. First, it offers existence, uniqueness and regularity or coherence results which contain, and in fact go much beyond the similar earlier results. Secondly, the method of proof is unusually simple and transparent, giving thus a particularly clear understanding of the basic mathematical phenomena involved, which - as previously in Chapters 2 and 3 - prove to be of an algebraic nature, related to properties of rings of sequences of continuous or smooth functions on Euclidean spaces.

§2. THE GENERAL EXISTENCE AND UNIQUENESS RESULT

First we start with the customary type of conditions on the semilinear hyperbolic system (4.1.1).

Concerning the $n \times n$ diagonal matrix of functions A, we assume that

$$(4.2.1) \qquad A \text{ or } D_x A \text{ is bounded on } \mathbb{R}^2$$

This condition is sufficient for the existence of the *characteristic curves* for all time $t \in \mathbb{R}$.

The nonlinear term F is assumed to satisfy the *bounded gradient* condition

$$(4.2.2) \qquad \begin{array}{l} \forall \;\; K \subset \mathbb{R}^2 \text{ compact :} \\ \exists \;\; C > 0 : \\ \forall \;\; (t,x) \in K, \; u = (u_1, \ldots, u_n) \in \mathbb{R}^n, \;\; 1 \leq i \leq n : \\ \quad |D_{u_i} F(t,x,u)| \leq C \end{array}$$

which guarantees that (4.1.1), (4.1.2) has a unique global solution $U \in (\mathcal{C}^\infty(\mathbb{R}^2))^n$, for every intial value $u \in (\mathcal{C}^\infty(\mathbb{R}))^n$.

Now we can turn to the construction of the suitable spaces of generalized functions, and to the appropriate concept of generalized solution.

For a convenient formulation of (4.1.1), let us define on $\mathbb{R}^2$ the following nonlinear partial differential operator

(4.2.3) $T(D)U(t,x) = U_t(t,x) + A(t,x)U_x(t,x) - F(t,x,U(t,x)),\ (t,x) \in \mathbb{R}^2$

Then obviously

(4.2.4) $T(D)(\mathcal{C}^{\ell+1}(\mathbb{R}^2))^n \subset (\mathcal{C}^{\ell}(\mathbb{R}^2))^n, \quad \ell \in \overline{\mathbb{N}}$

where we assume that $\infty + 1 = \infty$. In order to deal with the initial value problem (4.1.2), let us define the following linear operator

(4.2.5) $BU(t,x) = U(0,x), \quad (t,x) \in \mathbb{R}^2$

It follows that

(4.2.6) $B(\mathcal{C}^{\ell}(\mathbb{R}^2))^n \subset (\mathcal{C}^{\ell}(\mathbb{R}))^n, \quad \ell \in \overline{\mathbb{N}}$

Now (4.1.1), (4.1.2) can be written in the equivalent form

(4.2.7) $T(D)U = 0$

(4.2.8) $BU = u$

Concerning (4.2.7), we could try to follow the method for systems of nonlinear partial differential equations presented in Section 14, Chapter 1, noting that with the notation there, we would have

$$\Omega = \mathbb{R}^2, \quad a = b = n, \quad m = 1$$

Furthermore, as shown in Oberguggenberger [6], it will be convenient to take

(4.2.9) $\Lambda = (0,1) \subset \mathbb{R}$

However, as it stands, the method in Section 14, Chapter 1 would need to assume that F in (4.1.1) or (4.2.3) is polynomial. Fortunately, this and several other assumptions made in Section 14, Chapter 1 can be done away with, owing to the fact that the concept of generalized solution used in Oberguggenberger [6] is more general than that in Section 10, Chapter 1.

In this respect, instead of extensions of the type (1.14.14), we shall construct for $T(D)$ in (4.2.3) above extensions, see (3.4.70)

(4.2.10) $T(D) : E_1 \longrightarrow E_2$

where, see (1.6.7)

(4.2.11) $E_1 = \mathcal{S}_1/\mathcal{V}_1,\ E_2 = \mathcal{S}_2/\mathcal{V}_2 \in VS_{\mathcal{F},(0,1)}$

are suitable *quotient vector spaces*, while

$$(4.2.12) \qquad \mathcal{F} = \mathcal{C}^\infty(\mathbb{R}^2,\mathbb{R}^n)$$

that is, $\mathcal{F}$ is the *vector space* of all $\mathcal{C}^\infty$-smooth functions on $\mathbb{R}^2$ and with values in $\mathbb{R}^n$.

It should be noted that for $n \geq 2$, that is, for *nontrivial* systems, the usual, point wise operations on functions $f : \mathbb{R}^2 \to \mathbb{R}^n$ will only yield a vector space structure on $\mathcal{F}$, and *not* one of algebra. Therefore, we could *not* use $\mathcal{F}$, in case one of the two spaces of generalized functions E_1 or E_2 in (4.2.11) would have to be an algebra. In this respect, one of the advantages of the extension (4.2.10) used in Oberguggenberger [6] is precisely in the fact that *none* of the two spaces E_1 or E_2 need to be an algebra.

Now, in view of (1.6.5), (1.6.6), it follows that

$$(4.2.13) \qquad \begin{array}{ccccc} \mathcal{V}_i & \longrightarrow & \mathcal{S}_i & \longrightarrow & \mathcal{F}^{(0,1)} \\ \uparrow & & \uparrow & & \\ \mathcal{O} & \longrightarrow & \mathcal{U}_{\mathcal{F},(0,1)} & & \end{array}$$

with the neutrix property

$$(4.2.14) \qquad \mathcal{V}_i \cap \mathcal{U}_{\mathcal{F},(0,1)} = \mathcal{O}$$

for $i \in \{1,2\}$.

Turning to the initial value problem (4.2.8), we shall construct for B in (4.2.5) extensions of the type

$$(4.2.15) \qquad B : E_1 \longrightarrow E_0$$

where

$$(4.2.16) \qquad E_0 = \mathcal{S}_0/\mathcal{V}_0 \in VS_{\mathcal{K},(0,1)}$$

is a suitable *quotient vector space*, while

$$(4.2.17) \qquad \mathcal{K} = \mathcal{C}^\infty(\mathbb{R},\mathbb{R}^n)$$

in other words, $\mathcal{K}$ is the *vector space* of all $\mathcal{C}^\infty$-smooth functions on $\mathbb{R}$ and with values in $\mathbb{R}^n$. Again, therefore, if $n \geq 2$, then $\mathcal{K}$ is *not* an algebra with the usual pointwise operations on functions.

Similar to (4.2.13) and (4.2.14), we shall have

(4.2.18)
$$\begin{array}{ccccc} \mathcal{V}_0 & \longrightarrow & \mathcal{S}_0 & \longrightarrow & \mathcal{K}^{(0,1)} \\ \uparrow & & \uparrow & & \\ \mathcal{O} & \longrightarrow & \mathcal{U}_{\mathcal{K},(0,1)} & & \end{array}$$

as well as the neutrix property

(4.2.19)
$$\mathcal{V}_0 \cap \mathcal{U}_{\mathcal{K},(0,1)} = \mathcal{O}$$

We can proceed now with the details of the *construction* of extensions (4.2.10) and (4.2.15).

For that purpose, it is convenient to split the nonlinear operator $T(D)$ in (4.2.3) into its linear part

(4.2.20)
$$L(D)U(t,x) = U_t(t,x) + A(t,x)U_x(t,x), \quad (t,x) \in \mathbb{R}^2$$

and its remaining nonlinear part, which for simplicity will again be denoted by F, that is

(4.2.21)
$$FU(t,x) = F(t,x,U(t,x)), \quad (t,x) \in \mathbb{R}^2$$

Then, similar to (4.2.4) we obtain

(4.2.22)
$$L(D)(\mathcal{C}^{\ell+1}(\mathbb{R}^2))^n \subset (\mathcal{C}^{\ell}(\mathbb{R}^2))^n, \quad \ell \in \mathbb{N}$$

(4.2.23)
$$F(\mathcal{C}^{\ell}(\mathbb{R}^2))^n \subset (\mathcal{C}^{\ell}(\mathbb{R}^2))^n, \quad \ell \in \mathbb{N}$$

In particular, in view of (4.2.12), we have

(4.2.24)
$$L(D)\mathcal{F} \subset \mathcal{F}, \quad F\mathcal{F} \subset \mathcal{F}$$

On the other hand, (4.2.6), (4.2.12) and (4.2.17) yield

(4.2.25)
$$B\mathcal{F} \subset \mathcal{K}$$

Now, based on (4.2.24) and (4.2.25), we simply obtain the extensions

(4.2.26)
$$L(D) : \mathcal{F}^{(0,1)} \longrightarrow \mathcal{F}^{(0,1)}, \quad F : \mathcal{F}^{(0,1)} \longrightarrow \mathcal{F}^{(0,1)}$$

(4.2.27)
$$B : \mathcal{F}^{(0,1)} \longrightarrow \mathcal{K}^{(0,1)}$$

by defining termwise the respective mappings, that is, given

$$s = (\psi_\epsilon | \epsilon \in (0,1)) \in \mathcal{F}^{(0,1)}$$

we have

$$L(D)s = (L(D)\psi_\epsilon | \epsilon \in (0,1)), \quad Fs = (F\psi_\epsilon | \epsilon \in (0,1))$$

and

$$Bs = (B\psi_\epsilon | \epsilon \in (0,1))$$

Finally, we can come to the choice of the vector spaces of generalized functions E_1, E_2 and E_0 in (4.2.10) and (4.2.15).

For that, first, we shall take the vector subspaces

$$\mathcal{S}_1, \mathcal{S}_2 \subset \mathcal{F}^{(0,1)} \quad \text{and} \quad \mathcal{S}_0 \subset \mathcal{K}^{(0,1)}$$

so that the following *four conditions* are satisfied

$$(4.2.28) \quad \begin{aligned} &\mathcal{S}_1 \subset \mathcal{S}_2 \\ &L(D)\mathcal{S}_1 \subset \mathcal{S}_2 \\ &F\mathcal{S}_1 \subset \mathcal{S}_1 \\ &B\mathcal{S}_1 \subset \mathcal{S}_0 \end{aligned}$$

Next, we shall take the vector subspaces

$$\mathcal{V}_1 \subset \mathcal{S}_1, \ \mathcal{V}_2 \subset \mathcal{S}_2 \quad \text{and} \quad \mathcal{V}_0 \subset \mathcal{S}_0$$

in a way which satisfies the *four conditions*

$$(4.2.29) \quad \begin{aligned} &\mathcal{V}_1 \subset \mathcal{V}_2 \\ &L(D)\mathcal{V}_1 \subset \mathcal{V}_2 \\ &F(s+v) - F(s) \in \mathcal{V}_1, \quad \text{for} \quad s \in \mathcal{S}_1, \ v \in \mathcal{V}_1 \\ &B\mathcal{V}_1 \subset \mathcal{V}_0 \end{aligned}$$

In the examples in Section 3 next, which include the known results in literature obtained until recently, it will be shown how the above conditions (4.2.28) and (4.2.29) can be satisfied.

The point in these eight conditions (4.2.28) and (4.2.29), considered for the first time in Oberguggenberger [6], is that we can now define the following *four mappings* between the respective spaces of generalized functions. First, the two linear mapping

(4.2.30)
$$\begin{aligned} E_1 = \mathcal{S}_1/\mathcal{V}_1 &\xrightarrow{\quad i \quad} E_2 = \mathcal{S}_2/\mathcal{V}_2 \\ s + \mathcal{V}_1 &\longmapsto s + \mathcal{V}_2 \end{aligned}$$

and

(4.2.31)
$$\begin{aligned} E_1 = \mathcal{S}_1/\mathcal{V}_1 &\xrightarrow{\quad L(D) \quad} E_2 = \mathcal{S}_2/\mathcal{V}_2 \\ s + \mathcal{V}_1 &\longmapsto L(D) + \mathcal{V}_2 \end{aligned}$$

then the nonlinear mapping

(4.2.32)
$$\begin{aligned} E_1 = \mathcal{S}_1/\mathcal{V}_1 &\xrightarrow{\quad F \quad} E_1 = \mathcal{S}_1/\mathcal{V}_1 \\ s + \mathcal{V}_1 &\longmapsto Fs + \mathcal{V}_1 \end{aligned}$$

and finally, the linear mapping

(4.2.33)
$$\begin{aligned} E_1 = \mathcal{S}_1/\mathcal{V}_1 &\xrightarrow{\quad B \quad} E_0 = \mathcal{S}_0/\mathcal{V}_0 \\ s + \mathcal{V}_1 &\longmapsto Bs + \mathcal{V}_0 \end{aligned}$$

Consequently, the mappings (4.2.30), (4.2.31) and (4.2.32) allow us the definition of the *nonlinear* mapping

(4.2.34) $$T(D) : E_1 \longrightarrow E_2$$

by

(4.2.35) $$T(D) = L(D) - i \circ F$$

In this way, the problem of constructing the extensions in (4.2.10) and (4.2.15) got solved by (4.2.34) and (4.2.33) respectively.

We note that the condition

(4.2.36) $$\mathcal{S}_1 \cap \mathcal{V}_2 = \mathcal{V}_1$$

is necessary and sufficient for the canonical embedding i in (4.2.30) to be injective, in which case, we shall consider that the inclusion holds

(4.2.37) $$E_1 \subset E_2$$

We can now define the concept of *generalized solution* introduced in Oberguggenberger [6], once the framework (4.2.11)-(4.2.35) is given. Namely, a generalized function

(4.2.38) $U = s + \mathcal{V}_1 \in E_1 = \mathcal{S}_1/\mathcal{V}_1$

is called an $(E_1 \to E_2,\ E_0)$-*sequential solution* of the semilinear hyperbolic system

(4.2.39) $U_t(t,x) + A(t,x)U_x(t,x) = F(t,x,U(t,x)),\ (t,x) \in \mathbb{R}^2$

with the associated initial value problem

(4.2.40) $U = u$ at $t = 0$

if and only if the mappings (4.2.34) and (4.2.33) satisfy the conditions

(4.2.41) $T(D)U = 0$

and

(4.2.42) $BU = u$

where the initial value u is given such that

(4.2.43) $u \in E_0$

Remark 1

It is obvious from the above construction in (4.2.11)-(4.2.35) that the concept of $(E_1 \to E_2,\ E_0)$-*sequential solution* just defined is by *no means limited* to the particular partial differential equation in (4.1.1) or (4.2.39). Indeed, for a given *arbitrary* nonlinear partial differential operator $T(D)$, one can easily arrive at a corresponding concept of $(E_1 \to E_2,\ E_0)$-sequential solution, simply by adapting accordingly the conditions (4.2.28) and (4.2.29). Similarly, one can use infinite index sets Λ other than $(0,1) \subset \mathbb{R}$, chosen in (4.2.9). Finally, one can use vector subspaces $\mathcal{F}$ and $\mathcal{K}$ other than those given in (4.2.12) and (4.2.17) respectively.

At this stage, we can now turn to the question of *existence and uniqueness* of an $(E_1 \to E_2,\ E_0)$-sequential soluition for our rough initial value problem for the semilinear hyperbolic system (4.1.1), (4.1.2).

As is known, under the condition that

(4.2.44) A, F, u are $\mathcal{C}^\infty$-smooth

the problem (4.1.1), (4.1.2) *has a unique* classical solution

(4.2.45) $U \in \mathcal{C}^\infty(\mathbb{R}^2, \mathbb{R}^n)$

provided the (4.2.1) and (4.2.2) hold.

In order to obtain a *general existence and uniqueness* result for *rough initial values* such as in (4.2.43), the following *two conditions* are fundamental. First we assume that, given any $(\chi_\epsilon | \epsilon \in (0,1)) \in \mathcal{S}_0$, if ψ_ϵ, with $\epsilon \in (0,1)$, is the unique classical solution of (4.1.1) for the initial value problem $\psi_\epsilon(0,x) = \chi_\epsilon(x)$, $x \in \mathbb{R}$, then

(4.2.46) $\qquad s = (\psi_\epsilon | \epsilon \in (0,1)) \in \mathcal{S}_1$

Secondly, we assume that, given any $t, z \in \mathcal{S}_1$, such that $T(D)t, T(D)z \in \mathcal{V}_2$ and $Bt - Bz \in \mathcal{V}_0$, then

(4.2.47) $\qquad t - z \in \mathcal{V}_1$

Under the above conditions (4.2.1), (4.2.2), (4.2.44), (4.2.46) and (4.2.47), we obtain the following *general existence and uniqueness result.*

Theorem 1

Given an arbitrary initial value $u \in E_0$. Then the semilinear hyperbolic system

(4.2.48) $\qquad T(D)U = 0$

with the rough initial values

(4.2.49) $\qquad BU = u$

has a unique $(E_1 \longrightarrow E_2, E_0)$-sequential solution $U \in E_1$.

Proof

Assume that we have the representation

$$u = (\chi_\epsilon | \epsilon \in (0,1)) + \mathcal{V}_0 \in E_0 = \mathcal{S}_0/\mathcal{V}_0$$

with

$$(\chi_\epsilon | \epsilon \in (0,1)) \in \mathcal{S}_0$$

Then, with the respective construction preceeding (4.2.46), we obtain

$$s = (\psi_\epsilon | \epsilon \in (0,1)) \in \mathcal{S}_1$$

Now, from (4.2.34) and (4.2.33), it follows easily that

(4.2.50) $U = s + \mathcal{V}_1 \in E_1 = \mathcal{S}_1/\mathcal{V}_1$

is indeed an $(E_1 \rightarrow E_2, E_0)$-sequential solution of (4.2.48), (4.2.49).

The uniqueness of U in (4.2.50) follows at once from (4.2.47) □

Remark 2

In view of the significant generality of the existence and uniqueness result in Theorem 1 above, it is *particularly important* to establish the *coherence properties*, see Colombeau [1,2], of the unique generalized solutions given by this theorem. In other words, we have to establish the way in which these unique generalized solutions are related to the earlier known classical, distributional and generalized solutions. This coherence property will be illustrated next, in Sections 3 and 4, in the case of the earlier known $\mathcal{L}^1_{loc}$ and delta wave solutions.

Remark 3

It is important to note the fact that both the insight and the result in Theorem 1, gained by the general construction in this Section, are highly nontrivial. Indeed, on the one hand, they contain and unify in a clear and elegant manner the essential algebraic and analytic aspects of earlier known results. Here, to be more precise, we should mention that, at a closer study, the construction in this Section gives the obvious impression of requiring the *minimum-minimorum* of the algebraic and analytic conditions for bringing about the existence and uniqueness result in Theorem 1.

Let us be more specific, by noting the following. One of the *strong points* of Theorem 1 is that, as seen later in (4.4.1), the vector space E_0 of the *initial values* can be quite *large*, for instance, it can contain all the distributions in $\mathcal{D}'(\mathbb{R})$. Now, as is obvious, the essence of the construction in this Section is to choose the *six vector subspaces* $\mathcal{S}_1$, $\mathcal{S}_2$, $\mathcal{S}_0$, $\mathcal{V}_1$, $\mathcal{V}_2$ and $\mathcal{V}_0$ in such a way that the conditions (4.2.28) and (4.2.29) are satisfied. However, since $E_0 = \mathcal{S}_0/\mathcal{V}_0$, it follows that *large* E_0 means *large* $\mathcal{S}_0$ and *small* $\mathcal{V}_0$. And condition (4.2.28) does not prevent $\mathcal{S}_0$ from being large. On the other hand, conditon (4.2.29) may easily prevent $\mathcal{V}_0$ from being small. Which means that the construction of a large $E_0 = \mathcal{S}_0/\mathcal{V}_0$ is *not a trivial* matter.

It is precisely here that the mathematical difficulties involved in securing the result in Theorem 1 come to be manifested. And in view of (4.2.28), (4.2.29), these difficulties take the particularly simple, obvious and minimal form of *eight inclusions* involving vector spaces, in two of which linear partial differential operators are present.

In this way, the *enabling power* of the respective framework for solving systems of nonlinear partial differential equations is larger than that of customary functional analytic approaches which, owing to possible unnecessary topological ideosyncrasies, may require more stringent conditions.

On the other hand, the mentioned general construction opens the door to a remarkably large variety of spaces of generalized functions in which one can search for the solutions of large classes of systems of nonlinear partial differential equations, see the comment in Remark 1.

It should be noted that the use of the various Sobolev spaces had offered during the last decades a most impressive opening in the study of linear, and certain nonlinear partial differential equations. The difficulty however with this functional analytic method is in its near exclusive reliance on the *topologies* on the respective spaces of generalized functions. Indeed, as is known, Dacorogna, most of the even simplest *nonlinear* operations are *not continuous* in a large variety of such topologies. Therefore, a functional analytic approach to nonlinear partial differential equations often necessitates *stringent particularizations*, in order to be able to overcome such difficulties.

In more precise, technical terms, the enabling power of the framework in this Section comes from the *large variety* of the possiblities in the choice of the *vector subspaces* in (4.2.28) and (4.2.29). Indeed, as seen next in Sections 3 and 4, suitable choices of these vector subspaces make it easy to account for various *analytical properties* of generalized solutions of nonlinear partial differential equations.

At this stage of the ongoing research, with the opening given by the construction in this Section, one can proceed further and elaborate appropriate methods - in the basic algebraic and analytic spirit of this construction - which can be applied to various classes of systems of nonlinear partial differential equations.

As in Oberguggenberger [6], this Chapter presents one such method, specifically deviced for semilinear hyperbolic systems, see in particular Section 4 below.

§3. COHERENCE WITH $\mathcal{L}^1_{loc}$ SOLUTIONS

As is well known, for $u \in \mathcal{L}^1_{loc}(\mathbb{R},\mathbb{R}^n)$, the semilinear hyperbolic system (4.1.1), with the initial value problem (4.1.2) has a unique solution $U \in \mathcal{C}(\mathbb{R},\mathcal{L}^1_{loc}(\mathbb{R},\mathbb{R}^n))$.

We show now that these solutions are obtained by Theorem 1, Section 2 as well.

For that purpose, let

$$(4.3.1) \qquad \mathcal{F} = \mathcal{C}^\infty(\mathbb{R}^2,\mathbb{R}^n),\ \mathcal{K} = \mathcal{C}^\infty(\mathbb{R},\mathbb{R}^n)$$

Further, let us take

$$(4.3.2) \qquad \mathcal{V}_1 \subset \mathcal{S}_1 \subset \mathcal{F}^{(0,1)}$$

with $\mathcal{S}_1$ being the set of all convergent sequences in $\mathcal{C}(\mathbb{R},\mathcal{L}^1_{\ell oc}(\mathbb{R},\mathbb{R}^n))$, and $\mathcal{V}_1$ being the subset of those sequences which converge to zero. In other words, if for instance $v = (\psi_\epsilon \mid \epsilon \in (0,1)) \in \mathcal{V}_1$, then by convergence to zero of the sequence v we mean that $\psi_\epsilon \longrightarrow 0$ in $\mathcal{C}(\mathbb{R},\mathcal{L}^1_{\ell oc}(\mathbb{R},\mathbb{R}^n))$ when $\epsilon \longrightarrow 0$. Similarly for sequences $s \in \mathcal{S}_1$. Further, let us take

$$(4.3.3) \qquad \mathcal{V}_2 \subset \mathcal{S}_2 \subset \mathcal{F}^{(0,1)}$$

where $\mathcal{S}_2$ is the set of all convergent sequences in $\mathcal{D}'(\mathbb{R}^2,\mathbb{R}^n)$, while $\mathcal{V}_2$ is its subset of sequences convergent to zero. Finally, we take

$$(4.3.4) \qquad \mathcal{V}_0 \subset \mathcal{S}_0 \subset \mathcal{K}^{(0,1)}$$

Here $\mathcal{S}_0$ is the set of all convergent sequences in $\mathcal{L}^1_{\ell oc}(\mathbb{R},\mathbb{R}^n)$ and $\mathcal{V}_0$ is the subset of the sequences convergent to zero.

It follows easily that

$$(4.3.5) \qquad E_1 = \mathcal{C}(\mathbb{R},\mathcal{L}^1_{\ell oc}(\mathbb{R},\mathbb{R}^n)),\ E_2 = D'(\mathbb{R}^2,\mathbb{R}^n),\ E_0 = \mathcal{L}^1_{\ell oc}(\mathbb{R},\mathbb{R}^n)$$

Now, owing to the bounded gradient condition (4.2.2), we shall have (4.2.28) and (4.2.29) satisfied.

Further, as is known, the distributional solutions $U \in \mathcal{C}(\mathbb{R},\mathcal{L}^1_{\ell oc}(\mathbb{R},\mathbb{R}^n))$ depend continuously on the initial values $u \in \mathcal{L}^1_{\ell oc}(\mathbb{R},\mathbb{R}^n)$. From this, condition (4.2.46) follows easily.

Finally, condition (4.2.47) follows from the fact that the $\mathcal{L}^1_{\ell oc}$ solutions are unique.

In this way, Theorem 1 does indeed contain the unique, $\mathcal{L}^1_{\ell oc}$ solutions.

§4. THE DELTA WAVE SPACE

We construct spaces of generalized functions E_1, E_2, E_0 with the following properties:

(4.4.1) E_1, E_2, E_0 contain the $\mathcal{D}'$ distributions

(4.4.2) for every initial value $u \in E_0$, ther exists a unique $(E_1 \longrightarrow E_2, E_0)$-sequential solution $U \in E_1$ for the problem (4.1.1), (4.1.2)

(4.4.3) $\mathcal{L}^1_{\ell oc}$ solutions are $(E_1 \longrightarrow E_2, E_0)$-sequential solutions

(4.4.4) certain dela wave solutions are $(E_1 \longrightarrow E_2, E_0)$-sequential solutions

From (4.4.1) follows in particular that the initial value problem (4.1.2) for the semilinear hyperbolic system (4.1.1) admits solutions for as *rough initial values* as can be given by *arbitrary distributions*. However, as seen immediately, one can in fact use initial values which are quite a bit more rough.

In view of the above, the space of generalized functions E_1 is called the *delta wave space*, see Oberguggenberger [6].

Now let us proceed as follows. The space

(4.4.5) $$\mathcal{C} = \mathcal{C}(\mathbb{R}, \mathcal{L}^1_{\ell oc}(\mathbb{R}, \mathbb{R}^n))$$

is equipped with the seminorms

(4.4.6) $$p_k(\psi) = \sup_{|t| \leq k} \int_{-k}^{k} |\psi(t,x)| dx, \quad \psi \in \mathcal{C}, \; k \in \mathbb{N}$$

Then, we define the space, see (4.2.20)

(4.4.7) $$\mathcal{C}_{L(D)} = \left\{ S \in \mathcal{D}'(\mathbb{R}^2, \mathbb{R}^n) \;\middle|\; \begin{array}{l} \exists \;\; \psi \in \mathcal{C} : \\ S = L(D)\psi \end{array} \right\}$$

and define on this space the Fréchet topology given by the finest locally convex topology which makes continuous the mapping

(4.4.8) $$L(D) : \mathcal{C} \longrightarrow \mathcal{C}_{L(D)}$$

Obviously, this Fréchet space on $\mathcal{C}_{L(D)}$ is given by the seminorms

(4.4.9) $$q_k(S) = \inf\{p_k(\psi) \mid \psi \in \mathcal{C},\ L(D)\psi = S\},\ S \in \mathcal{C}_{L(D)},\ k \in \mathbb{N}$$

In order to follow the construction in Section 2, we take again

(4.4.10) $$\mathcal{F} = \mathcal{C}^\infty(\mathbb{R}^2,\mathbb{R}^n),\ \mathcal{K} = \mathcal{C}^\infty(\mathbb{R},\mathbb{R}^n)$$

while on the other hand, this time we take

(4.4.11) $$\mathcal{S}_1 = \mathcal{S}_2 = \mathcal{F}^{(0,1)},\ \mathcal{S}_0 = \mathcal{K}^{(0,1)}$$

Finally

(4.4.12) $$\mathcal{V}_1 \subset \mathcal{S}_1,\ \mathcal{V}_2 \subset \mathcal{S}_2,\ \mathcal{V}_0 \subset \mathcal{S}_0$$

will be taken as the respective sets of sequences which converge to zero in $\mathcal{C}$, $\mathcal{C}_{L(D)}$ and $\mathcal{L}^1_{loc}(\mathbb{R},\mathbb{R}^n)$.

In this way, we obtain the vector spaces of generalized functions

(4.4.13) $$E_1 = \mathcal{S}_1/\mathcal{V}_1,\ E_2 = \mathcal{S}_2/\mathcal{V}_2,\ E_0 = \mathcal{S}_0/\mathcal{V}_0$$

Next, we have to verify that the conditions (4.2.28), (4.2.29), (4.2.46) and (4.2.47) are satisfied.

First we note that in view of (4.4.11) and (4.4.12), the conditions (4.2.28) and (4.2.46) are trivially satisfied.

Now we turn to the verification of the remaining two conditions (4.2.29) and (4.2.47).

For that, for $1 \leq i \leq n$, let us denote by $a_i \in \mathcal{C}^\infty(\mathbb{R}^2,\mathbb{R})$ the i-th function on the diagonal of A in (4.1.1). Then, for given $(t,x) \in \mathbb{R}^2$, let us denote by

(4.4.14) $$\gamma_i(t,x,\cdot) : \mathbb{R} \to \mathbb{R}$$

the parametric representation of the i-th characteristic curve of the operator $L(D)$. In other words, $\gamma_i(t,x,\tau)$, with $\tau \in \mathbb{R}$, is precisely that integral curve of the vector field $D_t + a_i(t,x)D_x$ which passes through the point x at the time $\tau = t$.

Further, we note that the matrix differential operator $L(D)$ in (4.4.8) has a right inverse J given by

$$(4.4.15)\qquad (J\psi(t,x))_i = \int_0^t \psi(\tau,\gamma_i(t,x,\tau))d\tau,\quad 1 \le i \le n$$

for $\psi \in \mathcal{C}$ and $(t,x) \in \mathbb{R}^2$. Moreover, it is easy to verify that

$$(4.4.16)\qquad J : \mathcal{C} \to \mathcal{C}$$

is continuous.

We also have

Lemma 1

J maps $\mathcal{C}^\infty(\mathbb{R}^2,\mathbb{R}^n)$ continuously in $\mathcal{C}$, when the former space has the topology induced by $\mathcal{C}_{L(D)}$.

Proof

Consider the mapping $\xi : \mathcal{C} \to \mathcal{C}$ which translates initial values along characteristic curves, according to the relation

$$(\xi\psi(t,x))_i = \psi(0,\gamma_i(t,x,0)),\quad 1 \le i \le n$$

with $\psi \in \mathcal{C}$ and $(t,x) \in \mathbb{R}^2$. Then in view of (4.4.6), it is obvious that ξ is continuous. It follows that

$$\begin{aligned}&\forall\ k \in \mathbb{N} :\\ &\exists\ C > 0,\ \ \ell \in \mathbb{N},\ \ \ell \ge k :\\ &\forall\ \psi \in \mathcal{C} :\\ &\quad p_k(\xi\psi) \le Cp_\ell(\psi)\end{aligned}$$

We also note the inclusion

$$\mathcal{C}^\infty(\mathbb{R}^2,\mathbb{R}^n) \subset \mathcal{C}_{L(D)}$$

Therefore, given $\psi \in \mathcal{C}^\infty(\mathbb{R}^2,\mathbb{R}^n)$, $\epsilon > 0$ and $k \in \mathbb{N}$, one can find $\chi \in \mathcal{C}$, such that

$$L(D)\chi = \psi,\ p_\ell(\chi) \le q_\ell(\psi) + \epsilon$$

But J is the right inverse of $L(D)$, thus

$$L(D)(J\psi - \chi) = 0 \quad \text{in} \quad \mathcal{D}'(\mathbb{R}^2,\mathbb{R}^n)$$

This means that $J\psi - \chi$ is constant along the characteristic curves.

However, (4.4.15) implies that

$$J\psi = 0 \quad \text{at} \quad t = 0$$

Therefore

$$J\psi - \chi = \xi\chi$$

which means that

$$p_k(J\psi) = p_k(\chi+\xi\chi) \le p_k(\chi) + Cp_\ell(\chi) \le (C+1)(q_\ell(\psi)+\epsilon)$$

Since $\epsilon > 0$ is arbitrary, one obtains

$$p_k(J\psi) \le (C+1)q_\ell(\psi), \quad \psi \in \mathcal{C}^\infty(\mathbb{R}^2,\mathbb{R}^n)$$ □

We turn now to the verification of the remaining two conditions (4.2.29) and (4.2.47), which are essential for the application of Theorem 1 in Section 2.

The inclusion $\mathcal{V}_1 \subset \mathcal{V}_2$ in (4.2.29) follows from the continuity of J in (4.4.16), the continuity of $J : \mathcal{C}^\infty(\mathbb{R}^2,\mathbb{R}^n) \longrightarrow \mathcal{C}$ proved in Lemma 1, and the decomposition of the identity mapping on $\mathcal{C}^\infty(\mathbb{R}^2,\mathbb{R}^n)$ into $L(D)J$. The other inclusions in (4.2.29) follow easily from (4.4.11), (4.4.12) and (4.2.2).

Finally, the verification of condition (4.2.47) is a bit more involved.

Suppose given $s,z \in \mathcal{S}_1 = (\mathcal{C}^\infty(\mathbb{R}^2,\mathbb{R}^n))^{(0,1)}$, such that $T(D)s, T(D)z \in \mathcal{V}_2$ and also $Bs-Bz \in \mathcal{V}_0$. Then, by definition, there exist $v \in \mathcal{V}_0$ and $w \in \mathcal{V}_2$, such that

$$\text{(4.4.17)} \qquad \begin{aligned} &L(D)(s-z) - Fs + Fz = w \\ &B(s-z) = v \end{aligned}$$

Now, in order to verify condition (4.2.47), it only remains to show that

$$\text{(4.4.18)} \qquad s-z \in \mathcal{V}_1$$

Assume that

$$s = (\psi_\epsilon | \epsilon \in (0,1)), \quad z = (\chi_\epsilon | \epsilon \in (0,1))$$

$$v = (\rho_\epsilon | \epsilon \in (0,1)), \quad w = (\sigma_\epsilon | \epsilon \in (0,1))$$

$$F = (F_1,\ldots,F_n)$$

Let us fix $k \in \mathbb{N}$ and estimate the seminorm

$$p_k(\psi_\epsilon - \chi_\epsilon)$$

For a suitable $T > 0$, let $K_T \subset \mathbb{R}^2$ be the *domain of determinacy* of $L(D)$ which contains $[-k,k] \times [-k,k] \subset \mathbb{R}^2$ and is bounded by the lines $t = \pm T$, as well as the slowest and fastest characteristics respectively, passing through the endpoints of a sufficiently large x-interval $[\alpha_0, \beta_0] \subset \mathbb{R}$, at $t = 0$. Let $[\alpha_\tau, \beta_\tau]$ be the x-interval obtained by intersecting K_T with the line $t = \tau$, with $-T \leq \tau \leq T$. Then (4.4.17) gives, for $t \in [-T,T]$ and each coordinate $1 \leq i \leq n$, the inequalities

$$\int_{\alpha_t}^{\beta_t} |(\psi_\epsilon)_i(t,x) - (\chi_\epsilon)_i(t,x)|dx \leq$$

$$\leq \int_{\alpha_t}^{\beta_t} |(\rho_\epsilon)_i(\gamma_i(t,x,0))|dx +$$

$$+ \left|\int_{\alpha_t}^{\beta_t} \int_0^t ((\sigma_\epsilon)_i + F_i(\psi_\epsilon) - F_i(\chi_\epsilon))(\tau,\gamma_i(t,x,\tau))d\tau\, dx\right| \leq$$

$$\leq \int_{\gamma_i(t,\alpha_t,0)}^{\gamma_i(t,\beta_t,0)} |(\rho_\epsilon)_i(x)| \cdot \|D_x\gamma_i\|_{\mathcal{L}^\infty(K_T)} dx +$$

$$+ \int_{\alpha_t}^{\beta_t} |(J\sigma_\epsilon)_i(t,x)|dx +$$

$$+ \left|\int_0^t \int_{\gamma_i(t,\alpha_t,\tau)}^{\gamma_i(t,\beta_t,\tau)} C\, \|D_x\gamma_i\|_{\mathcal{L}^\infty(K_T)} \|\psi_\epsilon(t,x) - \chi_\epsilon(t,x)\| dx\, d\tau\right|$$

where C follows from (4.2.2) for K_T.

Now, in view of Lemma 1, we note that the integral involving J is bounded uniformly in $t \in [-T,T]$, by $q_\ell((\sigma_\epsilon)_i)$, for a suitable $\ell \geq k$. For the other two terms in the last inequality we note that

$$[\gamma_i(t,\alpha_t,\tau),\gamma_i(t,\beta_t,\tau)] \subset [\alpha_\tau,\beta_\tau]$$

Therefore, we obtain

$$\int_{\alpha_t}^{\beta_t} |(\psi_\epsilon)_i(t,x) - (\chi_\epsilon)_i(t,x)|dx \leq$$

$$\leq \int_{\alpha_0}^{\beta_0} |(\rho_\epsilon)_i(x)|\cdot\|D_x\gamma_i\|_{\mathcal{L}^\infty(K_T)} dx + q_\ell((\sigma_\epsilon)_i) +$$

$$+ |\int_0^t C \|D_x\gamma_i\|_{\mathcal{L}^\infty(K_T)} \int_{\alpha_\tau}^{\beta_\tau} \|\psi_\epsilon(t,x) - \chi_\epsilon(t,x)\|dx\, d\tau$$

Using Gronwall's inequality, it follows that

$$\int_{\alpha_t}^{\beta_t} \|\psi_\epsilon(t,x) - \chi_\epsilon(t,x)\|dx \leq$$

$$(C_1 \int_{\alpha_0}^{\beta_0} \|\rho_\epsilon(x)\|dx + q_\ell(\sigma_\epsilon))e^{C_1 T}$$

for all $t \in [-T,T]$, provided that $C_1 > 0$ is suitably chosen. In this way, we obtain indeed (4.4.18), which completes the proof of (4.2.47).

We conclude that the framework constructed in (4.4.10)-(4.4.13) does indeed satisfy the conditions in Theorem 1, Section 2.

In particular, we have, therefore, obtained the result claimed in (4.4.2).

We can now proceed with the proof of the remaining results claimed in (4.4.1), (4.4.3) and (4.4.4).

First we prove (4.4.1).

For that purpose, we take any fixed sequence $(\varphi_\epsilon | \epsilon \in (0,1)) \in (\mathcal{D}(\mathbb{R},\mathbb{R}))^{(0,1)}$ which converges in $\mathcal{D}'(\mathbb{R},\mathbb{R})$ to the Dirac delta distribution δ, when $\epsilon \to 0$. As is well known, we could take for instance

$$\varphi_\epsilon(x) = \frac{1}{\epsilon}\varphi(\frac{x}{\epsilon}), \quad \epsilon \in (0,1), \quad x \in \mathbb{R}$$

where $\varphi \in \mathcal{D}(\mathbb{R},\mathbb{R})$ is given in such a way that

$$\int_{\mathbb{R}} \varphi(x)dx = 1$$

Now, we define the mapping

$$(4.4.19) \qquad \mathcal{D}'(\mathbb{R},\mathbb{R}^n) \xrightarrow{\alpha} E_0 = \mathcal{S}_0/\mathcal{V}_0$$

with the help of the convolution *, as follows

$$(4.4.20) \qquad \alpha(S) = (S*\varphi_\epsilon \mid \epsilon \in (0,1)) + \mathcal{V}_0, \ S \in \mathcal{D}'(\mathbb{R},\mathbb{R}^n)$$

Similarly, we define the mapping

$$(4.4.21) \qquad \mathcal{D}'(\mathbb{R}^2,\mathbb{R}^n) \xrightarrow{\beta} E_1 = \mathcal{S}_1/\mathcal{V}_1$$

by

$$(4.4.22) \qquad \beta(Q) = (Q*\psi_\epsilon \mid \epsilon \in (0,1)) + \mathcal{V}_1, \quad Q \in \mathcal{D}'(\mathbb{R}^2,\mathbb{R}^n)$$

where

$$\psi_\epsilon(t,x) = \varphi_\epsilon(t)\varphi_\epsilon(x), \quad \epsilon \in (0,1), (t,x) \in \mathbb{R}^2$$

Then, in view of (4.4.12), it follows immediately that α and β above are *linear embeddings*. It follows that E_1 and E_0 do contain the $\mathcal{D}'$ distributions. The fact that E_2 also contains the $\mathcal{D}'$ distributions follows from (4.2.37) and Lemma 1. Therefore (4.4.1) holds indeed.

The proof of (4.4.3) follows from

Proposition 1

Given $u \in \mathcal{L}^1_{loc}(\mathbb{R},\mathbb{R}^n)$ and $U \in \mathcal{C}(\mathbb{R},\mathcal{L}^1_{loc}(\mathbb{R},\mathbb{R}^n))$ the unique solution of the semilinear hyperbolic system (4.1.1) with the initial value problem (4.1.2).

Then, for the initial value $v = \alpha(u) \in E_0$, there corresponds a unique $(E_1 \to E_2, E_0)$-sequential solution $V \in E_1$ for the semilinear hyperbolic system (4.1.1).

Moreover, we have the *coherence property*

$$(4.4.23) \qquad V = \beta(U)$$

Proof

In view of (4.4.20), we obtain

$$v = (u*\varphi_\epsilon | \epsilon \in (0,1)) + \mathcal{V}_0 \in E_0 = \mathcal{S}_0/\mathcal{V}_0$$

Then, according to the proof of Theorem 1 in Section 2, the unique solution $V \in E_1$ can be obtained as

$$V = (V_\epsilon | \epsilon \in (0,1)) + \mathcal{V}_1 \in E_1 = \mathcal{S}_1/\mathcal{V}_1$$

where V_ϵ is the unique classical, in fact, $\mathcal{C}^\infty$-smooth solution of (4.1.1) with the initial value $u*\varphi_\epsilon$.

But

$$V_\epsilon \longrightarrow U \text{ in } \mathcal{C}, \text{ when } \epsilon \longrightarrow 0$$

owing to the well known continuous dependence property of classical solutions of (4.1.1), (4.1.2).

Also we have obviously

$$U*\psi_\epsilon \longrightarrow U \text{ in } \mathcal{C}, \text{ when } \epsilon \longrightarrow 0$$

In this way, we obtain

$$(V_\epsilon - U*\psi_\epsilon | \epsilon \in (0,1)) \in \mathcal{V}_1$$

which, in view of (4.4.22), yields (4.4.23) □

Finally, we turn to the proof of property (4.4.4) concerning *delta waves.*

For that purpose we shall have to make the following two additional assumptions on the semilinear hyperbolic system (4.1.1), namely

(4.4.24) A is constant

and F is *bounded*, more precisely

(4.4.25)
$$\begin{aligned} &\forall \ K \subset \mathbb{R}^2 \text{ compact :} \\ &\exists \ C > 0 : \\ &\forall \ (t,x) \in K, \ u \in \mathbb{R}^n \\ &\quad |F(t,x,u)| \leq C \end{aligned}$$

Concerning the *rough initial values* u, we assume that

(4.4.26) $u \in \mathcal{D}'(\mathbb{R},\mathbb{R}^n)$ and supp u is finite.

We recall now the following result, Oberguggenberger [2,9], Rauch & Reed.

Let U_ϵ, with $\epsilon \in (0,1)$, be the classical C^∞-smooth unique solution of (4.1.1) corresponding to the initial value $u*\varphi_\epsilon$. Then, there exist $V \in \mathcal{D}'(\mathbb{R}^2,\mathbb{R}^n)$ and $W \in C^\infty(\mathbb{R}^2,\mathbb{R}^n)$, such that

(4.4.27) $$U_\epsilon \longrightarrow V + W \text{ in } \mathcal{D}'(\mathbb{R}^2,\mathbb{R}^n), \text{ when } \epsilon \longrightarrow 0$$

with V being the *distributional* solution of the *linear* hyperbolic initial value problem

(4.4.28) $$\begin{aligned} &L(D)V = 0 \\ &V = u \quad \text{at} \quad t = 0 \end{aligned}$$

while W is the *classical* solution of the *semilinear* hyperbolic initial value problem

(4.4.29) $$\begin{aligned} &T(D)W = 0 \\ &W = 0 \quad \text{at} \quad t = 0 \end{aligned}$$

Owing to (4.4.27), one calls $V + W$ the *delta wave solution* of the semilinear hyperbolic initial value problem, see (4.1.1), (4.1.2)

(4.4.30) $$\begin{aligned} &T(D)U = 0 \\ &U = u \quad \text{at} \quad t = 0 \end{aligned}$$

Its *decomposition* property (4.4.28) and (4.4.29) is quite surprising and remarkable, since the *rough initial value* u, see (4.4.26), only influences the *linear* part (4.4.28) of the semilinear hyperbolic system (4.4.30).

Furthermore, this delta wave solution $V + W$ has the following *coherence property* as well.

Proposition 2

The delta wave solution $V + W$ is precisely the unique $(E_1 \longrightarrow E_2, E_0)$-sequential solution of the semilinear hyperbolic initial value problem (4.4.30).

Proof

With the notations in (4.4.27)-(4.4.29), let us consider

$$V_\epsilon = V*\psi_\epsilon, \quad \epsilon \in (0,1)$$

Since A is constant and V is a solution of (4.4.28), we obviously obtain for $\epsilon \in (0,1)$

$$L(D)V_\epsilon = 0$$

$$V_\epsilon = u*\varphi_\epsilon \quad \text{at} \quad t = 0$$

Further, it follows, Oberguggenberger [2,9], that

$$U_\epsilon - V_\epsilon - W \longrightarrow 0 \quad \text{in} \quad \mathcal{C}, \quad \text{when} \quad \epsilon \longrightarrow 0$$

But W is $\mathcal{C}^\infty$-smooth, therefore

$$W - W*\psi_\epsilon \longrightarrow 0 \quad \text{in} \quad \mathcal{C}, \quad \text{when} \quad \epsilon \longrightarrow 0$$

In this way

$$(U_\epsilon - (V+W)*\psi_\epsilon | \epsilon \in (0,1)) \in \mathcal{V}_1$$

and then, according to (4.4.22), we obtain for $U = (U_\epsilon | \epsilon \in (0,1))$ the relation

(4.4.31) $\quad U = \beta(V+W) \in E_1 = \mathcal{S}_1/\mathcal{V}_1$ □

§5. A FEW REMARKS

The question of *coherence* with more general types of *delta wave solutions* developed in Oberguggenberger [2,9] and Rauch & Reed remains open.

As shown in Oberguggenberger [6], the choice of the very general framework in (4.2.10) and (4.2.15) where E_1, E_2 and E_0 are not necessarily algebras, has at least one *critically important advantage*, namely, it can offer *coherence properties* of generalized solutions, such as for instance in Propositions 1 and 2 in Section 4. Indeed, in case these three spaces are taken for instance as the differential algebras constructed in Colombeau [1,2], then the *coherence property* (4.4.31) will in general *fail*, even in the case *two dimensional*, *linear*, *variable coefficient* hyperbolic initial value problems.

The conditions (4.2.28) and (4.2.29) which define the essence of the framework in Section 2, are obviously related to the notions of stability, generality and exactness of generalized solutions, notions defined in Section 11, Chapter 1. Indeed, large $\mathcal{V}_1$ means high *stability* with respect to the possible perturbations of a given representative s defining the solution U, see (4.2.50). The *generality* property of solutions U increases with the size of $E_1 = \mathcal{S}_1/\mathcal{V}_1$. Therefore, it means large $\mathcal{S}_1$ and small $\mathcal{V}_1$. According to (4.4.1), E_1 is large enough to contain the $\mathcal{D}'$ distributions. At that point, one can already note that stability and generality are *conflicting*.

In case we omit condition (1.11.14) from the definition of exactness and also omit the requirement 'large $\mathcal{A}$' in condition (1.11.15), then this concept of exactness can be applied to the quotient vector space $E_2 = \mathcal{S}_2/\mathcal{V}_2$ as well. In this case, better *exactness* will mean smaller $\mathcal{V}_2$. Since however $\mathcal{V}_1 \subset \mathcal{V}_2$, see (4.2.29), we can note that stability will also *conflict* with exactness.

CHAPTER 5

DISCONTINUOUS, SHOCK, WEAK AND GENERALIZED SOLUTIONS OF BASIC NONLINEAR PARTIAL DIFFERENTIAL EQUATIONS

§1. THE NEED FOR NONCLASSICAL SOLUTIONS: THE EXAMPLE OF THE NONLINEAR SHOCK WAVE EQUATIONS

It is interesting to note that many of the *nonlinear* partial differential equations of physics

$$(5.1.1) \qquad F(t,x,U(t,x),\ldots,D_t^p D_x^q U(t,x),\ldots) = 0, \quad t > t_0, \quad x \in \Omega \subset \mathbb{R}^n$$

are defined by highly regular, in particular *analytic* functions F. In fact, in many cases, the order $|p| + |q|$ of the partial derivatives in (5.1.1) does not exceed 2, while the nonlinearities are polynomial, or even quadratic, as for instance in the equations of fluid dynamics, general relativity, etc. It can however happen that the initial and/or boundary value problems associated with (5.1.1) will no longer be given by analytic functions. Yet, under suitable well-posedness conditions satisfied by (5.1.1), such initial and/or boundary values may be replaced by analytic approximations. In this way, it may appear that we may restrict our attention to analytic partial differential equations and initial or boundary values. This of course would be a major advantage, as we could for instance use the classical Cauchy-Kovalevskaia theorm, which guarantees the *existence* of an analytic - therefore, *classical* - solution for every noncharacteristic analytic initial value problem.

Unfortunately, this and similar, Oleinik, Colton and the references mentioned there, existence results are of a *local* nature, i.e., they guarantee the existence of classical, in particular analytic, solutions only in a neighbourhood of the noncharacteristic hypersurface on which the initial values are given. And this situation is *twice* unsatisfactory: first, in many physical problems we are interested in solutions which exist on much *larger domains* than those granted by the above mentioned local existence results, then secondly, classical solutions will in general *fail* to exist on the larger domains of physical interest. A particularly relevant, simple, yet important example in this respect is given by the conservation law

$$(5.1.2) \qquad U_t + U_x U = 0, \quad t > 0, \quad x \in \mathbb{R}$$

with the initial value problem

$$(5.1.3) \qquad U(0,x) = u(x), \quad x \in \mathbb{R}$$

Obviously (5.1.2) is an analytic nonlinear partial differential equation which is of first order and has a polynomial, actually quadratic nonlinearity. Let us assume that the function u defining the initial value problem (5.1.3) is analytic on $\mathbb{R}$. It is easy to see that the classical, in fact analytic solution U of (5.1.2), (5.1.3) will be given by the implicit equation

$$(5.1.4) \qquad U(t,x) = u(x - tU(t,x)), \quad t \geq 0, \quad x \in \mathbb{R}$$

Hence, according to the implicit function theorem, if

$$(5.1.5) \qquad tu'(x - tU(t,x)) + 1 \neq 0$$

we can obtain $U(s,y)$ from (5.1.4), for s and y in suitable neighbourhoods of t and x respectively. Obviously (5.1.5) is satisfied for $t = 0$, hence, there exists a neighbourhood $\Omega \subset [0,\infty) \times \mathbb{R}$ of the x-axis $\mathbb{R}$, so that $U(t,x)$ exists for $(t,x) \in \Omega$. However, if for a no matter how small interval $I \subset \mathbb{R}$ we have

$$(5.1.6) \qquad u'(x) < 0, \quad x \in I$$

then the condition (5.1.5) may be violated for certain $t > 0$. This can happen irrespective of the extent of the domain of analyticity of u. Indeed, $u(x) = \sin x$ for instance is analytic not only for real but also for all complex x, yet, it satisfies (5.1.6) on every interval $I = ((2k+1)\pi,(2k+2)\pi) \subset \mathbb{R}$, with $k \in \mathbb{Z}$.

Now, it is well known, Lax, that the violation of (5.1.5) can mean that the classical solution U no longer exists for the respective t and x. In other words, we can have $\Omega \subsetneqq [0,\infty] \times \mathbb{R}$, i.e., for certain $x \in \mathbb{R}$, the classical solution $U(t,x)$ will cease to exist for sufficiently large $t > 0$. In particular, it follows that

$$(5.1.7) \qquad U \notin \mathcal{C}^1([0,\infty) \times \mathbb{R})$$

in other words, the equation (5.1.2) *fails* to have *classical solutions* on the *whole* of its domain of definition.

However, from physical point of view, it is precisely the points $(t,x) \in [0,\infty) \times \mathbb{R}\backslash\Omega$ which present interest in connection with the possible appearance and propagation of what are called *shock waves*. Fortunately, under rather general conditions, Lax, Schaeffer, one can define certain *generalized solutions* for all $t \geq 0$ and $x \in \mathbb{R}$

$$(5.1.8) \qquad U : [0,\infty) \times \mathbb{R} \longrightarrow \mathbb{R}$$

which are physically meaningful, and which are in fact classical solutions, except for points $(t,x) \in \Gamma$, where $\Gamma \subset [0,\infty) \times \mathbb{R}$ consists of certain families of curves called *shock fronts*. For clarity, let us consider the following example, when the initial value u in (5.1.3) is given by

$$(5.1.9) \qquad u(x) = \left| \begin{array}{lll} 1 & \text{if} & x \leq 0 \\ 1-x & \text{if} & 0 \leq x \leq 1 \\ 0 & \text{if} & x \geq 1 \end{array} \right.$$

in which case we have the *shock* front

$$(5.1.10)) \qquad \Gamma = \left\{ (t,x) \in [0,\infty) \times \mathbb{R} \;\middle|\; \begin{array}{ll} *) & t \geq 1 \\ **) & x = (t+1)/2 \end{array} \right\}$$

and for $0 \leq t < 1$, we have the classical solution

$$(5.1.11) \qquad U(t,x) = \left| \begin{array}{ll} 1 & \text{if } x \leq 0 \\ (x-1)/(t-1) & \text{if } 0 \leq x \leq 1 \\ 0 & \text{if } x \geq 1 \end{array} \right.$$

while for $t \geq 1$, we have a generalized solution

$$(5.1.12) \qquad U(t,x) = \left| \begin{array}{ll} 1 & \text{if } x < (t+1)/2 \\ 0 & \text{if } x > (t+1)/2 \end{array} \right.$$

with $U(t,x)$ defined at will for $(t,x) \in \Gamma$. It should be noted that in the above example (5.1.9)-(5.1.12), the failure of U to be a classical solution for all $(t,x) \in [0,\infty) \times \mathbb{R}$, does not come from the fact that u in (5.1.9) is not sufficiently smooth, for instance analytic, but from the fact that u in (5.1.9) satisfies (5.1.6) on $I = (0,1)$.

Before we go further and see the ways generalized solutions could be defined, it should be noted that within the *linear* theory of distributions, the above generalized solution (5.1.12) *cannot* be dealt with in a satisfactory way. Indeed, across the shock front Γ in (5.1.10), the generalized solution U in (5.1.12) has a jump discontinuity of the type the Heaviside function has at $x = 0$, i.e.,

$$(5.1.13) \qquad H(x) = \left| \begin{array}{ll} 0 & \text{if } x > 0 \\ 1 & \text{if } x > 0 \end{array} \right.$$

hence, its partial derivative U_x in (5.1.2) will have across Γ a singularity of the type the Dirac δ distribution has at $x = 0$. And then, the product $U_x U$ in (5.1.2) when simplified to one dimension, is of the type $H \cdot \delta$, which as is known, cannot be dealt with within the Schwartz distribution theory, since both the factors are singular at the same point $x = 0$.

Let us deal in some more detail with this difficulty.

For instance, one can naturally ask whether, nevertheless the *nonclassical solutions* U in (5.1.7) of (5.1.2) could perhaps be dealt with within the framework of the *distributions*. For instance, we could perhaps assume that, nevertheless, we may have

$$(5.1.14) \qquad U \in \mathcal{D}'((0,\infty) \times \mathbb{R})$$

Indeed, as is well known, Lax, Schaeffer, in many important cases, the nonclassical solutions U of (5.1.2), (5.1.3) are in fact smooth functions on

$(0,\infty) \times \mathbb{R}$, with the exception of certain smooth curves $\Gamma \subset (0,\infty) \times \mathbb{R}$. In other words, we often have

$$(5.1.15) \qquad U \in \mathcal{C}^{\infty}(((0,\infty) \times \mathbb{R}) \backslash \Gamma)$$

Furthermore, the nonclassical solutions U have finite jump discontinuities across the curves Γ. In other words, if we assume that

$$(5.1.16) \qquad \Gamma = \{(t,x) \in (0,\infty) \times \mathbb{R} \mid t = \gamma(x)\}, \quad \text{with} \quad \gamma \in \mathcal{C}^{\infty}(\mathbb{R})$$

then

$$(5.1.17) \qquad U(t,x) = U_-(t,x) + (U_+(t,x) - U_-(t,x)) \cdot H(t-\gamma(x)), \; t > 0, \; x \in \mathbb{R}$$

where

$$(5.1.18) \qquad U_-, U_+ \in \mathcal{C}^{\infty}((0,\infty) \times \mathbb{R})$$

and

$$(5.1.19) \qquad H(y) = \left| \begin{array}{ll} 0 & \text{if} \quad y < 0 \\ 1 & \text{if} \quad y > 0 \end{array} \right.$$

is the Heaviside function. In such a case, using the *distributional* derivatives, one obtains from (5.1.16)-(5.1.19) the following relations in $\mathcal{D}'((0,\infty) \times \mathbb{R})$

$$(5.1.20) \qquad \begin{aligned} U_t &= (U_-)_t + (U_+ - U_-)_t \cdot H + (U_+ - U_-) \cdot \delta \\ U_x &= (U_-)_x + (U_+ - U_-)_x \cdot H - \gamma'(U_+ - U_-) \cdot \delta \end{aligned}$$

where

$$(5.1.21) \qquad \delta \in \mathcal{D}'(\mathbb{R})$$

is the Dirac distribution, which is the distributional derivative of the Heaviside function, that is

$$(5.1.22) \qquad \delta = H' \quad \text{in} \quad \mathcal{D}'(\mathbb{R})$$

The inappropriateness of dealing with nonclassical solutions of *nonlinear* partial differential equations within the distributional framework becomes now obvious.

Indeed, if we try to *replace* (5.1.17), (5.1.20) into (5.1.2) in order to *check* whether or not U in (5.1.17) is indeed a solution, this simply *cannot* be done within $\mathcal{D}'((0,\infty) \times \mathbb{R})$, since the *nonlinear* term $U \cdot U_x$ in (5.1.2) would lead to the *singular* product

(5.1.23) $$H \cdot \delta$$

which, as we mentioned, is *not* defined within the distributions.

Needless to say, if instead of the first order, polynomial nonlinear partial differential equation in (5.1.2) we have a second order, polynomial nonlinear partial differential equation - a situation frequently occuring in physics - and we try to *check* whether or not a nonclassical solution U is a distributional solution, we can in addition to (5.1.23) end up with yet more singular products, such as for instance

(5.1.24) $$H \cdot \delta', \quad \delta \cdot \delta, \quad \delta \cdot \delta'$$

which are even less definable within the distributions $\mathcal{D}'$.

In some cases of nonlinear partial differential equations, such as for instance *conservative* ones, as is the case of (5.1.2) as well, the problem of having to deal with nonclassical solutions can be approached by replacing the respective nonlinear partial differential equations with more general, so called *weak* equations. For instance, in the case of (5.1.2), it is obvious that any function $U \in \mathcal{C}^1((0,\infty) \times \mathbb{R})$ which satisfies it, will also satisfy the *weak* equation

(5.1.25) $$\int_0^\infty \int_{-\infty}^\infty (U\psi_t + \tfrac{1}{2}U^2\psi_x)dt\ dx = 0, \quad \psi \in \mathcal{D}((0,\infty) \times \mathbb{R})$$

which is obviously more general than (5.1.2), since it can admit solutions

(5.1.26) $$U \in \mathcal{L}^2_{[\ni]}((0,\infty) \times \mathbb{R})$$

The obvious *advantage* of the *weak* equation (5.1.25) is that - unlike (5.1.2) - it only contains U but *none* of the partial derivatives of U. In this way, the *weak* equation (5.1.25) can accommodate *nonclassical* solutions U as well.

However, many important nonlinear equations of physics are *not* in a conservative form and therefore, they do not admit convenient weak generalizations. One of the simplest such examples is the following system which models the coupling between the velocity U and stress Σ in a one dimensional homogeneous medium of constant density

(5.1.27) $$\begin{aligned} U_t + U \cdot U_x &= \Sigma_x \\ \Sigma_t + U \cdot \Sigma_x &= k^2 U_x \end{aligned}, \quad t \geq 0,\ x \in \mathbb{R}$$

with $k > 0$ depending on the medium, and where the second equation - owing to the term $U \cdot \Sigma_x$ - is *not* in conservative form.

It follows that the *weak* equations have *two deficiencies* : first, they cannot always be used to replace nonlinear partial differential equations, and secondly, even when they can be used, they are more general than, and not necessarily equivalent with the nonlinear partial differential equations they replace.

A detailed approach to generalized solutions corresponding to *discontinuous* functions, such as for instance in (5.1.17), is presented in Chapter 7. This approach can deal with an arbitrary number of independent variables and with rather large classes of nonlinear partial differential equations, which include many of the equations of physics.

§2. INTEGRAL VERSUS PARTIAL DIFFERENTIAL EQUATIONS

There appears as well to exist deeper reasons for considering *nonclassical solutions* for linear and nonlinear partial differential equations.

We recall that most of the basic equations of physics which are *direct* expressions of physical laws, are *balance equations*, valid on sufficiently regular domains of space-time, and as such, they are written as *integro-differential* equations on the respective domains, Fung, Erigen, Peyret & Taylor. Since a *local*, space-time, point-wise description of the state of a physical system is often considered to be preferable from the point of view of satisfactory or hopefully sufficient information, the respective integro-differential equations are reduced - under suitable *additional* regularity assumptions - to *partial differential equations* whose classical, function solutions

$$(5.2.1) \qquad U : \Omega \longrightarrow \mathbb{R}, \quad \Omega \subset \mathbb{R}^n$$

are supposed to describe the state of the respective physical system. It follows that many of the basic partial differential equations of physics are consequences of physical laws *and* additional mathematical type regularity conditions needed in the reduction of the primal integro-differential equations to the mentioned partial differential equations. These additional assumptions or conditions can be seen as constituting a *localization principle*, Eringen, which under suitable forms, plays a crucial role in various notions of weak, distributional and generalized solutions. In fact, this localization principle determines an important *sheaf* structure, Seebach, et. al., on the respective spaces of distributions and generalized functions, see Appendex 2, Chapter 3.

A good example in connection with the above is given by *conservation laws*.

Suppose a scalar physical system occupying a fixed space domain $\Delta \subset R^m$ is such that the change in time in the total amount of that physical entity in any given sufficiently regular subdomain $G \subset \Delta$ is due to the flux of that physical entity across the boundary ∂G of G, and takes place according to the relation

(5.2.2) $$\frac{d}{dt} \int_G U(t,x)dx = - \int_{\partial G} \langle F(t,x),\ n(x)\rangle\ dS$$

where $U(t,x) \in \mathbb{R}$ is the density of the physical entity at time t and at the space point $x \in G$, while $F(t,x) \in \mathbb{R}^m$ is the flux of that physical entity at time t and at the space point $x \in \partial G$.

As is known, in case U and F are assumed to be sufficiently regular, for instance $\mathcal{C}^1$-smooth, the integro-differential equation (5.2.2) can be reduced to the partial differential equation

(5.2.3) $$\frac{\partial}{\partial t} U + \operatorname{div}_x F = 0, \quad t \in \mathbb{R},\ x \in \Delta$$

Indeed, in view of Gauss' formula and the regularity of G, (5.2.2) yields

(5.2.4) $$\int_G (\frac{\partial}{\partial t} U + \operatorname{div}_x F)dx = 0$$

and then, the arbitrariness of $G \subset \Delta$ will imply (5.2.3).

However, it is *important* to note that the integro-differential equation (5.2.2) which is the *direct* expression of the conservation law considered, is *more general* than the partial differential equation (5.2.3) which was obtained from (5.2.2) under the mentioned additional regularity assumptions on U and F, assumptions which are *not* required on the level of (5.2.2), a relation valid for nonsmooth but integrable U and F.

Nevertheless, if we make use of *test functions* $\psi \in \mathcal{C}^1(\mathbb{R} \times \Delta)$ with compact support, equation (5.2.3) yields after an integration by parts

(5.2.5) $$\int_{\mathbb{R}} \int_{\Delta} (U \frac{\partial}{\partial t} \psi + F \operatorname{div}_x \psi)dt\ dx = 0$$

This equation, when assumed to hold for every $\psi \in \mathcal{D}(\mathbb{R} \times \Delta)$, is the *weak* form of (5.2.3) and obviously, it is *more general* than (5.2.3), although not necessarily more general than (5.2.2). However, for many of the nonlinear integro-differential equations, one cannot obtain a corresponding convenient weak form, but only a nonlinear partial differential equation which, as mentioned, is obtained by assuming among others, suitable regularity conditions on the state function U of the respective physical system.

In this way, we can distinguish *two levels of localization* : first, localization on *compact* subsets of the space-time domain, and second, localization at *points* of the space-time domain.

The first localization leads to *weak* forms of the physical balance equations, as for instance in (5.2.5) and does not depend on additional regularity assumptions, but rather on specific features of the physical system itself, such as conservation properties, for instance.

The second localization leads to linear or nonlinear *partial differential equations* and does require additional regularity conditions, as for instance in the above example, where the partial differential equation (5.2.3) was obtained from a physical law *and* certain additional mathematical regularity conditions.

In general, the situation can be more complex, as it happens for instance in fluid dynamics, where in addition to the physical laws and certain mathematical regularity conditions, one also needs specific assumptions on the mechanical properties of the fluid, usually called *constitutive equations*, such as those concerning the stress-strain relationship, Fung. A typical and important example is that of the so called Newtonian viscous fluids described by the Navier-Stokes equations, where the balance or conservation of mass, momentum and energy give the respective *integro-differential* equations, Peyret & Taylor

$$(5.2.6) \qquad \frac{d}{dt}\int_{\Delta} \rho U \, dx + \int_{\partial\Delta} \rho \langle U,n\rangle dS = 0$$

$$(5.2.7) \qquad \frac{d}{dt}\int_{\Delta} \rho U \, dx + \int_{\partial\Delta} (\langle U,n\rangle \rho U - n\sigma) dS = \int_{\Delta} F \, dx$$

$$(5.2.8) \qquad \frac{d}{dt}\int_{\Delta} \rho E \, dx + \int_{\partial\Delta} \langle \rho E U - \sigma U + q, n\rangle dS = \int_{\Delta} \langle F,U\rangle \, dx$$

where $\Delta \subset \mathbb{R}^3$ is a domain occupied by the fluid, $U : \Delta \longrightarrow \mathbb{R}^3$ is the velocity of the fluid, while ρ, σ, F, E and q are the density, stress tensor, internal volume force, total energy and heat flux, respectively. If we assume now the suitable conditions of regularity on the above physical entities, as well as the usual constitutive equations, Fung, Peyret & Taylor, then, for an incompressible fluid, we obtain the *nonlinear partial differential system*

$$(5.2.9) \qquad \frac{\partial}{\partial t} U_i + \sum_{1\leq j\leq 3} U_j \frac{\partial}{\partial x_j} U_i = F_i - \frac{1}{\rho}\frac{\partial}{\partial x_i} P$$

$$+ \nu \sum_{1\leq j\leq 3} \frac{\partial^2}{\partial x_j^2} U_i, \quad 1 \leq i \leq 3$$

$$(5.2.10) \qquad \sum_{1\leq i\leq 3} \frac{\partial}{\partial x_i} U_i = 0$$

with $t \geq 0$ and $x \in \Delta$, where $U = (U_1, U_2, U_3)$, $F = (F_1, F_2, F_3)$, while $P : \Delta \longrightarrow \mathbb{R}$ is the pressure and ν is the kinematic viscosity.

Although, as seen above, many of the partial differential equations of physics are more particular than the initial, direct expressions of the respective physical laws, such as for instance the balance, integro-differential equations, we are often obliged to deal only with these linear or nonlinear partial differential equations, since the less restrictive weak forms may fail to exist under convenient form. Indeed, when looking for local, point-wise information in space-time on the state functions (5.2.1), partial differential equations seem to be irreplacable.

For the sake of completeness however, we should remember the following. The *continuous* - in particular, integro-differential-modelling of physical laws originated with Euler, while Newton's initial formulation of the second law of dynamics has been *discrete*, see Abbott, pp. 220-227, and the literature cited there. The difference between these two kind of formulations is that in certain cases - such as for instance irreversible physical processes - the discrete models are more general than the continuous ones.

As seen above, nonclassical, in particular generalized solutions appear in a necessary way in the study of some of the simplest nonlinear partial differential equations. In addition, we have to remember the following.

When replacing with partial differential equations the integro-differential balance equations which are the primary expressions of physical laws, we assumed certain additional regularity conditions on the state functions U. Usually these regularity conditions require a smoothness of U of an order which turns U into a classical solution of the resulting partial differential equations. Therefore, in order to avoid the possibility of eliminating physically meaningful nonclassical solutions when we reduce the original integro-differential equations to partial differential equations, we have to provide *sufficiently general* means allowing for the incorporation of nonclassical, in particular generalized solutions of these partial differential equations.

As we are mainly interested in *generalized solutions* for *nonlinear* partial differential equations, we shall in the next Section direct our attention towards two of the methods which have so far proved to be most suited, see for instance Sobolev [1,2], Lions [2].

However, in order to make clear the most basic ideas involved, it is useful to consider a few more, simple, linear examples. The equation

(5.2.11) $$U_t - U_x = 0, \quad t,x \in \mathbb{R}$$

has the classical solution

(5.2.12) $$U(t,x) = u(x+t), \quad t,x \in \mathbb{R}$$

where

(5.2.13) $$u \in \mathcal{C}^1(\mathbb{R})$$

and it describes the propagation of the space wave defined by u, along the characteristic lines $x + t$ = constant. It follows that from physical point of view, there is no any justification for the regularity condition (5.2.13), as the mentioned kind of wave propagation can as well make sense for functions $u \notin \mathcal{C}^1(\mathbb{R})$. In this way, we should be able to obtain for (5.2.11) solutions (5.2.12), when we no longer have (5.2.13). In such a case, the corresponding U in (5.2.12) will be a *generalized solution*. This indeed can be obtained if, similar to (5.2.5), we replace (5.2.11) by its more general *weak* form

(5.2.14) $$\iint_{\mathbb{R}^2} U\cdot(\psi_t - \psi_x)dt\,dx = 0, \quad \psi \in \mathcal{D}(\mathbb{R}^2)$$

Indeed, for every $u \in \mathcal{L}^1_{[\ni]}(\mathbb{R})$, the corresponding U in (5.2.11) will satisfy (5.2.14).

In addition to its generality, the weak form (5.2.14) draws attention upon another *important* property of generalized solutions, first pointed out and used in a systematic way in Sobolev [1,2]. Namely, if U_ν, with $\nu \in \mathbb{N}$, is a sequence of generalized solutions of (5.2.11) obtained from (5.2.14), and this sequence *converges* uniformly on compacts in $\mathbb{R}^2$ to a function U, then U will again satisfy (5.2.14), and hence, it will be a generalized solution of (5.2.11). In fact, it is obvious that much more general types of convergence will still exhibit the above *closure property* of generalized solutions. On the other hand, the *classical* solutions of (5.2.11) obviously *fail* to have this closure property. Indeed, if U_ν, with $\nu \in \mathbb{N}$, are $\mathcal{C}^1$-smooth classical solutions of (5.2.11), it may happen that $U \in \mathcal{C}^0 \backslash \mathcal{C}^1$, even if the convergence $U_\nu \longrightarrow U$ is uniform on compacts in $\mathbb{R}^2$.

The above situation generalizes entirely to the customary linear wave equation

(5.2.15) $$U_{t^2} - U_{x^2} = 0, \quad t,x \in \mathbb{R}$$

which has the classical solution

(5.2.16) $$U(t,x) = u(x-t) + v(x+t), \quad t,x \in \mathbb{R}$$

where

(5.2.17) $$u,v \in \mathcal{C}^2(\mathbb{R})$$

As the solution (5.2.16) describes the superposition of the propagation of the space waves u and v, along the characteristic lines $x - t$ = constant and $x + t$ = constant, respectively, it is obvious that from physical point of view, the regularity condition (5.2.17) is not necessary. Indeed, the *weak* form of (5.2.16) is

$$\iint_{\mathbb{R}^2} U\cdot(\psi_{t^2} - \psi_{x^2})dt = 0, \quad \psi \in \mathcal{D}(\mathbb{R}^2) \tag{5.2.18}$$

thus, for every $u,v \in \mathcal{L}^1_{[\![\ni]\!]}(\mathbb{R})$, the corresponding U in (5.2.16) will satisfy (5.2.18), hence it will be a *generalized solution* of (5.2.15). In view of (5.2.18), it is obvious that the respective generalized solutions will again have the above closure property. And as before, the classical solutions of (5.2.15) will fail to have the mentioned closure property.

In view of the above, it would appear that a proper way to proceed is to define the generalized solutions as solutions of the associated weak form of linear or nonlinear partial differential equations. However, such an approach proves to have several deficiencies. Indeed, in the case of *nonlinear* partial differential equations, the explicit expression of the associated weak form cannot be obtained, except for particular cases, such as conservation laws for instance. Furthermore, even when the weak form is available, it is not so easy to solve it in the unknown function U. But above all, even in the case of linear partial differential equations, the weak form proves to have an *insufficient generality* in order to handle the whole range of useful generalized solutions, such as for instance the *Green function* or *elementary solution* asasociated with a linear, constant coefficient partial differential equation. Indeed, let us consider the linear wave equation (5.2.15) in the following more general form

$$U_{t^2} - U_{x^2} = F, \quad t,x \in \mathbb{R} \tag{5.2.19}$$

where $F \in \mathcal{L}^1_{[\![\ni]\!]}(\mathbb{R}^2)$. Then, the associated weak form is

$$\iint_{\mathbb{R}^2} (U\cdot(\psi_{t^2} - \psi_{x^2}) - F\cdot\psi)dt\ dx = 0, \quad \psi \in \mathcal{D}(\mathbb{R}^2) \tag{5.2.20}$$

It is well known, Hörmander, that in case F has compact support, a generalized - in fact, distribution - solution of (5.2.19) can be obtained by

$$U = E*F \tag{5.2.21}$$

where $*$ is the convolution operator and E is the *elementary solution* of (5.2.15), i.e., it is the distribution solution of

$$U_{t^2} - U_{x^2} = \delta, \quad t,x \in \mathbb{R} \tag{5.2.22}$$

where δ is the Dirac delta distribution. Now, although (5.2.22) resembles (5.2.19), it *fails* to have an associated weak form (5.2.20), with F a function, owing to a basic property of the Dirac delta distribution,

Schwartz [1]. It follows that the elementary solution E of (5.2.22) *cannot* be obtained from the weak form (5.2.20) of (5.2.19), for any choice of the function F.

These examples can offer no more than a first illustration of the way generalized solutions will necessarily arise in the study of partial differential equations. For a rather impressive account of their utility both in linear and nonlinear partial differential equations, a large literature is available, among others, Hörmander and Lions [2], cited above.

§3. CONCEPTS OF GENERALIZED SOLUTIONS

Let us resume the main facts mentioned in the previous Section, with a view towards a suitable definition of *generalized solutions* for *nonlinear* partial differential equations.

Within the sphere of analytic - thus classical - solutions of analytic partial differential equations, we can perform *partial derivatives* of arbitrary order as well as sufficiently general *nonlinear operations*, in particular *multiplications*. A basic deficiency encountered is that we are quite often interested in solutions on *larger domains* than those on which the analytic solutions prove to exist, and the solutions on such larger domains may fail to be classical solutions.

It follows that a *desirable* concept of *generalized solution* should allow the following *three* things : partial derivability of sufficiently high order, sufficiently general nonlinear operations, in particular unrestricted multiplication, and finally, existence of such generalized solutions on sufficiently large domains.

It is particularly important to note that *indefinite* partial derivability of generalized functions, although highly desirable, is *not* absolutely necessary, as long as we deal with the usual linear or nonlinear partial differential equations which are of course of *finite* order. Indeed, all what is in fact required in these usual cases is that a generalized solution has partial derivatives up to, and including, the order of the respective partial differential equation.

This *relaxation* in the requirements on generalized solutions may be particularly welcome in the light of the conflict between insufficient smoothness, multiplication and differentiation, see Section 4, Chapter 1.

Historically, *two* basic ideas concerning the definition of generalized solutions have emerged. The first, in Sobolev [1,2], which may be called the *sequential approach*, has not known a sufficiently general and systematic theoretical development, yet, it led to a wide range of efficient, even if somewhat ad-hoc, solution methods especially for *nonlinear* partial differential equations, see for instance Lions [2]. The second idea, in Schwartz [1], which can be called the *linear functional analytic approach*, has been extensively developed from theoretical point of view, although its major power is restricted to linear partial differential equations.

Here we shall recall the main idea of the *sequential approach* which is at the basis of the *nonlinear* method presented in this volume.

Suppose given a nonlinear partial differential equation

$$(5.3.1) \qquad T(D)U(x) = 0, \quad x \in \Omega \subset \mathbb{R}^n$$

Then we construct an infinite seqauence of 'approximating' equations

$$(5.3.2) \qquad T_\nu(D)V_\nu(x) = 0, \quad x \in \Omega, \ \nu \in \mathbb{N}$$

in such a way that V_ν are *classical* solutions of (5.3.2) and they *converge* in a certain *weak* sense to U. For instance, we have

$$(5.3.3) \qquad \lim_{\nu,\mu\to\infty} \int_\Omega (V_\nu - V_\mu)\cdot\psi \, dx = 0, \quad \psi \in \mathcal{F}$$

where $\mathcal{F} \subset \mathcal{C}^0(\Omega)$ is a set of sufficiently smooth test functions. Then, U is *defined* as the '$\mathcal{F}$-weak limit' of V_ν, with $\nu \in \mathbb{N}$.

Here, the sequential and linear functional analytic approaches have a *common point*, as for many choices of $\mathcal{F}$, the relation (5.3.3) means that V_ν converges to U, when $\nu \to \infty$, within a certain space of distributions. However, as seen in this volume, the seqauential approach does possess a significant potential for *extentions* leading to systematic *nonlinear* theories, while the linear functional analytic approach does not seem to do so. Moreover, as follows from the *stability paradoxes* in Section 11, Chapter 1, a proper way for a systematic *nonlinear* extension of the sequential approach needs a careful reassessment of its above mentioned common point with the linear functional analytic approach.

§4. WHY USE DISTRIBUTIONS?

Although Schwartz's *linear* theory of distributions has severe limitation in even solving linear partial differential equations, see Section 5 next, there are significant advantages in using that theory. Here we shortly mention some of the more important ones.

It should also be mentioned that the *linear functional analytic approach* of Schwartz [1] had as *main priority* the *indefinite* partial derivability of generalized solutions, Hörmander, and then, in view of the so called impossibility result of Schwartz [2], had to suffer from certain *limitations* on its capability to accommodate unrestricted *nonlinear* operations, in particular unrestricted multiplication. It should however be pointed out that, as seen in this volume, there are *various* ways one can circumvent the restrictions which appear when we would like to have both indefinite partial derivability and unrestricted multiplication. In this respect, the classical, linear functional analytic approach in Schwartz [1] is but one of the many possible ones, and it seems to be less suited for a systematic study of nonlinear partial differential equations.

Nevertheless, from the point of view of *partial derivability*, the space $\mathcal{D}'$ of distributions possesses a *canonical* structure. Indeed, let us consider the chain of inclusions

$$(5.4.1) \qquad \mathcal{C}^\infty \subset \ldots \subset \mathcal{C}^\ell \subset \ldots \subset \mathcal{C}^0 \subset \mathcal{D}', \quad \ell \in \mathbb{N}$$

where only the elements of $\mathcal{C}^\infty$ are indefinitely partial derivable in the classical sense. As is known, Schwartz [1], $\mathcal{D}'$ is the set of all linear functionals $T : \mathcal{D} \longrightarrow \mathbb{C}$ which are continuous in the usual topology on $\mathcal{D}$. In particular, the embedding $\mathcal{C}^0 \ni f \longmapsto T_f \in \mathcal{D}'$ is defined by

$$(5.4.2) \qquad T_f(\psi) = \int_{\mathbb{R}^n} f(x)\psi(x)dx, \quad \psi \in \mathcal{D}(\mathbb{R}^n)$$

which in fact holds also for $f \in \mathcal{L}^1_{loc}$. In this way, the elements of $\mathcal{D}'$ will again be indefinitely partial derivable, although no longer in a classical sense, but in the following more general, *weak* sense : suppose given $T \in \mathcal{D}'(\mathbb{R}^n)$ and $p \in \mathbb{N}^n$, then $D^pT \in \mathcal{D}'(\mathbb{R}^n)$ is defined by

$$(5.4.3) \qquad (D^pT)(\psi) = (-1)^{|p|}T(D^p\psi), \quad \psi \in \mathcal{D}(\mathbb{R}^n)$$

Obviously, if $f \in \mathcal{C}^\ell$, with $\ell \in \mathbb{N}$, and $p \in \mathbb{N}^n$, $|p| \leq \ell$, then

$$(5.4.4) \qquad D^p(T_f) = T_{D^pf}$$

i.e., the weak and classical partial derivatives coincide for sufficiently smooth functions.

The canonical property of $\mathcal{D}'$ is the following

$$(5.4.5) \qquad \begin{array}{l} \forall \ T \in \mathcal{D}'(\mathbb{R}^n), \ K \Subset \mathbb{R}^n : \\ \exists \ f \in \mathcal{C}^0(K), \ p \in \mathbb{N}^n : \\ \quad T\big|_K = D^pf \end{array}$$

where D^p is the weak partial derivative in (5.4.3). In other words $\mathcal{D}'$

is a *minimal* extension of $\mathcal{C}^0$ in the sense that locally, every distribution is a weak partial derivative of a continuous function.

The above property makes $\mathcal{D}'$ sufficiently large in order to contain the Dirac δ distribution defined by

$$(5.4.6) \qquad \delta(\psi) = \psi(0), \quad \psi \in \mathcal{D}(\mathbb{R}^n)$$

which, among others, is essential in the study of *elementary solutions* of linear constant coefficient partial differential equations, see for instance (5.2.22). Indeed, for simplicity, let us consider the one dimensional case, when $n = 1$, and let us define $x_+ \in \mathcal{C}^0(\mathbb{R}^n)$ by

$$(5.4.7) \qquad x_+ = \begin{vmatrix} 0 & \text{if} & x \leq 0 \\ x & \text{if} & x \geq 0 \end{vmatrix}$$

then in view of (5.4.2) and (5.4.3), the weak derivative of x_+ is given by

$$(5.4.8) \qquad D^1 x_+ = H$$

where H is the Heaviside function

$$(5.4.9) \qquad H(x) = \begin{vmatrix} 0 & \text{if} & x < 0 \\ 1 & \text{if} & x > 0 \end{vmatrix}$$

Similarly, a further weak derivation yields

$$(5.4.10) \qquad \delta = D^1 H = D^2 x_+ \in \mathcal{D}'(\mathbb{R})$$

hence, δ is the *second* weak derivative of a *continuous* function, and (5.4.10) holds globally on $\mathbb{R}$ and not only locally.

The use of the space $\mathcal{D}'$ of distributions has a particularly important justification in the study of *linear*, *constant coefficient* partial differential equations. The basic result in this respect, first obtained in Ehrenpreis and Malgrange, concerns the proof of the *existence of an elementary solution* for *every* such equation, a result which has a wide range of useful consequences and applications, Hörmander, Treves [2], and which alone would fully justify the use of distributions. It should however be pointed out that the existence of generalized solutions for inhomogeneous linear constant coefficient partial differential equations can easily be obtained in *other* spaces of generalized functions as well. For instance, using elementary ring theoretical methods, Gutterman proved such existence results within the space of Mikusinski operators. In fact, similar simple algebraic arguments together with some from classical Fourier analysis can deliver the mentioned Ehrenpreis-Malgrange results, see

Struble. A whole range of other linear applications of the $\mathcal{D}'$ distributions can be found in the literature, among others, in the last three monographs cited above, as well as in Treves [3]. A recent useful, yet easy to read account of many of the more important linear applications can be found in Friedlander.

Although the space $\mathcal{D}'$ of distributions has the above mentioned canonical property, various linear extensions of it, given for instance by spaces of hyperfunctions, Sato et. al., have been studied in the literature.

§5. THE LEWY INEXISTENCE RESULT

Soon after the foundation of the modern linear theory of distributions. Schwartz [1], and the proof of the existence of an elementary solution for every linear constant coefficient partial differential equation, Malgrange, Ehrenpreis, a very simple example of a linear variable coefficient partial differential equation given by Lewy, showed that the linear theory of distributions is *not sufficient* even for the study of linear partial differential equations. Lewy's example is the following surprisingly simple equation

$$(5.5.1) \qquad \frac{\partial}{\partial x_1}U + i\frac{\partial}{\partial x_2}U - 2i(x_1+ix_2)\frac{\partial}{\partial x_3}U = f, \ x=(x_1,x_2,x_3) \in \mathbb{R}^3$$

which for a large class of $f \in \mathcal{C}^\infty(\mathbb{R}^3)$, *fails* to have distribution solutions $U \in \mathcal{D}'$ in any neighbourhood of any point $x \in \mathbb{R}^3$.

In this way, it follows that the solution of (5.5.1) requires spaces of *generalized functions* which are *larger* than the space $\mathcal{D}'$ of the Schwartz distributions.

The interesting thing about Lewy's equation (5.5.1) is that it is not a kind of artificial, counter-example type of equation, but it appears naturally in connection with certain studies in complex functions of several variables, Krantz.

The phenomenon of *insufficiency* of the $\mathcal{D}'$ distributional framework, pointed out by Lewy's example, became the object of several subsequent studies. We shall shortly relate the result of one of them. For that purpose we need several notations. Suppose given a domain $\Omega \subset \mathbb{R}^n$ and an m-th order linear variable coefficient partial differential operator

$$(5.5.2) \qquad P(x,D) = \sum_{\substack{p\in\mathbb{N}^n \\ |p|\leq m}} c_p(x)D^p , \quad x \in \Omega$$

with the coefficients $c_p \in \mathcal{C}^\infty(\Omega)$, for $p \in \mathbb{N}^n$, $|p| \leq m$. The *principal part* of $P(x,D)$ is by definition

(5.5.3) $$P_m(x,\xi) = \sum_{\substack{p\in\mathbb{N}^n \\ |p|=m}} c_p(x)\zeta^p, \quad x \in \Omega, \ \xi \in \mathbb{R}^n$$

and its complex conjugate is

(5.5.4) $$\dot{P}_m(x,\xi) = \sum_{\substack{p\in\mathbb{N}^n \\ |p|=m}} \overline{c_p(x)}\zeta^p, \quad x \in \Omega, \ \xi \in \mathbb{R}^n$$

finally, the *commutator* of $P(x,D)$ is defined by

(5.5.5) $$C_{2m-1}(x,\xi) = i \sum_{1\leq j\leq n} \left(\frac{\partial}{\partial \xi_j}P_m(x,\xi)\frac{\partial}{\partial x_j}\overline{P}_m(x,\xi) - \right.$$
$$\left. - \frac{\partial}{\partial x_j}P_m(x,\xi)\frac{\partial}{\partial \xi_j}\overline{P}_m(x,\xi)\right), \quad x \in \Omega, \ \xi \in \mathbb{R}^n$$

which obviously is a polynomial of degree $2m - 1$ in ξ and it has real, $\mathcal{C}^\infty$-smooth coefficients.

A basic *necessary condition* for solvability is the following.

<u>Theorem 1</u> (Hörmander)

Suppose the linear partial differential equation

(5.5.6) $$P(x,D)U = f, \quad x \in \Omega$$

has a solution $U \in \mathcal{D}'(\Omega)$, for every $f \in \mathcal{D}(\Omega)$, then

(5.5.7) $$\forall \ x \in \Omega, \ \xi \in \mathbb{R}^n$$
$$P_m(x,\xi) = 0 \Rightarrow C_{2m-1}(x,\xi) = 0$$ □

It is easy to see that in Lewy's example (5.5.1) we have $n = 3$, $\Omega = \mathbb{R}^3$, $m = 1$ and

(5.5.8) $$C_1(x,\xi) = 8\xi_3, \quad x,\xi \in \mathbb{R}^3$$

hence, for $x \in \mathbb{R}^3$ and $\xi_1 = -2x_2$, $\xi_2 = 2x_1$ and $\xi_3 = 1$, we have

(5.5.9) $$P_1(x,\xi) = 0, \quad C_1(x,\xi) \neq 0$$

which contradicts (5.5.7)

In general, if the coefficients of P_m, i.e., c_p, with $p \in \mathbb{N}^n$, $|p| = m$, are real, then obviously, C_{2m-1} is identically zero, hence (5.5.7) is satisfied. Unfortunately however, (5.5.7) is only a *necessary and not also a sufficient condition* for solvability, Trèves [1]. However, in the case of *first order* linear partial differential equations with $\mathcal{C}^\infty$-smooth coefficients, Nirenberg & Trèves could obtain a necessary and sufficient condition for solvability. In the general case of (5.5.2), a certain strengthened form of (5.5.7) proves to be both necessary and sufficient for local solvability, provided that multiple characteristics are not present, Hörmander.

It is interesting to note that in case we deal with the *easier* problem of solvability in the neighbourhood of a given, fixed point $x \in \mathbb{R}^n$, we can find still *simpler* linear partial differential equations *without* distribution solutions. Indeed, Grushin showed that the equation

$$(5.5.10) \qquad \frac{\partial}{\partial x_1}U + ix_1\frac{\partial}{\partial x_2}U = f, \quad x = (x_1, x_2) \in \mathbb{R}^2$$

with suitably chosen $f \in \mathcal{C}^\infty(\mathbb{R}^2)$, *fails* to have distribution solutions $U \in \mathcal{D}'(\Omega)$ in any neighbourhood $\Omega \subset \mathbb{R}^2$ of $x_0 = (0,0) \in \mathbb{R}^2$.

Finally, it should be mentioned that *linear variable* coefficient partial differential equations *fail* to have solutions in various *linear extensions* of the Schwartz $\mathcal{D}'$ distributions, such as for instance, spaces of hyperfunctions, Sato, et. al, Hörmander. A result in this respect can be found in Shapira.

In view of the above, it is important to note that in Colombeau [3] was for the *first time* proved the *existence* of generalized solutions for systems of *arbitrary* linear partial differential equations with $\mathcal{C}^\infty$-smooth coefficients, within the differential algebra of generalized functions, mentioned later in Chapter 8, see for details Rosinger [3, pp. 169-177].

APPENDIX 1

MULTIPLICATION, LOCALIZATION AND REGULARIZATION OF DISTRIBUTIONS

We shortly recall a few important properties of the $\mathcal{D}'$ distributions, which will be useful in the sequel, Schwartz [1], Friedlander.

As is well known, one can multiply every smooth function $\chi \in \mathcal{C}^\infty(\mathbb{R}^n)$ with every distribution $T \in \mathcal{D}'(\mathbb{R}^n)$ and obtain as product the distribution

(5.A1.1) $$S = \chi \cdot T \in \mathcal{D}'(\mathbb{R}^n)$$

defined by

(5.A1.2) $$S(\psi) = T(\chi \cdot \psi), \quad \psi \in \mathcal{D}(\mathbb{R}^n)$$

However, if $\chi \in \mathcal{D}'(\mathbb{R}^n) \backslash \mathcal{C}^\infty(\mathbb{R}^n)$, then we shall have $\chi \cdot \psi \notin \mathcal{D}(\mathbb{R}^n)$, for certain $\psi \in \mathcal{D}(\mathbb{R}^n)$. Hence, the right hand term in (5.A1.2) will no longer be defined, and then, we cannot use this relation in order to define the product in (5.A1.1).

As an application to the one dimensional case of the distributions in $\mathcal{D}'(\mathbb{R})$, we have

(5.A1.3) $$x \cdot \delta = 0 \in \mathcal{D}'(\mathbb{R})$$

Indeed, (5.A1.1) and (5.A1.2) will give for $\psi \in \mathcal{D}(\mathbb{R})$, the relations

$$(x \cdot \delta)(\psi) = \delta(x \cdot \psi) = 0 \cdot \psi(0) = 0 \in \mathbb{R}$$

Suppose now given $\Omega \subset \mathbb{R}^n$ open and let us define $\mathcal{D}^\ell(\Omega)$ as the set of all linear functionals $T : \mathcal{D}(\Omega) \longrightarrow \mathbb{C}$ which are continuous in the usual topology on $\mathcal{D}(\Omega)$.

The relation between $\mathcal{D}'(\mathbb{R})$ and $\mathcal{D}'(\Omega)$ is an example of a *localization principle* corresponding to a *sheaf* structure specified next.

Obviously we have an embedding of vector spaces

(5.A1.4) $$\mathcal{D}(\Omega) \subset \mathcal{D}(\mathbb{R}^n)$$

defined as follows : if $\psi \in \mathcal{D}(\Omega)$, then we can consider $\psi \in \mathcal{D}(\mathbb{R}^n)$, with ψ vanishing on $\mathbb{R}^n \backslash \Omega$. Now (5.A1.4) yields the embedding of vector spaces

(5.A1.5) $$\mathcal{D}'(\mathbb{R}^n) \subset \mathcal{D}'(\Omega)$$

defined by the restriction mapping

(5.A1.6) $$\mathcal{D}'(\mathbb{R}^n) \ni T \longmapsto T\Big|_{\mathcal{D}(\Omega)} \in \mathcal{D}'(\Omega)$$

However, we also have the following, less trivial, *converse* aspect of the localization principle.

Suppose given $\Omega_i \subset \Omega$ open, with $i \in I$, and $T_i \in \mathcal{D}'(\Omega_i)$, with $i \in I$, such that

(5.A1.7) $$T_i\Big|_{\Omega_i \cap \Omega_j} = T_j\Big|_{\Omega_j \cap \Omega_j}$$

whenever $\Omega_i \cap \Omega_j \neq \phi$, with $i,j \in I$. Further, let us suppose that

(5.A1.8) $$\bigcup_{i \in I} \Omega_i = \Omega$$

Then, there exists a *unique* $T \in \mathcal{D}'(\Omega)$, such that

(5.A1.9) $$T\Big|_{\Omega_i} = T_i, \quad i \in I$$

It follows in particular that a distribution $T \in \mathcal{D}'(\mathbb{R}^n)$ is uniquely determined if its restriction $T|_V$ to a neighbourhood V of an arbitrary point $x \in \mathbb{R}^n$ is known. Further we note that (5.A1.7)-(5.A1.9) allow us to define the *support* supp T of a distribution $T \in \mathcal{D}'(\Omega)$ as the closed subset in Ω which is the complementary of the largest open subset in Ω on which T vanishes.

As an important application of the above, we present the *regularization* of certain functions or distributions across singularities. This can help in better understanding the difficulties involved in the problems solved in Chapter 7.

Suppose given $\Gamma \subset \mathbb{R}^n$ nonvoid, closed, then $\Omega = \mathbb{R}^n \backslash \Gamma$ is open. It is easy to see that in general, the inclusion (5.A1.5) is strict. Therefore, let us define

(5.A1.10) $$\mathcal{D}'_\Gamma(\Omega) = \left\{ T \in \mathcal{D}'(\Omega) \;\middle|\; \begin{array}{l} \exists\ S \in \mathcal{D}'(\mathbb{R}^n) : \\ T = S\Big|_\Omega \end{array} \right\}$$

and call S the Γ-*regularization* of T. Obviously, even if Ω is dense in $\mathbb{R}^n$, i.e., Γ has no interior, two Γ-regularizations S_1 and S_2 of T can be different and their difference $S_2 - S_1 \in \mathcal{D}'(\mathbb{R}^n)$ will be a distribution with support contained in Γ. It follows that whenever it exists, the Γ-regularization of T is *unique modulo a distribution supported by* Γ.

Let us clarify the above by an example.

Suppose $\Gamma = \{0\} \subset \mathbb{R}$, then $\Omega = \mathbb{R} \setminus \Gamma = (-\infty, 0) \cup (0, \infty)$ is dense in $\mathbb{R}$. Further, suppose given $f \in \mathcal{C}^\infty(\Omega)$ defined by

(5.A1.11) $\quad f(x) = \ell n|x|, \quad x \in \Omega$

Obviously $f \in \mathcal{L}^1_{\ell oc}(\mathbb{R})$, hence (5.4.2) yields a distribution

(5.A1.12) $\quad T_f \in \mathcal{D}'(\mathbb{R})$

Now, with the usual derivative of smooth functions, we have

(5.A1.13) $\quad g(x) = Df(x) = 1/x, \quad x \in \Omega$

Hence $g \notin \mathcal{L}^1_{\ell oc}(\mathbb{R})$, therefore (5.4.2) cannot be applied to g on $\mathbb{R}$. Nevertheless, we have $g \in \mathcal{C}^\infty(\Omega) \subset \mathcal{L}^1_{\ell oc}(\Omega)$, hence (5.4.2) yields a distribution

(5.A1.14) $\quad T_g \in \mathcal{D}'(\Omega)$

We show now the *stronger property*

(5.A1.15) $\quad T_g \in \mathcal{D}'_\Gamma(\Omega)$

i.e., T_g admits a Γ-regularization $S \in \mathcal{D}'(\mathbb{R})$.

Indeed, in view of (5.A1.12), let us take

(5.A1.16) $\quad S = DT_f \in \mathcal{D}'(\mathbb{R})$

where D is the distributional derivative in $\mathcal{D}'(\mathbb{R})$. Then, (5.4.4) applied to $f \in \mathcal{C}^\infty(\Omega)$, together with (5.A1.13) will obviously give (5.A1.15).

With usual notations, the above can be sumarized by saying that the function $1/x$ which is singular at $x = 0$ and it is not locally integrable on $\mathbb{R}$, can nevertheless be regularized at $x = 0$ by the distribution $S = D(\ell n|x|) \in \mathcal{D}'(\mathbb{R})$. In view of this, we shall *identify* $1/x$ with S, and thus obtain

(5.A1.17) $\quad (1/x) = D(\ell n\ |x|) \in \mathcal{D}'(\mathbb{R}) \setminus \mathcal{L}^1_{\ell oc}(\mathbb{R})$

As mentioned above, the regularization (5.A1.17) of $1/x$ is *unique, modulo distributions with support* $\Gamma = \{0\} \subset \mathbb{R}$. These distributions are known to be of the form $\sum_{0 \le p \le \ell} c_p D^p \delta$, with $\ell \in \mathbb{N}$, $c_p \in \mathbb{C}$. Finally, in view of (5.A1.1) and (5.A1.17) we can define the product

$$(5.A1.18) \qquad (x) \cdot (1 \backslash x) \in \mathcal{D}'(\mathbb{R})$$

and we shall show that

$$(5.A1.19) \qquad (x) \cdot (1/x) = 1$$

Indeed, according to (5.4.2) and (5.4.3), we have for $\psi \in \mathcal{D}(\mathbb{R})$ the relations

$$((x) \cdot (1/x))(\psi) = (1/x)(x \cdot \psi) =$$

$$= -(\ell n|x|)(\psi + x \cdot \psi') = - \int_{\mathbb{R}} \ell n|x| \psi(x) dx - \int_{\mathbb{R}} x \ \ell n|x| \psi'(x) dx =$$

$$= - \int_{\mathbb{R}} \ell n|x| \psi(x) dx + \int_{\mathbb{R}} (\ell n|x| + 1) \psi(x) dx = \int_{\mathbb{R}} \psi(x) dx$$

which completes the proof of (5.A1.19)

It should be noted that the multiplication of two distributions which have *common singularities* does pose a considerable problem in the sense that, on the one hand, simple and natural definitions, such as for instance in (5.A1.1) and (5.A1.2) are no longer available, while on the other hand, there appears to be a large variety of other possible definitions with no natural or canonical candidate. Details in this connection can be found in Section 4, Chapter 1, as well as in Rosinger [1,2,3] and the literature cited there. Here, we should only like to recall the simple example of the product

$$(5.A1.20) \qquad \delta^2 = \delta \cdot \delta$$

which according to various interpretations, can be shown to be *no longer* a distribution, see details in Rosinger [1, pp. 11,29-31], Rosinger [2, pp. 66, 115-118], and Mikusinski [2].

Finally, for $\Omega \subset \mathbb{R}^n$ open and $\ell \in \mathbb{N}$, let us denote by

$$(5.A1.21) \qquad \mathcal{S}^{\ell}(\Omega), \quad \mathcal{V}^{\ell}(\Omega)$$

the set of all sequences $s \in (\mathcal{C}^{\ell}(\Omega))^{\mathbb{N}}$ which converge in $\mathcal{D}'(\Omega)$, respectively the set of all sequences $v \in \mathcal{S}^{\ell}(\Omega)$ which converge in $\mathcal{D}'(\Omega)$ to zero. As is well known, Schwartz [1], we have the *quotient vector space representation* given by the linear isomorphism

(5.A1.22)
$$\begin{array}{ccc} \mathcal{S}^{\ell}(\Omega)/\mathcal{V}^{\ell}(\Omega) & \longrightarrow & \mathcal{D}'(\Omega) \\ s + \mathcal{V}^{\ell}(\Omega) & \longmapsto & T \end{array}$$

where for $s = (\psi_\nu | \nu \in \mathbb{N}) \in \mathcal{S}^{\ell}(\Omega)$, we have

(5.A1.23)
$$T(\psi) = \lim_{\nu\to\infty} \int_\Omega \varphi_\nu(x)\psi(x)dx, \quad \psi \in \mathcal{D}(\Omega)$$

It follows that we can obtain the following *quotient vector space representation* of the L. Schwartz distributions

(5.A1.24)
$$\mathcal{D}'(\Omega) = \mathcal{S}^{\ell}(\Omega)/\mathcal{V}^{\ell}(\Omega), \quad \ell \in \mathbb{N}$$

CHAPTER 6

CHAINS OF ALGEBRAS OF GENERALIZED FUNCTIONS

§1. RESTRICTIONS ON EMBEDDINGS OF THE DISTRIBUTIONS INTO QUOTIENT ALGEBRAS

The aim of this Chapter is to present the basic results concerning the *necessary structure* of the nonlinear theories of generalized functions originated in Rosinger [7,8] and developed in Rosinger [1,2,3], as well as in their more particular form, in Colombeau [1,2].

The starting point of these theories are the *quotient algebras* of generalized functions as defined in (1.5.26) and (1.6.9), and subsequently used directly in Chapters 2 and 3.

The fundamental problem in connection with these quotient algebras $A = \mathcal{A}/\mathcal{I}$ is of course the clarification of their *deeper structure*, beyond that which is so simply given by their definition through the *two conditions* (1.6.10) and (1.6.11).

Surprisingly, to a good extent such a clarification can be *reduced* to the understanding of the structure of the *ideals* $\mathcal{I}$ alone.

A first set of results in this respect was already presented in Sections 5-8 in Chapter 3, in the nature of various *densely vanishing* conditions which *characterize* many of such ideals $\mathcal{I}$.

In this Chapter we clarify *three* further issues.

The first one, of a most general nature, is about the *detailed necessary structure* of the *inclusion diagrams* (1.6.10), and it is presented in Section 1, culminating in (6.1.27) and (6.1.33).

The second issue dealt with in this Chapter is an *alternative*, *most simple characterization* of a large class of the mentioned ideals $\mathcal{I}$. This characterization is obtained in Theorem 4, Section 3.

The third and last issue dealt with in this Chapter centers around the fact presented in Section 4, according to which a *natural* way to deal with the *conflict* between insufficient smoothness, multiplication and differentiation is to allow the *nonlinear* partial differential operators to act between possibly infinite *chains* of quotient algebras of generalized functions.

As seen in Chapter 1, in particular in Sections 1-4 and 8, a *nonlinear* theory of generalized functions has to face *two major problems*, namely:

- the conflict between insufficient smoothness, multiplication and differentiation, whose special case is the so called Schwartz impossibility result, and

- the nonlinear stability paradoxes.

In giving a solution to the latter problem, we have been led - see Sections 5, 6, 9-12 in Chapter 1 - to *quotient algebras* of generalized functions, introduced in (1.6.9)-(1.6.14), and to the respective interplay between the stability, generality and exactness properties of generalized solutions of nonlinear partial differential equations.

In this Chapter, a general and natural way is given in order to deal with the constraints imposed by the Schwartz impossibility result and in the same time to avoid the stability paradoxes. As seen, this is obtained from a detailed analysis of the way partial derivatives and certain products involving the Dirac δ distribution can be defined in algebras of generalized functions. One should not be surprised in this connection, since the derivative and a product involving the Dirac δ distribution are the central elements involved in the Schwartz impossibility results itself. After all, δ is a *highly nonsmooth* element, and therefore, its multiplication and derivatives are likely to provide aggravated instances of the *conflict* studied in Chapter 1.

In view of the already classical role played by the Schwartz distributisons in solving various linear or nonlinear partial differential equations, we shall require a level of *generality* not below the distributional one. In other words, we shall require that the spaces - in fact algebras - of generalized functions we deal with, contain the space $\mathcal{D}'$ of the Schwartz distributions, see for instance (1.11.6).

For the sake of simplicity, we shall only deal with the case when $\Omega = \mathbb{R}^n$ and therefore, no explicit mention of the domain will be made in the notation. The extension to general domains $\Omega \subset \mathbb{R}^n$ is immediate.

In order to reformulate for the mentioned level of *generality* the basic algebraic clarification and solution of the stability paradoxes - see in particular (1.8.54)-(1.8.58) - we shall use a slight extension of the representation of distributions

(6.1.1) $$\mathcal{D}' = \mathcal{S}^{\ell}/\mathcal{V}^{\ell}, \quad \ell \in \mathbb{N}$$

which was given in (5.A1.22).

Suppose given an arbitrary *infinite* index set Λ. Further, suppose given $\ell \in \mathbb{N}$. It is easy to see that there exist vector subspaces

(6.1.2) $$\mathcal{S}^{\ell}_{\Lambda} \subset (\mathcal{C}^{\ell})^{\Lambda}$$

with *linear surjections*

(6.1.3) $$\mathcal{S}^{\ell} \ni s \longmapsto T \in \mathcal{D}'$$

such that, see (1.6.4)

(6.1.4) $\qquad \mathcal{U}_{\mathcal{C}^{\ell},\Lambda} \subset \mathcal{S}^{\ell}_{\Lambda}$

and the mapping (6.1.3) is an extension of, see (5.4.2)

(6.1.5) $\qquad \mathcal{C}^{\ell} \ni \psi \longmapsto T_{\psi} \in \mathcal{D}'$

Denoting by

(6.1.6) $\qquad \mathcal{V}^{\ell}_{\Lambda}$

the kernel of the mapping (6.1.3), we obtain the following *vector space isomorphism*

(6.1.7) $\qquad \mathcal{S}^{\ell}_{\Lambda}/\mathcal{V}^{\ell}_{\Lambda} \ni s + \mathcal{V}^{\ell}_{\Lambda} \longmapsto T \in \mathcal{D}'$

hence the representation of distributions

(6.1.8) $\qquad \mathcal{D}' = \mathcal{S}^{\ell}_{\Lambda}/\mathcal{V}^{\ell}_{\Lambda}$

The existence of $\mathcal{V}^{\ell}_{\Lambda}$ and $\mathcal{S}^{\ell}_{\Lambda}$ which satisfy (6.1.2)-(6.1.8) is illustrated in Example 1 at the end of Section 4.

Now, it is obvious that a simple way for the embedding of $\mathcal{D}'$ into quotient algebras of generalized functions would be by constructing *inclusion diagrams*

(6.1.9)
$$\begin{array}{ccccc} \mathcal{I} & \longrightarrow & \mathcal{A} & \longrightarrow & (\mathcal{C}^{\ell})^{\Lambda} \\ \uparrow & & \uparrow & & \\ \mathcal{V}^{\ell}_{\Lambda} & \longrightarrow & \mathcal{S}^{\ell}_{\Lambda} & & \end{array}$$

with $\mathcal{A}$ subalgebra in $(\mathcal{C}^{\ell})^{\Lambda}$ and $\mathcal{I}$ ideal in $\mathcal{A}$, such that

(6.1.10) $\qquad \mathcal{I} \cap \mathcal{S}^{\ell}_{\Lambda} = \mathcal{V}^{\ell}_{\Lambda}$

Indeed, owing to (6.1.10), an inclusion diagram (6.1.9) would give a mapping

(6.1.11) $\qquad \mathcal{D}' = \mathcal{S}^{\ell}_{\Lambda}/\mathcal{V}^{\ell}_{\Lambda} \ni s + \mathcal{V}^{\ell}_{\Lambda} \longmapsto s + \mathcal{I} \in A = \mathcal{A}/\mathcal{I}$

which would be a *linear embedding* of $\mathcal{D}'$ into the algebra A.

Unfortunately, inclusion diagrams such as in (6.1.9), (6.1.10) *cannot* be constructed even when $\Lambda = \mathbb{N}$ and $\ell = \infty$, since in view of (1.8.57) we have

$$(6.1.12) \qquad (\mathcal{V}^\ell_\Lambda \cdot \mathcal{V}^\ell_\Lambda) \cap \mathcal{S}^\ell_\Lambda \not\subset \mathcal{V}^\ell_\Lambda$$

which contradicts (6.1.10).

The alternative simple way left is to turn $\mathcal{D}'$ into a quotient algebra by constructing inclusion diagrams

$$(6.1.13) \qquad \begin{array}{ccccc} \mathcal{I} & \longrightarrow & \mathcal{A} & \longrightarrow & (\mathcal{C}^\ell)^\Lambda \\ \downarrow & & \downarrow & & \\ \mathcal{V}^\ell_\Lambda & \longrightarrow & \mathcal{S}^\ell_\Lambda & & \end{array}$$

with $\mathcal{A}$ subalgebra in $(\mathcal{C}^\ell)^\Lambda$ and $\mathcal{I}$ ideal in $\mathcal{A}$, such that

$$(6.1.14) \qquad \mathcal{V}^\ell_\Lambda \cap \mathcal{A} = \mathcal{I}$$

$$(6.1.15) \qquad \mathcal{V}^\ell_\Lambda + \mathcal{A} = \mathcal{S}^\ell_\Lambda$$

in which case we would obtain the *linear injection* of the algebra $A = \mathcal{A}/\mathcal{I}$ *onto* $\mathcal{D}'$, given by

$$(6.1.16) \qquad A = \mathcal{A}/\mathcal{I} \ni s + \mathcal{I} \longmapsto s + \mathcal{V}^\ell_\Lambda \in \mathcal{D}' = \mathcal{S}^\ell_\Lambda/\mathcal{V}^\ell_\Lambda$$

Unfortunately again, inclusion diagrams such as in (6.1.13)-(6.1.15) *cannot* be constructed even in the case of $\Lambda = \mathbb{N}$ and $\ell = 0$. The difficulty is with condition (6.1.15) needed for the surjectivity of the linear injection (6.1.16). Indeed, this follows from the next classical result, see Rosinger [2, pp. 66, 115-118].

<u>Lemma 1</u>

Suppose given a sequence s of continuous functions on $\mathbb{R}$, such that

$$(6.1.17) \qquad \begin{array}{l} \text{supp } s_\nu \text{ shrinks to } 0 \in \mathbb{R}, \text{ when } \nu \longrightarrow \infty \\ s \in \mathcal{S}^0 \text{ and } \langle s, \cdot \rangle = \delta \end{array}$$

Then

$$(6.1.18) \qquad s^2 \notin \mathcal{S}^0$$

□

In view of the above, we are obliged to go to a next level of more involved inclusion diagrams of the form

$$
(6.1.19) \qquad
\begin{array}{ccccc}
\mathcal{I} & \longrightarrow & \mathcal{A} & \longrightarrow & (\mathcal{C}^{\ell})^{\Lambda} \\
\uparrow & & \uparrow & & \\
\mathcal{V} & \longrightarrow & \mathcal{S} & \longleftarrow & \mathcal{U}_{\mathcal{C}^{\ell},\Lambda} \\
\downarrow & & \downarrow & & \\
\mathcal{V}^{\ell}_{\Lambda} & \longrightarrow & \mathcal{S}^{\ell}_{\Lambda} & &
\end{array}
$$

with $\mathcal{A}$ subalgebra in $(\mathcal{C}^{\ell})^{\Lambda}$, $\mathcal{I}$ ideal in $\mathcal{A}$ and $\mathcal{V}$, $\mathcal{S}$ vector subspaces in $(\mathcal{C}^{\ell})^{\Lambda}$, such that

$$(6.1.20) \qquad \mathcal{I} \cap \mathcal{S} = \mathcal{V}$$

$$(6.1.21) \qquad \mathcal{V}^{\ell}_{\Lambda} \cap \mathcal{S} = \mathcal{V}$$

$$(6.1.22) \qquad \mathcal{V}^{\ell}_{\Lambda} + \mathcal{S} = \mathcal{S}^{\ell}_{\Lambda}$$

The interest in inclusion diagrams (6.1.19)-(6.1.22) is in the following. First we note that (6.1.4)-(6.1.6) and (6.1.20) yield

$$(6.1.23) \qquad \mathcal{I} \cap \mathcal{U}_{\mathcal{C}^{\ell},\Lambda} \subset \mathcal{V}^{\ell}_{\Lambda} \cap \mathcal{U}_{\mathcal{C}^{\ell},\Lambda} = 0$$

hence, in view of the *neutrix condition* (1.6.11), we have, see (1.6.9)

$$(6.1.24) \qquad A = \mathcal{A}/\mathcal{I} \in \mathrm{AL}_{\mathcal{C}^{\ell},\Lambda}$$

Further, it follows easily that the mappings

$$
(6.1.25) \qquad
\begin{array}{ccccc}
\mathcal{D}' = \mathcal{S}^{\ell}_{\Lambda}/\mathcal{V}^{\ell}_{\Lambda} & \longleftarrow & \mathcal{S}/\mathcal{V} & \longrightarrow & A = \mathcal{A}/\mathcal{I} \\
s+\mathcal{V}^{\ell}_{\Lambda} & \underset{\text{isom}}{\longleftarrow} & s+\mathcal{V} & \underset{\text{lin,inj}}{\longmapsto} & s + \mathcal{I}
\end{array}
$$

define a *linear embedding* of $\mathcal{D}'$ into the quotient algebra of generalized functions $A = \mathcal{A}/\mathcal{I}$.

The role of the inclusion

$$(6.1.26) \qquad \mathcal{U}_{\mathcal{C}^{\ell},\Lambda} \subset \mathcal{S}$$

in (6.1.19) is obvious. Indeed, in view of (6.1.25), it follows that the multiplication in $\Lambda = \mathcal{A}/\mathcal{I}$, when restricted to $\mathcal{C}^\ell$, will coincide with the usual multiplication of $\mathcal{C}^\ell$-smooth functions. A similar property concerning partial derivatives also follows, see Theorem 7 in Section 4.

However, in view of results such as those in Propositions 2, 3 and 4 in Section 4, Chapter 1, it will be appropriate to consider the following type of inclusion diagrams, which are more general than those in (6.1.19), since they do not require condition (6.1.26):

$$(6.1.27)\qquad \begin{array}{ccccc} \mathcal{I} & \longrightarrow & \mathcal{A} & \longrightarrow & (\mathcal{C}^\ell)^\Lambda \\ \uparrow & & \uparrow & & \\ \mathcal{V} & \longrightarrow & \mathcal{S} & & \\ \downarrow & & \downarrow & & \\ \mathcal{V}^\ell_\Lambda & \longrightarrow & \mathcal{S}^\ell_\Lambda & & \end{array}$$

where

$$(6.1.28)\qquad \Lambda = \mathcal{A}/\mathcal{I} \in \mathrm{AL}_{\mathcal{C}^\ell,\Lambda}$$

$$(6.1.29)\qquad \mathcal{I} \cap \mathcal{S} = \mathcal{V}$$

$$(6.1.30)\qquad \mathcal{V}^\ell_\Lambda \cap \mathcal{S} = \mathcal{V}$$

$$(6.1.31)\qquad \mathcal{V}^\ell_\Lambda + \mathcal{S} = \mathcal{S}^\ell_\Lambda$$

It is obvious that the *linear embedding* (6.1.25) of $\mathcal{D}'$ into the quotient algebra of generalized functions $\Lambda = \mathcal{A}/\mathcal{I}$ will still hold for the above more general inclusion diagrams in (6.1.27)-(6.1.31).

Remark 1

1. The intermediate quotient space $\mathcal{S}/\mathcal{V}$ in (6.1.25) obviously plays the role of a *regularization* of the representation of the distributions $\mathcal{D}'$ given in (6.1.8).

2. It is important to note that the form of the inclusion diagram (6.1.27) is *necessary* in the following sense. Suppose that for a quotient algebra $\Lambda = \mathcal{A}/\mathcal{I} \in \mathrm{AL}_{\mathcal{C}^\ell,\Lambda}$ the inclusion

$$(6.1.32)\qquad \mathcal{D}' \subset \Lambda$$

holds in the sense of the existence of a *commutative diagram*

(6.1.33)
$$\begin{array}{ccc} \mathcal{S} \ni s & \longmapsto & \langle s,\cdot\rangle \in \mathcal{D}' \\ & & \Big\downarrow \text{inj} \\ & & s + \mathcal{I} \in A \end{array}$$

with a suitably chosen vector subspace $\mathcal{S}$, such that

(6.1.34) $$\mathcal{S} \subset \mathcal{A} \cap \mathcal{S}_\Lambda^\ell$$

Then, if we take

(6.1.35) $$\mathcal{V} = \mathcal{I} \cap \mathcal{S}$$

we obtain an inclusion diagram (6.1.27)-(6.1.31)

§2. REGULARIZATIONS

The inclusion diagram (6.1.27)-(6.1.31) contains *four* undetermined spaces, that is, $\mathcal{V}$, $\mathcal{S}$, $\mathcal{I}$ and $\mathcal{A}$. In view of (6.1.29), $\mathcal{V}$ can be obtained from $\mathcal{I}$ and $\mathcal{S}$, thus only the *three spaces* $\mathcal{S}$, $\mathcal{I}$ and $\mathcal{A}$ are arbitrary.

It is particularly important to note however that as shown in the sequel, the construction of such inclusion diagrams (6.1.27)-(6.1.31) can often be *reduced* to the choice of the ideals $\mathcal{I}$ alone, see also Rosinger [1,2,3].

For convenience, we shall only deal with the case when $\ell = \infty$, however, it is obvious that the theory in this Section remains valid for all $\ell \in \mathbb{N}$. In that case (6.1.8) becomes

(6.2.1) $$\mathcal{D}' = \mathcal{S}_\Lambda^\infty / \mathcal{V}_\Lambda^\infty$$

The general case of $\ell \in \mathbb{N}$ is presented in detail in Rosinger [2].

We start with certain $\mathcal{V}_\Lambda^\infty \subset \mathcal{S}_\Lambda^\infty \subset (\mathcal{C}^\infty)^\Lambda$ as given in (6.1.2)-(6.1.6). Further, we assume given a subalgebra $\mathcal{A} \subset (\mathcal{C}^\infty)^\Lambda$, such that

(6.2.2) $$\mathcal{U}_{\mathcal{C}^\infty,\Lambda} \subset \mathcal{A}$$

(6.2.3) $$\mathcal{S}_\Lambda^\infty \subset \mathcal{V}_\Lambda^\infty + \mathcal{A}$$

conditions which are obviously satisfied in the particular case when $\mathcal{A} = (\mathcal{C}^\infty)^\Lambda$.

The *basic notion* in the general theory of embeddings (6.1.25) is presented in:

Definition 1

Given vector subspaces $\mathcal{V}$ and $\mathcal{S}$ in $\mathcal{A} \cap \mathcal{S}^{\infty}_{\Lambda}$ and an ideal $\mathcal{I}$ in $\mathcal{A}$, we call

(6.2.4) $\qquad (\mathcal{V},\mathcal{S},\mathcal{I},\mathcal{A})$

a *regularization* of the representation (6.2.1) of the $\mathcal{D}'$ distributions, if and only if (6.1.27)-(6.1.31) are satisfied. For convenience, the representation (6.2.1) being assumed given, it will no longer be mentioned when dealing with regularizations (6.2.4).

In case we also have satisfied the condition

(6.2.5) $\qquad \mathcal{U}_{\mathcal{C}^{\infty},\Lambda} \subset \mathcal{S}$

then $(\mathcal{V},\mathcal{S},\mathcal{I},\mathcal{A})$ is called a $\mathcal{C}^{\infty}$-*smooth regularization*.

It is possible to introduced the following *simplification*:

Definition 2

An ideal $\mathcal{I}$ in $\mathcal{A}$ is called *regular*, if and only if there exist vector subspaces $\mathcal{V}$ and $\mathcal{S}$ in $\mathcal{A} \cap \mathcal{S}^{\infty}_{\Lambda}$ such that $(\mathcal{V},\mathcal{S},\mathcal{I},\mathcal{A})$ is a regularization.

The ideal $\mathcal{I}$ is called $\mathcal{C}^{\infty}$-*smooth regular*, if and only if $(\mathcal{V},\mathcal{S},\mathcal{I},\mathcal{A})$ is a $\mathcal{C}^{\infty}$-smooth regularization.

The first basic result which is an extension of Theorem 4 in Rosinger [2, p. 75] and gives a useful structural *characterization* of regularizations and $\mathcal{C}^{\infty}$-smooth regularizations is the following.

Theorem 1

Suppose $\mathcal{I}$ is an ideal in $\mathcal{A}$ and there exist vector subspaces $\mathcal{S}'$ and $\mathcal{T}$ in $\mathcal{A} \cap \mathcal{S}^{\infty}_{\Lambda}$ such that

(6.2.6) $\qquad \mathcal{I} \cap \mathcal{T} = \mathcal{V}^{\infty}_{\Lambda} \cap \mathcal{T} = \mathcal{O}$

(6.2.7) $\qquad \mathcal{I} \cap \mathcal{S}^{\infty}_{\Lambda} \subset \mathcal{V}^{\infty}_{\Lambda} \oplus \mathcal{T}$

(6.2.8) $\mathcal{S}_\Lambda^\infty \subset \mathcal{V}_\Lambda^\infty \oplus \mathcal{S}' \oplus \mathcal{T}$

Then, for every vector subspace $\mathcal{V}$ in $\mathcal{I} \cap \mathcal{V}_\Lambda^\infty$ we have that

(6.2.9) $(\mathcal{V},\mathcal{V} \oplus \mathcal{S}' \oplus \mathcal{T},\mathcal{I},\mathcal{A})$ is a regularization

thus $\mathcal{I}$ is regular.

Conversely, every regularization $(\mathcal{V},\mathcal{S},\mathcal{I},\mathcal{A})$ has the form (6.2.6)-(6.2.9).

If in addition to (6.2.6)-(6.2.8), we also have

(6.2.10) $\mathcal{U}_{\mathcal{C}^\infty,\Lambda} \subset \mathcal{S}' \oplus \mathcal{T}$

then

(6.2.11) $(\mathcal{V},\mathcal{V} \oplus \mathcal{S}' \oplus \mathcal{T},\mathcal{I},\mathcal{A})$ is a $\mathcal{C}^\infty$-smooth regularization

thus $\mathcal{I}$ is $\mathcal{C}^\infty$-smooth regular.

Conversely again, every $\mathcal{C}^\infty$-smooth regularization $(\mathcal{V},\mathcal{S},\mathcal{I},\mathcal{A})$ has the form in (6.2.6)-(6.2.8), (6.2.10), (6.2.11).

Proof

Let us denote

(6.2.12) $\mathcal{S} = \mathcal{V} \oplus \mathcal{S}' \oplus \mathcal{T}$

and prove that $(\mathcal{V},\mathcal{S},\mathcal{I},\mathcal{A})$ is indeed a regularization.

First we note that (6.1.27) is obviously satisfied. Further, (6.1.30) and (6.1.31) follow easily from (6.2.8). Now we prove (6.1.29). For that we note that the above choice of $\mathcal{S}$ in (6.2.12) yields the inclusions

$$\mathcal{I} \cap \mathcal{S} \subset \mathcal{I} \cap (\mathcal{V} \oplus \mathcal{S}' \oplus \mathcal{T}) \subset$$

$$\subset (\mathcal{I} \cap \mathcal{S}_\Lambda^\infty) \cap (\mathcal{V} \oplus \mathcal{S}' \oplus \mathcal{I}) \subset$$

$$\subset (\mathcal{V}_\Lambda^\infty \oplus \mathcal{I}) \cap (\mathcal{V} \oplus \mathcal{S}' \oplus \mathcal{I})$$

the last inclusion being implied by (6.2.7). Now (6.2.8) yields

$$\mathcal{I} \cap \mathcal{S} \subset \mathcal{V} \oplus \mathcal{T}$$

thus (6.1.29) follows from (6.2.6) and the inclusion $\mathcal{V} \subset \mathcal{I}$.

Conversely, let us assume that, $(\mathcal{V},\mathcal{S},\mathcal{I},\mathcal{A})$ is a given regularization. Let us then take $\mathcal{S}' = \mathcal{O}$ and $\mathcal{T}$ a vector subspace in $\mathcal{S}$, such that

$$(6.2.13) \qquad \mathcal{S} = \mathcal{V} \oplus \mathcal{T}$$

Then obviously

$$\mathcal{V}^{\infty}_{\Lambda} \cap \mathcal{T} \subset \mathcal{V}^{\infty}_{\Lambda} \cap \mathcal{S} \subset \mathcal{V}$$

the last inclusion being implied by (6.1.30). In this way we obtain

$$\mathcal{V}^{\infty}_{\Lambda} \cap \mathcal{T} \subset \mathcal{V} \cap \mathcal{T} = 0$$

which gives the second equality in (6.2.6). Further we note that

$$\mathcal{I} \cap \mathcal{T} \subset \mathcal{I} \cap \mathcal{S} \subset \mathcal{V}$$

the last inclusion resulting from (6.1.29). Thus

$$\mathcal{I} \cap \mathcal{T} \subset \mathcal{V} \cap \mathcal{T} \subset \mathcal{O}$$

which proves (6.2.6).

We also have (6.2.8), since obviously

$$\mathcal{V}^{\infty}_{\Lambda} \oplus \mathcal{S}' \oplus \mathcal{T} \supset \mathcal{V}^{\infty}_{\Lambda} + \mathcal{T} \supset \mathcal{V}^{\infty}_{\Lambda} + \mathcal{V} + \mathcal{T} \supset \mathcal{V}^{\infty}_{\Lambda} + \mathcal{S} \subset \mathcal{S}^{\infty}_{\Lambda}$$

the last inclusion being obtained from (6.1.31).

Further, (6.1.31) yields

$$\mathcal{I} \cap \mathcal{S}^{\infty}_{\Lambda} \subset \mathcal{S}^{\infty}_{\Lambda} \subset \mathcal{V}^{\infty}_{\Lambda} + \mathcal{S}$$

hence (6.2.13) gives

$$\mathcal{I} \cap \mathcal{S}^{\infty}_{\Lambda} \subset \mathcal{V}^{\infty}_{\Lambda} + (\mathcal{V} \oplus \mathcal{T}) \subset \mathcal{V}^{\infty}_{\Lambda} \oplus \mathcal{T}$$

which proves (6.2.7).

Let us now assume that (6.2.10) also holds, then (6.2.11) will obviously satisfy (6.1.19).

Conversely, if we assume that $(\mathcal{V},\mathcal{S},\mathcal{I},\mathcal{A})$ is a $\mathcal{C}^{\infty}$-smooth regularization, then we can again take $\mathcal{S}' = \mathcal{O}$, while $\mathcal{T}$ will be taken as a vector subspace in $\mathcal{S}$, such that, see (6.2.13)

$$(6.2.14) \qquad \mathcal{S} = \mathcal{V} \oplus \mathcal{T}, \quad \mathcal{U}_{\mathcal{C}^{\infty},\Lambda} \subset \mathcal{T}$$

which is possible, since (6.1.19) gives

$$\mathcal{U}_{\mathcal{C}^\infty,\Lambda} \subset \mathcal{S}$$

while in view of (6.1.21) and (6.1.23), we have

$$\mathcal{V} \cap \mathcal{U}_{\mathcal{C}^\infty,\Lambda} \subset \mathcal{V}^\infty_\Lambda \cap \mathcal{U}_{\mathcal{C}^\infty,\Lambda} = \mathcal{O}$$

Now, we have (6.2.10) since by the above choice of $\mathcal{S}'$ and $\mathcal{T}$ in (6.2.14), it follows that

$$\mathcal{S}' \oplus \mathcal{T} = \mathcal{O} \oplus \mathcal{T} = \mathcal{T} \supset \mathcal{U}_{\mathcal{C}^\infty,\Lambda} \qquad \square$$

Based on the above structural characterization of regularizations, we can identify a class of regular ideals $\mathcal{I}$ given in the following:

Definition 3

An ideal $\mathcal{I}$ in $\mathcal{A}$ is called *small*, if and only if

(6.2.15) $$\underset{\mathcal{I}\cap\mathcal{S}^\infty_\Lambda}{\operatorname{codim}} \mathcal{I} \cap \mathcal{V}^\infty_\Lambda \leq \underset{\mathcal{V}^\infty_\Lambda}{\operatorname{codim}} \mathcal{I} \cap \mathcal{V}^\infty_\Lambda$$

A useful property of small ideals, which is an extension of Proposition 1 in Rosinger [2, p. 77], is presented now.

Proposition 1

Suppose the ideal $\mathcal{I}$ in $\mathcal{A}$ is small. Then there exist vector subspaces $\mathcal{T} \subset \mathcal{A} \cap \mathcal{S}^\infty_\Lambda$ such that

(6.2.16) $$\mathcal{I} \cap \mathcal{T} = \mathcal{V}^\infty_\Lambda \cap \mathcal{T} = \mathcal{O}$$

(6.2.17) $$\mathcal{V}^\infty_\Lambda + (\mathcal{I} \cap \mathcal{S}^\infty_\Lambda) = \mathcal{V}^\infty_\Lambda \oplus \mathcal{T}$$

Proof

Let us take

$$\mathrm{E} = \mathcal{S}^\infty_\Lambda, \ \mathrm{A} = \mathcal{V}^\infty_\Lambda, \ \mathrm{B} = \mathcal{I} \cap \mathcal{S}^\infty_\Lambda$$

in Lemma 2 below. Then taking

$$\mathcal{T} = C$$

the proof is completed □

Lemma 2

Suppose A and B are vector subspaces in E and

(6.2.18) $$\underset{B}{\text{codim}}\ A \cap B \leq \underset{A}{\text{codim}}\ A \cap B$$

Then there exist vector subspaces C in E, such that

(6.2.19) $$A \cap C = B \cap C = 0 \quad \text{(the null subspace)}$$

(6.2.20) $$A + B = A \oplus C$$

Proof

Assume that

$$(a_i \mid i \in I), \quad (b_j \mid j \in J)$$

are algebraic vector space bases in A and B respectively, such that

$$(c_k \mid \in K)$$

with

$$K = I \cap J, \quad c_k = a_k = b_k, \quad \text{for } k \in K$$

is an algebraic vector space base in $A \cap B$.

In view of the hypothesis, there exists an *injective* mapping

$$\alpha : (J\backslash K) \longrightarrow (I\backslash K)$$

and then it follows easily that

$$(a_{\alpha(j)} + b_j \mid j \in J\backslash K) \quad \text{is linear independent}$$

We show that we can take

$$C$$

as the vector subspace generared by the above family of linear independent *vectors*. Indeed if $x \in A \cap C$, then

$$x = \sum_{i \in I} \lambda_i a_i = \sum_{j \in J\backslash K} \mu_j (a_{\alpha(j)} + b_j)$$

hence

$$\sum_{i\in I} \lambda_i a_i - \sum_{j\in J\setminus K} \mu_j a_{\alpha(j)} = \sum_{j\in J\setminus K} \mu_j b_j$$

which implies

$$\mu_j = 0, \quad j \in J\setminus K$$

thus $x = 0$.

Now if $x \in B \cap C$, then

$$x = \sum_{j\in J} \lambda_j b_j = \sum_{j\in J\setminus K} \mu_j (a_{\alpha(j)} + b_j)$$

hence

$$\sum_{j\in J\setminus K} \mu_j a_{\alpha(j)} = \sum_{j\in J} \lambda_j b_j - \sum_{j\in J\setminus K} \mu_j b_j$$

implying that

$$\mu_j = 0, \quad j \in J\setminus K$$

since α is injective. Thus again $x = 0$. Finally, it suffices to show that

$$B \subset A + C$$

Assume $x \in B$. Obviously

$$x = \sum_{k\in K} \lambda_k c_k + \sum_{j\in J\setminus K} \mu_j b_j$$

hence

$$x = - \sum_{j\in J\setminus K} \mu_j a_{\alpha(j)} + \sum_{k\in K} \lambda_k c_k + \sum_{j\in J\setminus K} \mu_j (a_{\alpha(j)} + b_j)$$

therefore, indeed $x = A + C$ □

It can be seen that if E is finite-dimensional, the condition in the above inequality of codimensions is essential. Indeed, if $E = A + B$ and $\dim A < \dim B$, then $A \oplus C = E$ implies $B \cap C \neq 0$, as otherwise we have the contradiction

$$\dim E \geq \dim B + \dim C = \dim B + \dim E - \dim A > \dim E$$

The basic property concerning the existence of regular ideals is given in:

Theorem 2

Every small ideal $\mathcal{I}$ in $\mathcal{A}$ is regular.

Proof

Let us take $\mathcal{T}$ given by Proposition 1. Then (6.2.6) and (6.2.7) are satisfied in view of (6.2.16) and (6.2.17). Furthermore, owing to (6.2.3), we can choose vector subspaces $\mathcal{S}'$ in $\mathcal{A} \cap \mathcal{S}^{\infty}_{\Lambda}$ such that (6.2.8) will also hold.

In this way, Theorem 1 implies that $\mathcal{I}$ is regular □

Remark 2

1. The existence of large classes of *small* and therefore *regular ideals* $\mathcal{I}$ which are nontrivial, that is $\mathcal{I} \neq \mathcal{O}$, is proved in Rosinger [2, pp. 81-88], in the case of $\Lambda = \mathbb{N}$ and $\ell \in \mathbb{N}$. For instance, with the notation in (2.2.3), $\mathcal{I}_{\mathrm{nd}}(\mathbb{R}^n) \cap (\mathcal{C}^{\infty}(\mathbb{R}^n))^{\mathbb{N}}$, is a small and thus regular ideal.

2. The regular ideal mentioned in pct. 1. above is also $\mathcal{C}^{\infty}$-smooth regular.

§3. NEUTRIX CHARACTERIZATION OF REGULAR IDEALS

As seen in (1.6.11) and (6.1.23), the *neutrix* or *off diagonality* condition

$$(6.3.1) \qquad \mathcal{I} \cap \mathcal{U}_{\mathcal{C}^{\ell},\Lambda} = \mathcal{O}$$

is a *necessary condition* for every ideal $\mathcal{I}$ of an inclusion diagram (6.1.19) used in the construction of the quotient algebras of generalized functions, see (6.1.24)

$$(6.3.2) \qquad A = \mathcal{A}/\mathcal{I} \in \mathrm{AL}_{\mathcal{C}^{\ell},\Lambda}$$

As a *fundamental algebraic characterization* of these algebras of generalized functions, we prove in this Section that, for a large class of such algebras, the neutrix condition does in fact characterize the *regular ideals* $\mathcal{I}$ in the respective quotient algebras $A = \mathcal{A}/\mathcal{I}$.

Indeed, in the case of the index set $\Lambda = \mathbb{N}$ and for $\ell = \infty$, if we take

(6.3.3) $\mathcal{A} = (\mathcal{C}^\infty(\mathbb{R}^n))^{\mathbb{N}}$

then the *neutrix condition* (6.3.1) will *characterize* the *regular* and $\mathcal{C}^\infty$-*smooth regular ideals* $\mathcal{I}$, within a large class of so called co-final invariant ideals. It is easy to see that the neutrix characterization presented in Theorem 4 below, extends in the obvious way to all $\ell \in \mathbb{N}$. This neutrix characterization of regular and $\mathcal{C}^\infty$-smooth regular ideals, first obtained in Rosinger [2], is presented here in its main features.

For convenience, we shall use the framework in (6.1.1) with $\ell = \infty$, that is, the distributions are given by the quotient representation

(6.3.4) $\mathcal{D}'(\mathbb{R}^n) = \mathcal{S}^\infty/\mathcal{V}^\infty$

A useful class of ideals in specified in:

Definition 4

An ideal $\mathcal{I}$ in $(\mathcal{C}^\infty(\mathbb{R}^n))^{\mathbb{N}}$ is called vanishing, if and only if

$$(6.3.5) \quad \begin{aligned} &\forall \ w \in \mathcal{I}, \ \mu \in \mathbb{N} : \\ &\exists \ \nu \in \mathbb{N}, \ \nu \geq \mu, \ x \in \mathbb{R}^n : \\ &\quad w_\nu(x) = 0 \end{aligned}$$

With the definition in (2.2.4), it is obvious that

(6.3.6) $\mathcal{I}_{nd} \cap (\mathcal{C}^\infty(\mathbb{R}^n))^{\mathbb{N}}$ is a vanishing ideal

The basic property of vanishing ideals is presented now.

Theorem 3

Every vanishing ideasl $\mathcal{I}$ is small and therefore regular.

Proof

In view of (6.2.15), we only have to show that

(6.3.7) $\underset{\mathcal{I}\cap\mathcal{S}^\infty}{\text{codim}}\ \mathcal{I} \cap \mathcal{V}^\infty \leq \underset{\mathcal{V}^\infty}{\text{codim}}\ \mathcal{I} \cap \mathcal{V}^\infty$

For that, we note that obvious inequalities

$$\underset{\mathcal{I} \cap \mathcal{S}^\infty}{\text{codim}}\ \mathcal{I} \cap \mathcal{V}^\infty \leq \dim \mathcal{I} \cap \mathcal{S}^\infty \leq \text{car}\ \mathcal{I} \cap \mathcal{S}^\infty$$

But

$$\mathcal{I} \cap \mathcal{S}^\infty \subset (\mathcal{C}^0(\mathbb{R}^n))^{\mathbb{N}}$$

and

$$\text{car}\ \mathcal{C}^0(\mathbb{R}^n) = \text{car}\ \mathbb{R}$$

Therefore

$$\underset{\mathcal{I} \cap \mathcal{S}^\infty}{\text{codim}}\ \mathcal{I} \cap \mathcal{V}^\infty \leq (\text{car}\ \mathbb{R})^{\text{car}\ \mathbb{N}} = \text{car}\ \mathbb{R}$$

Now in order to prove (6.3.7), it suffices to show that

(6.3.8) $$\underset{\mathcal{V}^\infty}{\text{codim}}\ \mathcal{I} \cap \mathcal{V}^\infty \geq \text{car}\ \mathbb{R}$$

For that, we define $v_\alpha \in \mathcal{V}^\infty$, with $\alpha \in (0,1)$, by

$$(v_\alpha)_\nu(x) = \alpha^\nu, \quad \nu \in \mathbb{N}, \quad x \in \mathbb{R}^n$$

Then it follows that

(6.3.9) $$(v_\alpha | \alpha \in (0,1))$$

is linear independent in $\mathcal{V}^\infty$. Let us denote by

$$\mathcal{V}$$

the vector subspace in $\mathcal{V}^\infty$ generated by (6.3.9). Then

(6.3.10) $$\mathcal{I} \cap \mathcal{V} = \mathcal{O}$$

Indeed, assume $w \in \mathcal{I} \cap \mathcal{V}$. Then $w \in \mathcal{V}$ implies

(6.3.11) $$w = \sum_{1 \leq i \leq h} \lambda_i v_{\alpha_i}$$

with suitable $h \in N$, $\lambda_i \in \mathbb{C}$ and $\alpha_i \in (0,1)$.

Hence

(6.3.12) $$w_\nu(x) = \sum_{1\leq i\leq h} \lambda_i(\alpha_i)^\nu, \quad \nu \in \mathbb{N}, \quad x \in \mathbb{R}^n$$

But $w \in \mathcal{I}$, hence (6.3.5), (6.3.12) imply

(6.3.13) $$\sum_{1\leq i\leq h} \lambda_i(\alpha_i)^\nu = 0, \quad \text{for infinitely many} \quad \nu \in \mathbb{N}$$

Therefore

(6.3.14) $$\lambda_1 = \ldots = \lambda_h = 0$$

Indeed, assume $h = 1$. Then $\lambda_1 = 0$, since $\alpha_1 \in (0,1)$. Assume now $h \geq 2$. We can further aassume that

$$0 < \alpha_1 < \ldots < \alpha_h < 1$$

and

(6.3.15) $$\lambda_i \neq 0, \quad 1 \leq i \leq h$$

Dividing (6.3.13) by α_h, we obtain

(6.3.16) $$\lambda_1(\alpha_1/\alpha_h)^\nu + \ldots + \lambda_{h-1}(\alpha_{h-1}/\alpha_h)^\nu + \lambda_h = 0$$

for infinitely many $\nu \in \mathbb{N}$. Since

$$0 < \alpha_i/\alpha_h < 1, \quad 1 \leq i \leq h-1$$

the relation (6.3.16) yields

$$\lambda_h = 0$$

which contradicts (6.3.15) and thus the proof of (6.3.14) is completed. Obviously (6.3.11) and (6.3.14) yield (6.3.10).

In view of (6.3.9), we have

$$\dim \mathcal{V} = \text{car}\ \mathbb{R}$$

hence (6.3.10) implies (6.3.8) and the proof of (6.3.7) is completed.

It follows that $\mathcal{I}$ is a small ideal in $(\mathcal{C}^\infty(\mathbb{R}^n))^{\mathbb{N}}$, hence in view of Theorem 2 in Section 2, it is regular.

A simple and useful *characterization* of *vanishing ideals* can be obtained within the following large class of ideals.

Definition 5

An ideal $\mathcal{I}$ in $(\mathcal{C}^\infty(\mathbb{R}^n))^{\mathbb{N}}$ is called *cofinal invariant*, if and only if

$$(6.3.17) \qquad \left(\begin{array}{l} \forall \quad w \in (\mathcal{C}^\infty(\mathbb{R}^n))^{\mathbb{N}} : \\ \exists \quad w' \in \mathcal{I},\ \mu \in \mathbb{N} : \\ \forall \quad \nu \in \mathbb{N},\ \nu \geq \mu : \\ \qquad w_\nu = w'_\nu \end{array}\right) \Rightarrow w \in \mathcal{I}$$

In view of (2.2.4), it is easy to see that

$$(6.3.18) \qquad \mathcal{I}_{nd} \cap (\mathcal{C}^\infty(\mathbb{R}^n))^{\mathbb{N}} \text{ is a cofinal invariant ideal}$$

Proposition 2

A cofinal invariant ideal $\mathcal{I}$ is vanishing if and only if it is a proper ideal in $(\mathcal{C}^\infty(\mathbb{R}^n))^{\mathbb{N}}$.

Proof

Assume $\mathcal{I}$ is not vanishing. Then (6.3.5) yields $w \in \mathcal{I}$ and $\mu \in \mathbb{N}$, such that

$$(6.3.19) \qquad w_\nu(x) \neq 0, \quad \nu \in \mathbb{N}, \quad \nu \geq \mu, \quad x \in \mathbb{R}^n$$

Let us define $w' \in (\mathcal{C}^\infty(\mathbb{R}^n))^{\mathbb{N}}$ by

$$(6.3.20) \qquad w'_\nu(x) = \left| \begin{array}{ll} 1 & \text{if } \nu \in \mathbb{N},\ \nu < \mu \\ w_\nu(x) & \text{if } \nu \in \mathbb{N},\ \nu \geq \mu \end{array}\right.$$

for $\nu \in \mathbb{N}$ and $x \in \mathbb{R}^n$. Since $\mathcal{I}$ was assumed cofinal invariant, (6.3.17) and (6.3.20) yield

$$(6.3.21) \qquad w' \in \mathcal{I}$$

But (6.3.19), (6.3.20) obviously imply

$$(6.3.22) \qquad 1/w' \in (\mathcal{C}^\infty(\mathbb{R}^n))^{\mathbb{N}}$$

Now (6.3.21), (6.3.22) give

$$u(1) = w' \cdot (1/w') \in \mathcal{I} \cdot (\mathcal{C}^\infty(\mathbb{R}^n))^{\mathbb{N}} \subset \mathcal{I}$$

which means that $\mathcal{I}$ cannot be a proper ideal in $(\mathcal{C}^\infty(\mathbb{R}^n))^{\mathbb{N}}$. The converse is obvious. □

The fundamental result given by the *neutrix characterization* of *regular* and *$\mathcal{C}^\infty$-smooth regular ideals* which are cofinal invariant is as follows.

Theorem 4

For a cofinal invariant ideal $\mathcal{I}$ in $(\mathcal{C}^\infty(\mathbb{R}^n))^{\mathbb{N}}$ the following *three* conditions are *equivalent* :

(6.3.23) $\mathcal{I}$ is regular

(6.3.24) $\mathcal{I}$ is $\mathcal{C}^\infty$-smooth regular

(6.3.25) $\mathcal{I} \cap \mathcal{U}_{\mathcal{C}^\infty,\mathbb{N}} = \mathcal{O}$

in which case $\mathcal{I}$ will also be a vanishing ideal.

Proof

We obviously have the implications

$$(6.3.24) \Rightarrow (6.3.23) \Rightarrow (6.3.25)$$

with the last implication resulting from (6.1.28) and (6.3.1).

Therefore, we only have to prove the implication

$$(6.3.25) \Rightarrow (6.3.24)$$

Assume (6.3.25) holds, then obviously

$$\mathcal{I} \text{ is a proper ideal in } (\mathcal{C}^\infty(\mathbb{R}^n))^{\mathbb{N}}$$

Hence, in view of Proposition 2, $\mathcal{I}$ is a vanishing ideal, therefore, according to Theorem 3, $\mathcal{I}$ is a small ideal.

Now, Proposition 1 in Section 2 will yield vector subspaces

$$\mathcal{T} \subset \mathcal{S}^\infty$$

which satisfy (6.2.16) and (6.2.17).

Let us assume for the time being that $\mathcal{T}$ can be chosen so as also to satisfy

(6.3.26) $$(\mathcal{V}^\infty \oplus \mathcal{T}) \cap \mathcal{U}_{\mathcal{C}^\infty,\mathbb{N}} = \mathcal{T} \cap \mathcal{U}_{\mathcal{C}^\infty,\mathbb{N}}$$

If we take a vector subspace

$$\mathcal{U}' \subset \mathcal{U}_{\mathcal{C}^\infty,\mathbb{N}}$$

such that

(6.3.27) $$\mathcal{U}_{\mathcal{C}^\infty,\mathbb{N}} = (\mathcal{T} \cap \mathcal{U}_{\mathcal{C}^\infty,\mathbb{N}}) \oplus \mathcal{U}'$$

then we obtain

(6.3.28) $$(\mathcal{V}^\infty \oplus \mathcal{T}) \cap \mathcal{U}' = \mathcal{O}$$

since

$$(\mathcal{V}^\infty \oplus \mathcal{T}) \cap \mathcal{U}' \subset ((\mathcal{V}^\infty \oplus \mathcal{T}) \cap \mathcal{U}_{\mathcal{C}^\infty,\mathbb{N}}) \cap \mathcal{U}' =$$
$$= (\mathcal{T} \cap \mathcal{U}_{\mathcal{C}^\infty,\mathbb{N}}) \cap \mathcal{U}' = \mathcal{O}$$

the latter two equalities being implied by (6.3.26) and (6.3.27).

In view of (6.3.27), (6.3.28), we can choose vector subspaces

$$\mathcal{S}' \subset \mathcal{S}^\infty$$

which will satisfy (6.2.8) and

(6.3.29) $$\mathcal{U}_{\mathcal{C}^\infty,\mathbb{N}} \subset \mathcal{S}' \oplus \mathcal{T}$$

Then, according to Theorem 1 in Section 2, $\mathcal{I}$ is indeed a $\mathcal{C}^\infty$-smooth regular ideal.

Now all that is left, is to prove (6.3.26).

For that, we shall use Lemma 2 in Section 2, as well as Lemma 3 below. Let us denote

(6.3.30) $$\mathrm{E} = \mathcal{S}^\infty, \quad \mathrm{A} = \mathcal{V}^\infty, \; \mathrm{B} = \mathcal{U}_{\mathcal{C}^\infty,\mathbb{N}} \quad \text{and} \quad \mathrm{C} = \mathcal{I} \cap \mathcal{S}^\infty$$

Then we have

(6.3.31) $$\mathrm{A} \cap \mathrm{B} = \mathrm{B} \cap \mathrm{C} = 0 \quad \text{(the null space)}$$

the latter equality being implied by (6.3.25). Now, with the notation in Lemma 3 below, we have

$$\text{(6.3.32)}\qquad \begin{aligned} \bar{A} &= \mathcal{V}^{\infty} \oplus (\mathcal{U}_{\mathcal{C}^{\infty},\mathbb{N}} \cap (\mathcal{V}^{\infty} + (\mathcal{I} \cap \mathcal{S}^{\infty}))) \\ \bar{B} &= (\mathcal{I} \cap \mathcal{S}^{\infty}) \oplus (\mathcal{U}_{\mathcal{C}^{\infty},\mathbb{N}} \cap (\mathcal{V}^{\infty} + (\mathcal{I} \cap \mathcal{S}^{\infty}))) \end{aligned}$$

Our aim is to apply Lemma 2 to $\bar{A}, \bar{B} \subset E$. To do so, we have first to prove the relation

$$\text{(6.3.33)}\qquad \operatorname{codim}_{\bar{B}} \bar{A} \cap \bar{B} \leq \operatorname{codim}_{\bar{A}} \bar{A} \cap \bar{B}$$

Indeed, an argument similar to that used in the proof of Theorem 3 in order to establish (6.3.7), yields

$$\operatorname{codim}_{\bar{B}} \bar{A} \cap \bar{B} \leq \dim \bar{B} \leq \operatorname{car} \mathbb{R}$$

Now in order to obtain (6.3.33) it suffices to show that

$$\text{(6.3.34)}\qquad \operatorname{codim}_{\bar{A}} \bar{A} \cap \bar{B} \geq \operatorname{car} \mathbb{R}$$

To do so, we shall again use the sequences of functions $v_\alpha \in \mathcal{V}^{\infty}$, with $\alpha \in (0,1)$, defined in the proof of Theorem 3, as well as the vector subspace $\mathcal{V}$ generated by them, see (6.3.9), and prove the relation

$$\text{(6.3.35)}\qquad \bar{B} \cap \mathcal{V} = \mathcal{O}$$

Indeed, assume $\bar{w} \in \bar{B} \cap \mathcal{V}$. Then $\bar{w} \in \mathcal{V}$ implies that

$$\bar{w} = \sum_{1 \leq i \leq h} \lambda_i v_{\alpha_i}$$

with $h \in \mathbb{N}$, $\lambda_i \in \mathbb{R}$ and $\alpha_i \in (0,1)$, hence

$$\text{(6.3.36)}\qquad \bar{w}_\nu(x) = \sum_{1 \leq i \leq h} \lambda_i (\alpha_i)^\nu, \quad \nu \in \mathbb{N}, \quad x \in \mathbb{R}^n$$

But $\bar{w} \in \bar{B}$ and (6.3.32) yield

$$\text{(6.3.37)}\qquad \bar{w} = w + u(\psi)$$

with $w \in \mathcal{I} \cap \mathcal{S}^\infty$ and $\psi \in \mathcal{C}^\infty(\mathbb{R}^n)$.

The point is that

$$(6.3.38) \qquad \psi(x) = 0, \quad x \in \mathbb{R}^n$$

Indeed, assume (6.3.38) is false and that $\epsilon > 0$ and $\Omega' \in \mathbb{R}^n$, with Ω' non-void open, are such that

$$\psi(x) < -2\epsilon, \quad x \in \Omega'$$

Then in view of (6.3.36) and (6.3.37) we have

$$(6.3.39) \qquad \begin{array}{l} \exists \ \mu \in \mathbb{N} \ ; \\ \forall \ \nu \in \mathbb{N}, \ \nu \geq \mu, \ x \in \Omega' : \\ \quad w_\nu(x) = \bar{w}_\nu(x) - \psi(x) > \epsilon \end{array}$$

Now define $w' \in (\mathcal{C}^\infty(\mathbb{R}^n))^{\mathbb{N}}$ by

$$w'_\nu = \begin{vmatrix} 1 & \text{if } \nu < \mu \\ w_\nu & \text{if } \nu \geq \mu \end{vmatrix}$$

Then obviously

$$(6.3.40) \qquad w' \in \mathcal{I}$$

since $w \in \mathcal{I}$, and $\mathcal{I}$ is cofinal invariant. Take then any

$$(6.3.41) \qquad \chi \in \mathcal{C}^\infty(\mathbb{R}^n), \text{ with } \sup \chi \subset \Omega'$$

and in view of (6.3.39), define $t \in (\mathcal{C}^\infty(\mathbb{R}^n))^{\mathbb{N}}$ by

$$t_\nu = \begin{vmatrix} \chi & \text{if } \nu < \mu \\ \chi/w_\nu & \text{if } \nu \geq \mu \end{vmatrix}$$

Then (6.3.40) gives

$$w' \cdot t = u(\chi) \in \mathcal{I} \cap \mathcal{U}_{\mathcal{C}^\infty, \mathbb{N}}$$

Therefore, in view of (6.3.25), we have

$$\chi(x) = 0, \quad x \in \mathbb{R}^n$$

which is absurd, since χ can be chosen arbitrarily within the condition

(6.3.41). This completes the proof of (6.3.28).

Now (6.3.37) and (6.3.38) imply that

$$\bar{w} = w \in \bar{B} \cap \mathcal{V} \cap \mathcal{I} \cap \mathcal{S}^\infty \subset \mathcal{I} \cap \mathcal{V} = \mathcal{O}$$

the latter equality resulting from (6.3.10) in the proof of Theorem 3, as well as the fact that - as we have noticed - $\mathcal{I}$ is vanishing. This completes the proof of (6.3.35).

We can therefore conclude that (6.3.34) holds, since $\mathcal{V} \subset \mathcal{V}^\infty \subset \bar{A}$ and $(v_\alpha \mid \alpha \in (0,1))$ is an algebraic base in $\mathcal{V}$.

Since (6.3.34) implies (6.3.33), we can finally apply Lemma 2 to $\bar{A}$ and $\bar{B}$ given in (6.3.32) and obtain the existence of vector subspaces $\bar{C}$ in $E = \mathcal{S}^\infty$, such that

$$\bar{A} \cap \bar{C} = \bar{B} \cap \bar{C} = 0 \quad \text{and}$$

$$\bar{A} + \bar{B} = \bar{A} \oplus \bar{C}$$

Then (6.3.31) and Lemma 3 below impply the existence of vector subspaces $D \subset E = \mathcal{S}^\infty$, such that

$$\begin{aligned} & A \cap D = C \cap D = 0 \\ & A + C = A \oplus D \\ & B \cap (A+C) = B \cap D \end{aligned} \tag{6.3.42}$$

Taking, finally, $\mathcal{T} = D$, the latter two relations in (6.3.42) will yield (6.3.26) □

<u>Lemma 3</u>

If A, B and C are vector subspaces in E and

$$A \cap B = B \cap C = 0 \quad \text{(the null space)},$$

then the following two properties are equivalent

$$\begin{aligned} & \exists \; D \subset E \text{ vector subspace :} \\ & A \cap D = C \cap D = 0 \\ & A + C = A \oplus D \\ & B \cap (A+C) = B \cap D \end{aligned} \tag{6.3.43}$$

and

$$
(6.3.44) \qquad \begin{array}{l} \exists \ \bar{C} \subset E \ \text{ vector subspace :} \\ \bar{A} \cap \bar{C} = \bar{B} \cap \bar{C} = 0 \\ \bar{A} + \bar{B} = \bar{A} \oplus \bar{C} \end{array}
$$

where

$$\bar{A} = A \oplus (B \cap (A+C))$$

$$\bar{B} = C \oplus (B \cap (A+C))$$

Proof

Assume (6.3.44) holds. Then (6.3.43) results by direct verification, if one takes $D = (B \cap (A+C)) \oplus \bar{C}$. Assuming (6.3.43) and taking any vector subspace $\bar{C} \subset E$ such that $D = (B \cap D) \oplus \bar{C}$, direct verification will yield (6.3.44) □

The following consequence will be particularly useful in Chapter 7 in the construction of chains of algebras of generalized functions used in the resolution of singularities of solutions of nonlinear partial differential equations.

Corollary 1

$\mathcal{I}_{nd} \cap (\mathcal{C}^\infty(\mathbb{R}^n))^{\mathbb{N}}$ is a $\mathcal{C}^\infty$-smooth regular ideal in $(\mathcal{C}^\infty(\mathbb{R}^n))^{\mathbb{N}}$.

Proof

It follows from (6.3.18), Theorem 4 and (2.2.4) □

§4. THE UTILITY OF CHAINS OF ALGEBRAS OF GENERALIZED FUNCTIONS

We come now to the *high point* of the theoretical argument in this Chapter, concerning the *necessary structure* for *nonlinear* theories of generalized functions.

From the very beginning, in (1.1.5) it was already noted that in order to avoid certain aspects of the conflict between discontinuity, multiplication and differentiation, we may have to allow that the derivative operators act between *two different algebras*

$$(6.4.1) \qquad D : A \longrightarrow \bar{A}$$

That, of course, would imply the same for the nonlinear partial differential operators $T(D)$ containing such derivative operators, see for instance (2.3.6).

However, in Chapter 8, we shall see a more particular situation within the nonlinear theory given by the *coupled calculus* on Colombeau's differential algebra of generalized functions $\mathcal{G}(\mathbb{R}^n)$. When dealing in this theory with linear or polynomial nonlinear partial differential equations

(6.4.2) $$T(D)U(x) = f(x), \quad x \in \Omega \subset \mathbb{R}^n$$

one is in fact dealing with *one single* differential algebra $\mathcal{G}(\mathbb{R}^n)$ which is *both* the domain *and* the range of the respective partial differential operator, that is

(6.4.3) $$T(D) : \mathcal{G}(\mathbb{R}^n) \to \mathcal{G}(\mathbb{R}^n)$$

In other words, all the algebraic and differential operations connected with $T(D)$ are performed in the same, one single differential algebra $\mathcal{G}(\mathbb{R}^n)$.

That situation can obviously present advantages in so far that it allows the maximum economy in the number of spaces of generalized functions involved. However, it may as well present some disadvantages. Indeed, Colombeau's coupled calculus on the algebra $\mathcal{G}(\mathbb{R}^n)$, although precludes the stability paradoxes of type (1.8.1), nevertheless, it can exhibit them in the milder forms, see Rosinger [3, pp. 197, 199]. And obviously, a reason for the presence of the milder stability paradoxes is the fact that in (6.4.3), one single space of generalized functions is involved. Indeed, as seen in (1.9.1), if (6.4.3) is replaced with the more general framework

(6.4.4) $$T(D) : E \to A$$

involving two *different* spaces E and A of generalized functions, the need for a coupled calculus disappears, and so do the various forms of stability paradoxes, see Section 12, Chapter 1.

In addition, as seen in Chapters 2 and 3, when solving linear or nonlinear partial differential equations, it is often convenient to use the general framework (6.4.4) with two different spaces of generalized functions.

A further disadvantage of a framework such as in (6.4.3) which involves one single differential algebra may come from the limitations it can impose on the interplay between the stability, generality and exactness properties of generalized solutions, see Section 11, Chapter 1.

Finally, we also have to face the fact that a differential algebra, that is an algebra with *arbitrary* order differentiation, will by necessity have a *peculiar* multiplication. Indeed, as an example of that we have for instance the fact that *none* of the inclusions

$$(6.4.5) \qquad \mathcal{C}^m(\mathbb{R}^n) \subset \mathcal{G}(\mathbb{R}^n), \; m \in \mathbb{N}$$

is an inclusion of algebras. More precisely, the multiplications in the two algebras in (6.4.5) are different, and we have an inclusion of algebras only in the case of

$$(6.4.6) \qquad \mathcal{C}^\infty(\mathbb{R}^n) \subset \mathcal{G}(\mathbb{R}^n)$$

The above difficulty in (6.4.5) is in fact the very reason for the need of the coupled calculus on $\mathcal{G}(\mathbb{R}^n)$.

It is *particularly important* to note that the difficulty in (6.4.5) is to a large extent *unavoidable*, and it is not only a particular accident, specific to Colombeau's coupled calculus. Indeed, as seen in Sections 1-4 in Chapter 1, the difficulty in (6.4.5) is but one expression of the conflict between insufficient smoothness, multiplication and differentiation.

The above then leads us to the idea of using *more than one* space of generalized functions, and possibly *not* both of them being differential algebras, that is, with arbitrary order differentiation. The hope is that in such a way we could avoid the above mentioned difficulties.

Fortunately, as seen next, that hope can be achieved with the help of *chains of algebras of generalized functions*

$$(6.4.7) \qquad \mathcal{D}'(\mathbb{R}^n) \subset A^\infty \rightarrow \ldots \rightarrow A^\ell \rightarrow \ldots \rightarrow A^0, \quad \ell \in \mathbb{N}$$

where A^∞ is a differential algebra, while the algebras A^ℓ, with $\ell \in \mathbb{N}$, satisfy

$$(6.4.8) \qquad A^{\ell+m} \overset{m}{\leq} A^\ell, \quad \ell, m \in \mathbb{N}$$

with the notation in (1.9.11). The arrows $\rightarrow$ in (6.4.7) are algebra homomorphisms, which have a number of convenient properties extending the classical situation of the chain of inclusions

$$(6.4.9) \qquad \mathcal{C}^\infty(\mathbb{R}^n) \subset \ldots \subset \mathcal{C}^\ell(\mathbb{R}^n) \subset \ldots \subset \mathcal{C}^0(\mathbb{R}^n), \quad \ell \in \mathbb{N}$$

The question as to what extent can inclusions of algebras such as in (6.4.5) be achieved within the chains of algebras (6.4.7) is not completely answered. A variety of partial positive ansers can be found in Rosinger [2, pp. 88-104, 110-112]. See also Section 6 in the sequel.

Given an m-th order linear or polynomial nonlinear partial differential operator $T(D)$, it will be possible to consider it in the following frameworks

$$(6.4.10) \qquad T(D) : A^\infty \rightarrow A^\infty$$

which is similar with (6.4.3), or more generally

(6.4.11) $\quad T(D) : A^{\ell} \rightarrow A^{k}, \quad \ell, k \in \mathbb{N}, \quad k+m, \leq \ell$

which corresponds to (6.4.4).

Concerning the multiplication of *smooth functions*, we have the *inclusion of algebras*

(6.4.12) $\quad \mathcal{C}^{\infty}(\mathbb{R}^n) \subset A^{\ell}, \quad \ell \in \mathbb{N}$

therefore, the algebras A^{ℓ}, with $\ell \in \mathbb{N}$, extend the multiplication of $\mathcal{C}$-smooth functions.

Finally, concerning the multiplication of *nonsmooth* functions or distributions, we can have

(6.4.13) $\quad x^p \cdot D^q \delta = 0 \quad \text{in} \quad A^{\ell}$

for $\ell \in \mathbb{N}$, $p, q \in \mathbb{N}^n$, $|p| > \ell$, if $p \leq q$ does *not* hold, see Rosinger [2, pp. 229, 230].

We proceed now to the construction of a large class of chains of algebras (6.4.7). Suppose given an arbitrary infinite index set Λ, together with vector spaces

(6.4.14) $\quad \mathcal{V}^{\infty}_{\Lambda} \subset \mathcal{S}^{\infty}_{\Lambda} \subset (\mathcal{C}^{\infty}(\mathbb{R}^n))^{\Lambda}$

satisfying (6.1.2)-(6.1.6), which means the commutativity of the diagram

(6.4.15)
$$\begin{array}{ccc} \mathcal{S}^{\infty}_{\Lambda} \ni s & \xrightarrow{\quad \theta \text{ lin, sur} \quad} & T \in \mathcal{D}'(\mathbb{R}^n) \\ \subset \uparrow & & \uparrow \text{id} \\ \mathcal{U}_{\mathcal{C}^{\infty}_{\Lambda}} \ni u(\psi) & \longmapsto & T_{\psi} \in \mathcal{D}'(\mathbb{R}^n) \end{array}$$

together with the relation

(6.4.16) $\quad \mathcal{V}^{\infty}_{\Lambda} = \ker \theta$

We shall also assume that

(6.4.17) $\quad D^p \mathcal{V}^{\infty}_{\Lambda} \subset \mathcal{V}^{\infty}_{\Lambda}, \quad D^p \mathcal{S}^{\infty}_{\Lambda} \subset \mathcal{S}^{\infty}_{\Lambda}, \quad p \in \mathbb{N}^n$

Finally, suppose given

(6.4.18) $\mathcal{I}$ a $\mathcal{C}^\infty$-smooth regular ideal in $(\mathcal{C}^\infty(\mathbb{R}^n))^\Lambda$

In view of Example 1 at the end of this Section, conditions (6.4.14)-(6.4.18) can be satisfied.

Our aim is to *construct* for each $\ell \in \mathbb{N}$, a $\mathcal{C}^\infty$-smooth regularization

(6.4.19) $(\mathcal{V}_\ell, \mathcal{S}_\ell, \mathcal{I}_\ell, \mathcal{A}_\ell)$

of

(6.4.20) $\mathcal{D}'(\mathbb{R}^n) = \mathcal{S}^\infty{}_\Lambda/\mathcal{V}^\infty_\Lambda$

and then obtain the algebras in the chain (6.4.7) by the method given in (6.1.24), that is

(6.4.21) $A^\ell = \mathcal{A}_\ell/\mathcal{I}_\ell, \quad \ell \in \mathbb{N}$

We now proceed to construct the $\mathcal{C}^\infty$-smooth regularizations (6.4.19) from the given *$\mathcal{C}^\infty$-smooth regular ideal* $\mathcal{I}$ in (6.4.18).

In view of (6.4.18) and Theorem 1 in Section 2, we can further assume given

(6.4.22) $(\mathcal{V},\mathcal{V}\oplus\mathcal{S},\mathcal{I},(\mathcal{C}^\infty(\mathbb{R}^n))^{\mathbb{N}})$ a $\mathcal{C}^\infty$-smooth regularization

of (6.4.20), such that

(6.4.23) $\mathcal{I} \cap (\mathcal{V} \oplus \mathcal{S}) = \mathcal{V}, \quad \mathcal{U}_{\mathcal{C}^\infty,\Lambda} \subset \mathcal{S}$

see (6.1.20), (6.2.10) and (6.2.11).

Now, several auxiliary definitions and notations are first needed.

A subset $\mathcal{X} \subset (\mathcal{C}^\infty(\mathbb{R}^n))^\Lambda$ is called *derivative invariant*, if and only if

(6.4.24) $D^p\mathcal{X} \subset \mathcal{X}, \quad p \in \mathbb{N}^n$

Obviously $\mathcal{X} = (\mathcal{C}^\infty(\mathbb{R}^n))^\Lambda$ is derivative invariant. Also the intersection of any family of derivative invariant subsets is again derivative invariant.

Given the vector subspaces $\mathcal{V},\mathcal{S} \subset (\mathcal{C}^\infty(\mathbb{R}^n))^\Lambda$ and $\ell \in \mathbb{N}$, we denote

(6.4.25) $$\mathcal{V}_\ell = \left\{ v \in \mathcal{V} \,\middle|\, \begin{array}{l} \forall \ \ p \in \mathbb{N}^n, \ \ |p| \le \ell : \\ D^p v \in \mathcal{V} \end{array} \right\}$$

Further, we denote by

(6.4.26) $$\mathcal{A}_\ell(\mathcal{V},\mathcal{S})$$

the derivative invariant subalgebra generated by $\mathcal{V}_\ell + \mathcal{S}$ in $(\mathcal{C}^\infty(\mathbb{R}^n))^\Lambda$.

Finally, we denote by

(6.4.27) $$\mathcal{I}_\ell(\mathcal{V},\mathcal{S})$$

the ideal generated by $\mathcal{V}_\ell$ in $\mathcal{A}_\ell(\mathcal{V},\mathcal{S})$. Obviously, owing to (6.4.23), $\mathcal{A}_\ell(\mathcal{V},\mathcal{S})$ has a unit element, therefore we have

(6.4.28) $(\mathcal{I}_\ell(\mathcal{V},\mathcal{S})$ is the vector space generated by $\mathcal{V}_\ell \cdot \mathcal{A}_\ell(\mathcal{V},\mathcal{S})$

With the above notations, we shall take (6.4.19) as given by

(6.4.29)
$$\begin{array}{l} \mathcal{V}_\ell \ \text{ in (6.4.25)} \\ \mathcal{S}_\ell = \mathcal{V}_\ell \oplus \mathcal{S} \\ \mathcal{I}_\ell = \mathcal{I}_\ell(\mathcal{V},\mathcal{S}) \\ \mathcal{A}_\ell = \mathcal{A}_\ell(\mathcal{V},\mathcal{S}) \end{array}$$

for $\ell \in \mathbb{N}$.

<u>Theorem 5</u>

Given $\mathcal{I}$ in (6.4.18) and $\mathcal{V}$, $\mathcal{S}$ in (6.4.22), (6.4.23), then, with the construction in (6.4.29), it follows for $\ell \in \mathbb{N}$, that

(6.4.30) $(\mathcal{V}_\ell,\mathcal{S}_\ell,\mathcal{I}_\ell,\mathcal{A}_\ell)$ is a $\mathcal{C}^\infty$-smooth regularization

of

(6.4.31) $$\mathcal{D}'(\mathbb{R}^n) = \mathcal{S}^\infty_\Lambda/\mathcal{V}^\infty_\Lambda$$

Proof

The inclusions in (6.1.19) are obvious.

For the relation (6.1.20) we note that

(6.4.32) $$\mathcal{I}_\ell \subset \mathcal{I}, \quad \ell \in \mathbb{N}$$

Indeed, (6.4.25) and (6.4.22) yield

(6.4.33) $$\mathcal{V}_\ell \subset \mathcal{V} \subset \mathcal{I}, \quad \ell \in \mathbb{N}$$

But $\mathcal{I}$ is an ideal in $(\mathcal{C}^\infty(\mathbb{R}^n))^\Lambda$, hence (6.4.33) and (6.4.27) will indeed give (6.4.32), if we recall the notation in (6.4.29).

Now obviously

$$\mathcal{I}_\ell \cap \mathcal{S}_\ell \subset \mathcal{I} \cap (\mathcal{V}_\ell \oplus \mathcal{S}) \subset \mathcal{V}_\ell$$

the last inclusion being implied by (6.4.23).

The relations (6.1.21), (6.1.22) are direct consequences of (6.4.22) □

In view of Theorem 5, we indeed obtain *quotient algebras of generalized functions*

(6.4.34) $$A^\ell = \mathcal{A}_\ell / \mathcal{I}_\ell \in AL_{\mathcal{C}^\infty, \Lambda}, \quad \ell \in \mathbb{N}$$

In the rest of this Section, we shall present some of the more important properties of the algebras (6.4.34) needed in order to establish (6.4.7), (6.4.8) and (6.4.10)-(6.4.13). Further details can be found in Rosinger [2].

Theorem 6

Given $\mathcal{I}$ in (6.4.18) and $\mathcal{V}$, $\mathcal{S}$ in (6.4.22), (6.4.23), then, with the construction in (6.4.29), we have the following:

(1) $A^\ell = \mathcal{A}_\ell / \mathcal{I}_\ell$, with $\ell \in \mathbb{N}$, is an associative and commutative algebra with unit element.

(2) For $\ell \in \mathbb{N}$, the multiplication in A^ℓ, when restricted to $\mathcal{C}^\infty(\mathbb{R}^n)$, coincides with the usual multiplication of functions, that is, we have the *inclusion of algebras*

(6.4.35) $\quad \mathcal{C}^\infty(\mathbb{R}^n) \subset A^\ell, \quad \ell \in \mathbb{N}$

(3) For $h, k, \ell \in \mathbb{N}, \quad h \le k \le \ell$, the following diagram is commutative

(6.4.36)
$$\begin{array}{ccccc}
 & & \gamma_{\ell h}\ (A^\ell \to A^h) & & \\
A^\ell & \xrightarrow{\ \gamma_{\ell k}\ } & A^k & \xrightarrow{\ \gamma_{kh}\ } & A^h \\
\uparrow \epsilon_\ell & & \uparrow \epsilon_k & & \uparrow \epsilon_h \\
\mathcal{D}'(\mathbb{R}^n) & \xleftrightarrow[\ \text{id}\ ]{} & \mathcal{D}'(\mathbb{R}^n) & \xleftrightarrow[\ \text{id}\ ]{} & \mathcal{D}'(\mathbb{R}^n)
\end{array}$$

where $\gamma_{\ell k}, \gamma_{kh}, \gamma_{\ell h}$ are *algebra homomorphisms*, defined as follows

(6.4.37) $\quad \gamma_{\ell k}(s + \mathcal{I}_\ell) = s + \mathcal{I}_k, \quad s \in \mathcal{A}_\ell$

and similarly for $\gamma_{kh}, \gamma_{\ell h}$, while $\epsilon_\ell, \epsilon_k, \epsilon_h$ are *linear injective*, defined by (6.1.25).

Proof

It is useful to note that (6.4.25) yields

(6.4.38) $\quad \mathcal{V}_\ell \subset \mathcal{V}_k, \quad \ell, k \in \mathbb{N}, \quad k \le \ell$

hence in view of (6.4.29) we obtasin

(6.4.39) $\quad \mathcal{S}_\ell \subset \mathcal{S}_k, \quad \ell, k \in \mathbb{N}, \quad k \le \ell$

Now (6.4.26)-(6.4.29) yield

(6.4.40) $\quad \mathcal{A}_\ell \subset \mathcal{A}_k, \quad \mathcal{I}_\ell \subset \mathcal{I}_k, \quad \ell, k \in \mathbb{N}, \quad k \le \ell$

In this way (3) is immediate.

(1) follows easily

(2) follows from (6.4.23) □

In view of Theorem 6 above, the chain of algebras (6.4.7) obtains a detailed validation.

It is easy to see that the *algebra homomorphis*

(6.4.41) $\qquad \gamma_{\ell k} : A^{\ell} \longrightarrow A^{k}, \quad \ell,k \in \mathbb{N}, \quad k \leq \ell$

is *injective*, if and only if

(6.4.42) $\qquad \mathcal{I}_k \cap \mathcal{A}_\ell \subset \mathcal{I}_\ell$

The basic result, presented next, concerns the way partial derivatives and therefore, linear or polynomial nonlinear partial differential operators can be defined on the quotient algebras of generalized functions A^{ℓ}, with $\ell \in \mathbb{N}$.

Theorem 7

With the assumptions in Theorem 6, the following hold:

(1) $A^{\ell} \overset{m}{\leq} A^{k}, \quad \ell,k,m \in \mathbb{N}, \quad k+m \leq \ell$ (see (1.9.11))

(2) The *partial derivatives* are the *linear* mappings

(6.4.43) $\qquad D^p : A^{\ell} \longrightarrow A^{k}, \quad \ell,k \in \mathbb{N}, \ p \in \mathbb{N}^n, \ k + |p| \leq \ell$

defined by

(6.4.44) $\qquad D^p(s+\mathcal{I}_\ell) = D^p s + \mathcal{I}_k, \quad s \in \mathcal{A}_\ell$

and when restricted to $\mathcal{C}^{\infty}(\mathbb{R}^n)$, they coincide with the usual partial derivatives of functions.

(3) The partial derivaties (6.4.43) satisfy the Leibnitz rule of product derivative, that is

(6.4.45) $\qquad D^p(S\cdot T) = \sum_{\substack{q \in \mathbb{N}^n \\ q \leq p}} \binom{p}{q} (D^q S)\cdot(D^{p-q}T), \quad S,T \in A^{\ell}$

where

$$D^p, D^q, D^{p-q} : A^{\ell} \longrightarrow A^{k}$$

and $\ell,k \in \mathbb{N}, \quad p \in \mathbb{N}^n, \quad k + |p| \leq \ell.$

(4) For $\ell,\ell',k,k' \in \mathbb{N}, \quad \ell' \leq \ell, \quad k' \leq k, \quad p \in \mathbb{N}^n, \quad k + |p| \leq \ell,$ $k' + |p| \leq \ell'$, the following diagram is commutative

(6.4.46)
$$\begin{array}{ccc} A^{\ell} & \xrightarrow{\;D^p\;} & A^{k} \\ \gamma_{\ell\ell'}\downarrow & & \downarrow\gamma_{kk'} \\ A^{\ell'} & \xrightarrow[\;D^p\;]{} & A^{k'} \end{array}$$

Proof

In view (6.4.25), it is easy to see that we have

(6.4.47) $$D^p \mathcal{V}_\ell \subset \mathcal{V}_k, \quad \ell, k \in \mathbb{N}, \quad p \in \mathbb{N}^n, \quad k + |p| \leq \ell$$

But (6.4.26) and (6.4.38) yield

$$D^p \mathcal{A}_\ell(\mathcal{V},\mathcal{S}) \subset \mathcal{A}_\ell(\mathcal{V},\mathcal{S}) \subset \mathcal{A}_k(\mathcal{V},\mathcal{S}), \quad \ell, k \in \mathbb{N}, \quad k \leq \ell, \quad p \in \mathbb{N}^n$$

hence, in view of (6.4.29), we obtain

(6.4.48) $$D^p \mathcal{A}_\ell \subset \mathcal{A}_k, \quad \ell, k \in \mathbb{N}, \quad k \leq \ell, \quad p \in \mathbb{N}^n$$

therefore (6.4.47) and (6.4.48) yield

(6.4.49) $$D^p \mathcal{I}_\ell \subset \mathcal{I}_k, \quad \ell, k \in \mathbb{N}, \ p \in \mathbb{N}^n, \ k + |p| \leq \ell$$

Now (1) follows from (6.4.48), (6.4.49) and (1.9.11).

(2)

In view of (6.4.48), (6.4.49), it is obvious that (6.4.43), (6.4.44) is a correct definition. The inclusion

$$\mathcal{U}_{\mathcal{C}^\infty,\Lambda} \subset \mathcal{S}$$

in (6.4.23) and (6.4.29) complete the proof of (2).

(3) and (4) follow from (6.4.44) by direct verification □

The result in pct. (2) in Theorem 7 above can be improved in suitable circumstances as specified next.

Suppose given

(6.4.50) $\mathcal{A}$ a derivative invariant subalgebra in $(\mathcal{C}^\infty(\mathbb{R}^n))^\Lambda$ which satisfies (6.2.2), (6.2.3), and suppose that

(6.4.51) $\mathcal{I}$ is a derivative invariant, $\mathcal{C}^\infty$-smooth regular ideal in $\mathcal{A}$

Then in view of Theorem 1 in Section 2, we can assume given

(6.4.52) $(\mathcal{V}, \mathcal{V} \oplus \mathcal{S}, \mathcal{I}, \mathcal{A})$ a $\mathcal{C}^\infty$-smooth regularization

of (6.4.20), such that (6.4.23) holds.

It is easy to see that the constructions in (6.4.25)-(6.4.29) remain valid and we shall have

(6.4.53) $\mathcal{A}_\ell(\mathcal{V},\mathcal{S}) \subset \mathcal{A}, \quad \ell \in \mathbb{N}$

Furthermore, Theorems 5, 6 and 7 remain also valid.

Let us now assume that

(6.4.54) $\mathcal{V}$ is derivative invariant

Then obviously

(6.4.55) $\mathcal{V}_\ell = \mathcal{V}, \mathcal{S}_\ell = \mathcal{V} \oplus \mathcal{S}, \quad \ell \in \mathbb{N}$

hence

(6.4.56) $\mathcal{A}_\ell(\mathcal{V},\mathcal{S}) = \mathcal{A}_\infty(\mathcal{V},\mathcal{S}) \subset \mathcal{A}, \quad \ell \in \mathbb{N}$

which means that

(6.4.57) $\mathcal{I}_\ell(\mathcal{V},\mathcal{S}) = \mathcal{I}_\infty(\mathcal{V},\mathcal{S}) \subset \mathcal{I}, \quad \ell \in \mathbb{N}$

It follows from (6.4.29) that

(6.4.58) $(\mathcal{V}_\ell, \mathcal{S}_\ell, \mathcal{I}_\ell, \mathcal{A}_\ell) = (\mathcal{V}, \mathcal{V} \oplus \mathcal{S}, \mathcal{I}_\infty, \mathcal{A}_\infty), \quad \ell \in \mathbb{N}$

In this way the*chain of algebras* in (6.4.7) will *collapse* into the one single *differential algebra* in the chain, which is A^∞. Indeed, (6.4.34) yields

(6.4.59) $A^\ell = A^\infty \in AL_{\mathcal{C}^\infty, \Lambda}, \quad \ell \in \mathbb{N}$

Then (6.4.36) becomes the *linear injective* mapping

(6.4.60) $\mathcal{D}'(\mathbb{R}^n) \xrightarrow{\epsilon_\infty} A^\infty$

and (6.4.43), (6.4.44) and (6.4.46) become

(6.4.61) $D^p : A^\infty \to A^\infty, \quad p \in \mathbb{N}^n$

with

(6.4.62) $D^p(s+\mathcal{I}_\infty) = D^p_s + \mathcal{I}_\infty, \quad s \in \mathcal{A}_\infty$

Finally, in addition to (6.4.14)-(6.4.17), let us assume that the representation of distributions in (6.4.20), that is

(6.4.63) $\mathcal{D}'(\mathbb{R}^n) = \mathcal{S}^\infty_\Lambda/\mathcal{V}^\infty_\Lambda$

is such that, given any distribution $T \in \mathcal{D}'(\mathbb{R}^n)$ and its representation

(6.4.64) $T = s + \mathcal{V}^\infty_\Lambda, \quad s \in \mathcal{S}^\infty_\Lambda$

then the distributional partial derivatives $D^pT \in \mathcal{D}'(\mathbb{R}^n)$, with $p \in \mathbb{N}^n$, will have the representation

(6.4.65) $D^pT = D^ps + \mathcal{V}^\infty_\Lambda, \quad s \in \mathcal{S}^\infty_\Lambda, \quad p \in \mathbb{N}^n$

We note that in view of (6.4.17), the relation (6.4.65) is well defined.

Theorem 8

If in addition to (6.4.50)-(6.4.52), (6.4.54) and (6.4.63)-(6.4.65) we also have

(6.4.66) $\mathcal{V} \oplus \mathcal{S}$ is derivative invariant

then the *partial derivatives*

(6.4.67) $D^p : A^\infty \to A^\infty, \quad p \in \mathbb{N}^n$

defined in (6.4.62), coincide with the distributional partial derivatives, when restricted to $\mathcal{D}'(\mathbb{R}^n)$, according to the embedding (6.4.60)

Proof

We recall that in view of (6.1.25), ϵ_∞ in (6.4.60) is defined by

(6.4.68)
$$\begin{array}{ccccc} \mathcal{D}'(\mathbb{R}^n) = \mathcal{S}^\infty_\Lambda/\mathcal{V}^\infty_\Lambda & \longleftarrow & (\mathcal{V}\oplus\mathcal{S})/\mathcal{V} & \longrightarrow & A^\infty = \mathcal{A}_\infty/\mathcal{I}_\infty \\ s + \mathcal{V}^\infty_\Lambda & \xleftarrow[\text{isom}]{} & s + \mathcal{V} & \xmapsto[\text{lin,inj}]{} & s + \mathcal{I}_\infty \end{array}$$

that is

(6.4.69) $$\epsilon_\infty(s+\mathcal{V}^\infty_\Lambda) = s + \mathcal{I}_\infty, \quad s \in \mathcal{V} \oplus \mathcal{S}$$

Let us take now $T \in \mathcal{D}'(\mathbb{R}^n)$, then in view of (6.4.68), we obtain

(6.4.70) $$T = s + \mathcal{V}^\infty_\Lambda, \quad s \in \mathcal{V} \oplus \mathcal{S}$$

Thus in view of (6.4.65), we obtain

(6.4.71) $$D^pT = D^ps + \mathcal{V}^\infty_\Lambda, \quad s \in \mathcal{V} \oplus \mathcal{S}, \quad p \in \mathbb{N}^n$$

since according to (6.4.58)

$$\mathcal{V} \oplus \mathcal{S} = \mathcal{S}_\ell \subset \mathcal{S}^\infty_\Lambda, \; \ell \in \mathbb{N}$$

But (6.4.66) applied to (6.4.70) yields

(6.4.72) $$D^ps \in \mathcal{V} \oplus \mathcal{S}, \quad p \in \mathbb{N}^n$$

Finally, (6.4.69) applied to (6.4.70)-(6.4.72) gives in view of (6.4.42) the relation

(6.4.73) $$\epsilon_\infty(D^pT) = D^p(\epsilon_\infty(T)), \quad T \in \mathcal{D}'(\mathbb{R}^n), \quad p \in \mathbb{N}^n$$

which completes the proof □

Example 1

If $\operatorname{car} \Lambda = \operatorname{car} \mathbb{N}$, then with the notation in (6.6.1), we can take $\mathcal{V}^\ell_\Lambda = \mathcal{V}^\ell$ and $\mathcal{S}^\ell_\Lambda = \mathcal{S}^\ell$, with $\ell \in \mathbb{N}$, and the conditions (6.1.2)-(6.1.6) will obviously be satisfied.

If car $\Lambda >$ car $\mathbb{N}$, we take

(6.4.74) $\sigma : \mathbb{N} \longrightarrow \Lambda$ injective

and define the *injective algebra homomorphism*

(6.4.75) $\sigma : (\mathcal{C}^0(\mathbb{R}^n))^{\mathbb{N}} \longrightarrow (\mathcal{C}^0(\mathbb{R}^n))^{\Lambda}$

by

(6.4.76) $(\sigma s)_\lambda = \begin{vmatrix} s_n & \text{if } \lambda = \sigma(n) \text{ for some } n \in \mathbb{N} \\ 0 & \text{if } \lambda \in \Lambda \backslash \sigma(\mathbb{N}) \end{vmatrix}$

then obviously

(6.4.77) $(\sigma \mathcal{S}^\ell) \cap \mathcal{U}_{\mathcal{C}^\ell,\Lambda} = \mathcal{O}, \quad \ell \in \mathbb{N}$

Now we define

(6.4.78) $\mathcal{S}^\ell_\Lambda = (\sigma \mathcal{S}^\ell) \oplus \mathcal{U}_{\mathcal{C}^\ell,\Lambda}, \quad \ell \in \mathbb{N}$

as well as the *linear surjection*

(6.4.79) $\mathcal{S}^\ell \ni s \longmapsto T \in \mathcal{D}'(\mathbb{R}^n)$

where, for

(6.4.80) $s = \sigma s^\ell + u(\psi), \quad s^\ell \in \mathcal{S}^\ell, \quad \psi \in \mathcal{C}^\ell(\mathbb{R}^n)$

we have

(6.4.81) $T = T^\ell + T_\psi$

with T_ψ given in (5.4.2) and T^ℓ being the weak limit of s^ℓ in $\mathcal{D}'(\mathbb{R}^n)$. In view of (6.4.78), the definition (6.4.79) is coorrect.

By taking $\mathcal{V}^\ell$ as the kernel of the mapping (6.4.79), it is obvious that $\mathcal{V}^\ell_\Lambda$ and $\mathcal{S}^\ell_\Lambda$ satisfy (6.1.2)-(6.1.6).

In view of (6.4.78), it is easy to see that

(6.4.82) $D^p \mathcal{S}^\ell_\Lambda \subset \mathcal{S}^k_\Lambda, \quad \ell, k \in \mathbb{N}, \quad p \in \mathbb{N}^n, \quad k + |p| \leq \ell$

in particular

$$(6.4.83) \qquad D^p \mathcal{S}_\Lambda^\infty \subset \mathcal{S}_\Lambda^\infty, \quad p \in \mathbb{N}^n$$

Similarly we obtain

$$(6.4.84) \qquad D^p \mathcal{V}_\Lambda^\ell \subset \mathcal{V}_\Lambda^k, \quad \ell, k \in \mathbb{N}, \quad p \in \mathbb{N}^n, \quad k + |p| \leq \ell$$

$$(6.4.85) \qquad D^p \mathcal{V}_\Lambda^\infty \subset \mathcal{V}_\Lambda^\infty, \quad p \in \mathbb{N}^n$$

When considering regularizations of

$$(6.4.86) \qquad \mathcal{D}'(\mathbb{R}^n) = \mathcal{S}_\Lambda^\ell / \mathcal{V}_\Lambda^\ell$$

with arbitrary $\ell \in \mathbb{N}$, the following construction may be useful.

<u>Example 2</u>

Suppose car $\Lambda >$ car $\mathbb{N}$ and we are given

$$(6.4.87) \qquad \mathcal{V}_\Lambda^\infty \subset \mathcal{S}_\Lambda^\infty \subset (\mathcal{C}^\infty(\mathbb{R}^n))^\Lambda$$

satisfying (6.1.2)-(6.1.6), that is, we have the commutative diagram

$$(6.4.88) \qquad \begin{array}{ccc} \mathcal{S}_\Lambda^\infty \ni s & \xrightarrow{\eta \text{ lin,sur}} & T \in \mathcal{D}'(\mathbb{R}^n) \\ \subset \uparrow & & \uparrow \text{id} \\ \mathcal{U}_{\mathcal{C}^\infty,\Lambda} \ni u(\psi) & \longmapsto & T_\psi \in \mathcal{D}'(\mathbb{R}^n) \end{array}$$

and

$$(6.4.89) \qquad \mathcal{V}_\Lambda^\infty = \ker \eta$$

In view of Example 1, we can also assume that $\mathcal{V}_\Lambda^\infty$, $\mathcal{S}_\Lambda^\infty$ satisfy the relations (6.4.83), (6.4.85)

Let us take $\lambda_0 \in \Lambda$ and

$$(6.4.90) \qquad \pi : \Lambda \longrightarrow \Lambda \setminus \{\lambda_0\}$$

and define the *injective algebra homomorphism*

(6.4.91) $$\pi : (\mathcal{C}^0(\mathbb{R}^n))^\Lambda \to (\mathcal{C}^0(\mathbb{R}^n))^\Lambda$$

by

(6.4.92) $$(\pi s)_\lambda = \begin{vmatrix} s_\mu & \text{if} \quad \lambda = \pi(\mu) \quad \text{for some} \quad \mu \in \Lambda \\ 0 & \text{if} \quad \lambda = \lambda_0 \end{vmatrix}$$

Then obviously

(6.4.93) $$(\pi \mathcal{S}_\Lambda^\infty) \cap \mathcal{U}_{\mathcal{C}^0,\Lambda} = \mathcal{O}$$

We can now take

(6.4.94) $$\mathcal{S}_\Lambda^0 = (\pi \mathcal{S}_\Lambda^\infty) \oplus \mathcal{U}_{\mathcal{C}^0,\Lambda}$$

and define the *linear surjection*

(6.4.95) $$\mathcal{S}_\Lambda^0 \ni s \overset{\theta}{\longmapsto} T \in \mathcal{D}'(\mathbb{R}^n)$$

in the following way: if

(6.4.96) $$s = \pi s^\infty + u(\psi), \quad s^\infty \in \mathcal{S}_\Lambda^\infty, \quad \psi \in \mathcal{C}^0(\mathbb{R}^n)$$

then

(6.4.97) $$\theta(s) = \eta(s^\infty) + T_\psi$$

Finally, we take

(6.4.98) $$\mathcal{V}_\Lambda^0 = \ker \theta$$

Obviously, $\mathcal{V}_\Lambda^0$ and $\mathcal{S}_\Lambda^0$ will satisfy (6.1.2)-(6.1.6)

§5. NONLINEAR PARTIAL DIFFERENTIAL OPERATORS IN CHAINS OF ALGEBRAS

Suppose given the polynomial nonlinear partial differntial operator

$$(6.5.1) \qquad T(D) = \sum_{1\leq i\leq h} c_i(x) \prod_{1\leq j\leq k_i} D^{p_{ij}}, \quad x \in \mathbb{R}^n$$

where $h, k_i \in \mathbb{N}$, $p_{ij} \in \mathbb{N}^n$ and $c_i \in \mathcal{C}^\infty(\mathbb{R}^n)$. We recall that the *order* of $T(D)$ is by definition

$$(6.5.2) \qquad m = \max\{|p_{ij}| \,|\, 1 \leq i \leq h, \ 1 \leq j \leq k_i\}$$

provided that none of the c_i, with $1 \leq i \leq h$, vanishes identically on $\mathbb{R}^n$.

If we assume the conditions in Theorem 6, Section 4, then it follows easily that $T(D)$ will operate according to

$$(6.5.3) \qquad T(D) : A^\ell \longrightarrow A^k, \quad \ell, k \in \mathbb{N}, \quad k+m \leq \ell$$

see pct. (1) and (2) in Theorem 6, Section 4, as well as Section 9, Chapter 1.

In the next Chapter, we shall be interested in the following type of situation:
We are given the polynomial nonlinear partial differential equation

$$(6.5.4) \qquad T(D)U(x) = f(x), \quad x \in \mathbb{R}^n$$

where $f \in \mathcal{C}^\infty(\mathbb{R}^n)$. We can construct a sequence of smooth functions

$$(6.5.5) \qquad s = (\psi_\lambda | \lambda \in \Lambda) \in (\mathcal{C}^\infty(\mathbb{R}^n))^\Lambda$$

such that, see (6.4.29)

$$(6.5.6) \qquad s \in \mathcal{A}_\ell, \quad \ell \in \mathbb{N}$$

hence

$$(6.5.7) \qquad U_\ell = s + \mathcal{I}_\ell \in A^\ell = \mathcal{A}_\ell/\mathcal{I}_\ell, \quad \ell \in \mathbb{N}$$

Moreover, in the sense of (6.5.3), we have

$$(6.5.8)\qquad T(D)U_\ell = f \in A^k, \quad \ell,k \in \mathbb{N}, \quad k+m \le \ell$$

In such a case, the sequence of smooth functions s in (6.5.5) will be called a *chain generalized solution* of (6.5.4).

In the particular case when $\Lambda = \mathbb{N}$ and

$$(6.5.9)\qquad s = (\psi_\nu | \nu \in \mathbb{N}) \in \mathcal{S}^\infty$$

s will be called a *chain weak solution* of (6.5.4).

This latter case will be the one encountered in the next Chapter.

In view of Section 13, Chapter 1, it is easy to see that the chains of algebras (6.4.7) can be adapted in such a way that (6.5.3) still holds for *general* nonlinear partial differential operators. Similar adaptations can be made in order to accommodate *systems* of nonlinear partial differential equations. Details can be found in Rosinger [2].

§6. LIMITATIONS ON THE EMBEDDING OF SMOOTH FUNCTIONS INTO CHAINS OF ALGEBRAS

We have seen in (6.4.6) and pct. (2) in Theorems 6 and 7 in Section 4, that within the *chains of algebras* containing the distributions, see (6.4.7)

$$(6.6.1)\qquad \mathcal{D}'(\mathbb{R}^n) \subset A^\infty \to \dots \to A^\ell \to \dots \to A^0, \quad \ell \in \mathbb{N}$$

we can have *inclusions*

$$(6.6.2)\qquad \mathcal{C}^\infty(\mathbb{R}^n) \subset A^\ell, \quad \ell \in \mathbb{N}$$

which preserve *both* the *algebra* and *differential* structure of $\mathcal{C}^\infty(\mathbb{R}^n)$. In other words, the multiplication and partial derivatives in A^ℓ, when restricted to $\mathcal{C}^\infty(\mathbb{R}^n)$, will coincide with the usual multiplication and partial derivatives of $\mathcal{C}^\infty$-smooth functions. For short, we shall say that the inclusions (6.6.2) *preserve* the algebra and differential structures, or that the spaces involved in the inclusions have *compatible* algebra and differential structures.

There exists a natural interest in *extending* the left hand term in the inclusions (6.6.2) in such a way that the algebra and/or differential structures involved will still be compatible. Therefore, we are led to the problem of finding the *maximal* extensions of such kind.

First we note that we may have to consider separtely the algebra and differential structures involved.

Indeed, concerning *differentiation*, the natural candidate is the space of the Schwartz distributions which is closed under the differentiation of distributions and is contained in the algebras of the chains (6.6.1), that is, we have

$$(6.6.3) \qquad \mathcal{D}'(\mathbb{R}^n) \subset A^\ell, \quad \ell \in \mathbb{N}$$

However, as $\mathcal{D}'(\mathbb{R}^n)$ does not have a natural algebra structure, the inclusion (6.6.3) can only be considered from the point of view of the preservation of the differential structure. A positive answer is obtained in (6.4.67), and in particular, in the case of the inclusion, see Chapter 8

$$(6.6.4) \qquad \mathcal{D}'(\mathbb{R}^n) \subset \mathcal{G}(\mathbb{R}^n)$$

However this happens when the chain of algebras (6.6.1) is *collapsed* into one single differential algebra, see (6.4.59) and Chapter 8. Therefore, the question arises whether inclusions (6.6.3) with the preservation of the differential structures can be obtained without collapsing the chain of algebras (6.6.1). This question is so far open.

Concerning the *algebra structure*, there are two natural candidates

$$(6.6.5) \qquad \mathcal{C}^0(\mathbb{R}^n) \subset A^\ell, \quad \ell \in \mathbb{N}$$

or

$$(6.6.6) \qquad \mathcal{L}^\infty_{\ell oc}(\mathbb{R}^n) \subset A^\ell, \quad \ell \in \mathbb{N}$$

since we obviously have the inclusions

$$(6.6.7) \qquad \mathcal{C}^0(\mathbb{R}^n) \subset \mathcal{L}^\infty_{\ell oc}(\mathbb{R}^n) \subset \mathcal{L}^1_{\ell oc}(\mathbb{R}^n) \subset \mathcal{D}'(\mathbb{R}^n)$$

and $\mathcal{L}^1_{\ell oc}(\mathbb{R}^n)$ is not an algebra. However, neither $\mathcal{C}^0(\mathbb{R}^n)$ nor $\mathcal{L}^\infty_{\ell oc}(\mathbb{R}^n)$ have a natural differentiation. Nevertheless, $\mathcal{C}^0(\mathbb{R}^n)$ is connected to the usual differentiation of smooth functions through

$$(6.6.8) \qquad D^p : \mathcal{C}^\ell(\mathbb{R}^n) \rightarrow \mathcal{C}^k(\mathbb{R}^n), \quad \ell, k \in \mathbb{N}, \quad p \in \mathbb{N}^n, \ k+|p| \leq \ell$$

that is, $\mathcal{C}^0(\mathbb{R}^n)$ is the range space of the usual partial derivatives of smooth functions.

In view of the above, we shall concentrate on inclusions

$$(6.6.9) \qquad \mathcal{C}^k(\mathbb{R}^n) \subset A^\ell, \quad \ell \in \mathbb{N}, \quad k \in \mathbb{N}$$

and try to find out to what extent they can preserve both the algebra and differential structures involved.

The *main answer* will be that it is *not possible* to obtain inclusions

$$(6.6.10) \qquad \mathcal{C}^{\ell-4}(\mathbb{R}^n) \subset A^{\ell}, \quad \ell \in \mathbb{N}, \quad \ell \geq 4$$

which would preserve *both* the algebra and differential structure. As seen below, this limitation is of such a simple and general nature that its validity goes beyond the framework of the chains of algebras (6.6.1). In other words, this limitation is not specific to these chains, as it does not arise from the way such chains are constructed in Section 4.

The result in (6.6.10) follows from an extension of Proposition 2 in Section 4, Chapter 1, an extension which goes along similar lines with the extension of the Schwartz impossiblity result presented in Rosinger [3, p. 31].

Indeed, suppose given the algebras and algebra homomorphisms

$$(6.6.11) \qquad A^4 \longrightarrow A^3 \longrightarrow A^2 \longrightarrow A^1 \longrightarrow A^0$$

together with the derivatives, see (1.1.15)

$$(6.6.12) \qquad A^4 \xrightarrow{D} A^3 \xrightarrow{D} A^2 \xrightarrow{D} A^1 \xrightarrow{D} A^0$$

such that

$$(6.6.13) \qquad \begin{array}{l} A^3,\ A^2,\ A^1,\ A^0 \ \text{ are commutative} \\ \text{and } \ A^1,\ A^0 \ \text{ are associative} \end{array}$$

and the following diagrams are commutative

$$(6.6.14) \qquad \begin{array}{ccc} A^{i+1} & \longrightarrow & A^{i} \\ D \big\downarrow & & \big\downarrow D, \quad 1 \leq i \leq 3 \\ A^{i} & \xrightarrow[D^p]{} & A^{i-1} \end{array}$$

Suppose that

$$(6.6.15) \qquad x, x_+, x_- \in \bigcap_{1 \leq i \leq 4} A^i$$

$$(6.6.16) \qquad 1 \in \bigcap_{1 \leq i \leq 3} A^i$$

and 1 is the unit element in A^i, with $1 \leq i \leq 3$, while $1,\ x,\ x_+,\ x_-$ are invariants of the respective algebra homomorphisms in (6.6.11).

Further, suppose that

(6.6.17) $x_+ + x_- = x$ in A^i, with $2 \leq i \leq 3$

(6.6.18) $x \cdot x_+ = (x_+)^2$, $x \cdot x_- = (x_-)^2$ in A^4

(6.6.19) $Dx = 1$ in A^i, with $1 \leq i \leq 3$

(6.6.20) $D(x_+)^2 = 2 \cdot x_+$, $D(x_-)^2 = 2 \cdot x_-$ in A^3

We note that

(6.6.21) $1, x \in \mathcal{C}^\infty(\mathbb{R}), x_+, x_- \in \mathcal{C}^0(\mathbb{R}), (x_+)^2, (x_-)^2 \in \mathcal{C}^1(\mathbb{R})$

and the algebraic operations in (6.6.17), (6.6.18) coincide with those in $\mathcal{C}^0(\mathbb{R})$, while the derivatives in (6.6.19), (6.6.20) coincide with those in $\mathcal{C}^1(\mathbb{R})$.

We can also recall that

(6.6.22) $Dx_+ = H$, $D^2 x_+ = \delta$

in the sense of the distributional derivaties in $\mathcal{D}'(\mathbb{R})$.

<u>Proposition 3</u>

Within the conditions (6.6.11)-(6.6.20), it follows that

(6.6.23) $D^2(x_+) = D^2(x_-) = 0$ in A^0

<u>Proof</u>

For convenience, let us again denote

(6.6.24) $a = x_+$, $b = x_-$

thus (6.6.18) yields

(6.6.25) $x \cdot a = a^2$, $x \cdot b = b^2$ in A^4

which by derivation gives

$$a + x \cdot Da = 2 \cdot a, \quad b + x \cdot Db = 2 \cdot b \quad \text{in } A^3$$

if we take into account (6.6.16), (6.6.19) and (6.6.20), hence

(6.6.26) $x \cdot Da = a, \quad x \cdot Db = b \quad \text{in} \quad A^3$

In view of (6.6.19), (6.6.16), a derivation of (6.6.26) gives

$$Da + x \cdot D^2 a = Da, \quad Db + x \cdot D^2 b = Db \quad \text{in} \quad A^2$$

hence

(6.6.27) $x \cdot D^2 a = x \cdot D^2 b = 0 \quad \text{in} \quad A^2$

Now the derivation of (6.6.25) also yields

$$a + x \cdot Da = 2 \cdot a \cdot Da, \quad b + x \cdot Db = 2 \cdot b \cdot Db \quad \text{in} \quad A^3$$

owing to (6.6.13), (6.6.16) and (6.6.19). Hence, in view of (6.6.26), we obtain

(6.6.28) $a \cdot Da = a, \quad b \cdot Db = b \quad \text{in} \quad A^3$

In view of (6.6.17), (6.6.26) and (6.6.28), we have

$$a \cdot Db = (x - b) \cdot Db = x \cdot Db = b \cdot Db = 0 \quad \text{in} \quad A^3$$

$$b \cdot Da = (x - a) \cdot Da = x \cdot Da - a \cdot Da = 0 \quad \text{in} \quad A^3$$

that is

(6.6.29) $a \cdot Db = b \cdot Da = 0 \quad \text{in} \quad A^3$

which by derivation and in view of (6.6.13), (6.6.16) yields

(6.6.30) $Da \cdot Db + a \cdot D^2 b = Da \cdot Db + b \cdot D^2 a = 0 \quad \text{in} \quad A^2$

In view of (6.6.17) and (6.6.27) we have

(6.6.31) $a \cdot D^2 b = -b \cdot D^2 b, \quad b \cdot D^2 a = -a \cdot D^2 a \quad \text{in} \quad A^2$

Then (6.6.17) and (6.6.19) give

$$Da + Db = 1 \quad \text{in} \quad A^2$$

hence (6.6.16) yields

$$Da \cdot Db = Da \cdot (1 - Da) = Da - (Da)^2$$

$$= (1 - Db) \cdot Db = Db - (Db)^2 \quad \text{in} \quad A^2$$

which together with (6.6.30), (6.6.31) give

$$(6.6.32)\qquad \begin{aligned} &Da \cdot Db = -a \cdot D^2 b = -b \cdot D^2 a = a \cdot Da^2 = \\ &= b \cdot D^2 b = Da - (Da)^2 = Db - (Db)^2 = c \in A^2 \end{aligned}$$

We show that

$$(6.6.33)\qquad A^2 \ni c \longmapsto 0 \in A^1$$

with the algebra homomorphism $A^2 \rightarrow A^1$ in (6.6.11). Indeed, the derivation of (6.6.29) gives through (6.6.13) the relations

$$Da \cdot Db + a \cdot D^2 b = Da \cdot Db + b \cdot D^2 a = 0 \quad \text{in} \quad A^2$$

and then, by one more derivation, we obtain

$$(6.6.34)\qquad \begin{aligned} &D^2 a \cdot Db + Da \cdot D^2 b + Da \cdot D^2 b + a \cdot D^3 b = \\ &= D^2 a \cdot Db + Da \cdot D^2 b + Db \cdot D^2 a + b \cdot D^3 a = 0 \quad \text{in} \quad A^1 \end{aligned}$$

Now we note that applying to (6.6.27) the algebra homomorphism $A^2 \rightarrow A^1$ in (6.6.11), and in view of (6.6.14)-(6.6.16), we obtain

$$(6.6.35)\qquad x \cdot D^2 a = x \cdot D^2 b = 0 \quad \text{in} \quad A^1$$

Multiplying (6.6.34) by x and taking into account (6.6.13) and (6.6.35), we obtain

$$(6.6.36)\qquad x \cdot a \cdot D^3 b = x \cdot b \cdot D^3 a = 0 \quad \text{in} \quad A^1$$

On the other hand, the derivation of (6.6.27) gives through (6.6.19) the relations

$$D^2 a + x \cdot D^3 a = 0 \quad \text{in} \quad A^1$$

$$D^2 b + x \cdot D^3 b = 0 \quad \text{in} \quad A^1$$

which multiplied by b and a respectively, give in view of (6.6.36) and (6.6.13) the relations

$$(6.6.37)\qquad a \cdot D^2 b = b \cdot D^2 a = 0 \quad \text{in} \quad A^1$$

Now again, if we apply to (6.6.32) the algebra homomorphism $A^2 \rightarrow A^1$ in (6.6.11) and take into account (6.6.14)-(6.6.16), then (6.6.37) implies

$$\text{(6.6.38)} \quad \begin{aligned} Da \cdot Db &= -a \cdot D^2 b = -b \cdot D^2 a = a \cdot Da^2 \\ &= b \cdot D^2 b = Da - (Da)^2 = Db - (Db)^2 = 0 \quad \text{in} \quad A^1 \end{aligned}$$

hence in particular (6.6.33).

From (6.6.38) it follows that

$$(Da)^2 = Da, \quad (Db)^2 = Db \quad \text{in} \quad A^1$$

hence (6.6.13) yields for $p \in \mathbb{N}$, $p \geq 1$

$$\text{(6.6.39)} \qquad (Da)^p = Da, \quad (Db)^p = Db \quad \text{in} \quad A^1$$

One more derivation applied to (6.6.39) gives through (6.6.13) the relations

$$p(Da)^{p-1} \cdot D^2 a = D^2 a, \quad p(Db)^{p-1} \cdot D^2 b = D^2 b \quad \text{in} \quad A^0$$

for $p \in \mathbb{N}$, $p \geq 2$, thus in view of (6.6.39) we obtain

$$p \cdot Da \cdot D^2 a = D^2 a, \quad p \cdot Db \cdot D^2 b = D^2 b \quad \text{in} \quad A^0$$

hence

$$\frac{1}{p} \cdot D^2 a = Da \cdot D^2 a, \quad \frac{1}{p} \cdot D^2 b = Db \cdot D^2 b \quad \text{in} \quad A^0$$

It follows that for $p,q \in \mathbb{N}$, $p,q \geq 2$, we have

$$\frac{1}{p} \cdot D^2 a = \frac{1}{q} \cdot D^2 a, \quad \frac{1}{p} \cdot D^2 b = \frac{1}{q} \cdot D^2 b \quad \text{in} \quad A^0$$

which completes the proof of (6.6.23) □

Now the result mentioned in (6.6.10) can be specified within the following framework, which is *more general* than that of the chains of algebras constructed in Section 4. For simplicity, we shall only formulate it in the one dimensional case.

Suppose given the commutative and associative algebras and algebra homomorphisms

$$\text{(6.6.40)} \qquad A^{\ell+1} \longrightarrow A^{\ell}, \quad \ell \in \mathbb{N}$$

and derivatives, see (1.1.15)

$$\text{(6.6.41)} \qquad A^{\ell+1} \xrightarrow{\ D\ } A^{\ell}, \quad \ell \in \mathbb{N}$$

with the following commutative diagrams

(6.6.42)
$$\begin{array}{ccc} A^{\ell+2} & \longrightarrow & A^{\ell+1} \\ {\scriptstyle D}\downarrow & & \downarrow{\scriptstyle D}, \quad \ell \in \mathbb{N} \\ A^{\ell+1} & \longrightarrow & A^{\ell} \end{array}$$

The algebras (6.6.40)-(6.6.42) are called a *differential chain of algebras.*

Obviously, the chains of algebras constructed in Section 4 are n-dimensional versions of differential chains of algebras, when considered for $\ell \in \mathbb{N}$ only.

Given $\ell \in \mathbb{N}$, $k \in \overline{\mathbb{N}}$, $\ell, k \geq 1$, we say that the inclusion

(6.6.43) $$\mathcal{C}^k(\mathbb{R}) \subset A^{\ell}$$

preserves the algebra and differential structures, if and only if for $h, p \in \mathbb{N}$, $p \leq h \leq \ell$, we have the commutative diagrams

(6.6.44)
$$\begin{array}{ccc} A^{\ell} & \longrightarrow & A^{\ell-h} \\ {\scriptstyle \subset}\downarrow & & \downarrow{\scriptstyle \subset} \\ \mathcal{C}^r(\mathbb{R}) & \longrightarrow & \mathcal{C}^{r-h}(\mathbb{R}) \end{array}$$

with $r = \max\{k, h\}$, where $A^{\ell} \xrightarrow{D^p} A^{\ell-h}$ is defined in the obvious way by (6.6.40)-(6.6.42), while $\mathcal{C}^r(\mathbb{R}) \subset A^{\ell}$ and $\mathcal{C}^{r-h}(\mathbb{R}) \subset A^{\ell-h}$ are algebra embeddings with the constant function 1 being the unit element in the respective algebras, finally, $\mathcal{C}^r(\mathbb{R}) \xrightarrow{D^p} \mathcal{C}^{r-h}(\mathbb{R})$ is the usual p-th order derivative of $\mathcal{C}^r$-smooth functions.

<u>Theorem 9</u>

Within differential chains of algebras (6.6.40)-(6.6.42), one *cannot* have for any $\ell \in \mathbb{N}$, $\ell \geq 4$, an inclusion

(6.6.45) $$\mathcal{C}^{\ell-1}(\mathbb{R}) \subset A^{\ell}$$

which preserves the algebra and differential structures, *unless*

(6.6.46) $$D^2(x_+) = 0 \quad \text{in} \quad A^0$$

Further, one *cannot* have for any $\ell \in \mathbb{N}$, $\ell \geq 2$, an inclusion

(6.6.47) $$\mathcal{C}^{\ell-2}(\mathbb{R}) \subset A^{\ell}$$

which preserves the algebra and differential structures, *unless* for $a \in A^0$ we have

$$(6.6.48) \qquad x \cdot a = 0 \in A^0 \Longrightarrow a = 0 \in A^0$$

Proof

It follows easily from Proposition 3 above and Theorem 2 in Rosinger [3, p. 31] □

Remark 3

In view of (6.6.22), which by its simplicity and important role would be desirable in a nonlinear theory of generalized functions, the relation (6.6.46) does *not* seem to be desirable. Indeed, if we have any vector space embedding

$$(6.6.49) \qquad \mathcal{D}'(\mathbb{R}) \subset A^0$$

then

$$(6.6.50) \qquad \delta \in \mathcal{D}'(\mathbb{R}), \quad \delta \neq 0 \in \mathcal{D}'(\mathbb{R})$$

will necessarily yield

$$(6.6.51) \qquad \delta \in A^0, \ \delta \neq 0 \in A^0$$

Therefore, (6.6.22) and (6.6.46) can only mean that there is a rather sharp *difference* between the distributional derivatives and the derivatives (6.6.41), (6.6.42) in the differential chain of algebras to which A^0 belongs.

This is precisely the meaning of the impossiblity in (6.6.10) or (6.6.45).

The implication in (6.6.48), which is an extension of the Schwartz impossibility result (1.2.11), seems to be of a lesser concern. Indeed, just as in (6.6.49)-(6.6.51), it *cannot* mean that

$$(6.6.52) \qquad \delta = 0 \in A^0$$

and it can only mean that in A^0, the singularity of δ at $x = 0$ is *higher* than that of $1/x$, see (6.4.13).

For further details see Sections 2 and 4 in Chapter 1, as well as Rosinger [2, pp. 88-104, 100-112].

CHAPTER 7

RESOLUTION OF SINGULARITIES OF WEAK SOLUTIONS FOR POLYNOMIAL NONLINEAR PARTIAL DIFFERENTIAL EQUATIONS

§1. INTRODUCTION

In this Chapter the general method developed in Chapter 2 will be applied to the *resolution of closed, nowhere dense singularities* of sequential solutions for polynomial nonlinear partial differential equations, within the framework of the *chains of algebras* of generalized functions constructed in Chapter 6, Section 4.

We shall deal with the important particular case of sequential solutions given by usual *weak solutions*. The interest in these kind of solutions comes from a wide range of applications, see for instance Sections 3-8 in the sequel.

It should be mentioned that the results in this Chapter can easily be extended to larger classes of sequential solutions, as well as to general nonlinear partial differential equations and systems. An illustration of that was given in Chapter 2, with the *global* version of the Cauchy-Kovalevskaia theorem. Another illustration which presents a strengthening of that global result can be seen in Section 9 in this Chapter. Futther details concerning possible extensions of the classes of weak solutions and nonlinear partial differential equations dealt with in this Chapter can be found in Rosinger [2, pp. 121- 162].

Stated simply, the *resolution of singularities* means that the weak solutions considered will satisfy the respective polynomial *nonlinear* partial differential equations in the *usual algebraic sense*, that is, with the *multiplication* and *partial derivatives* as they are defined within the *chains of algebras* of generalized functions in Chapter 6, Section 4. In particular, the *stability paradoxes* will be *avoided* in this way. One of the effects of the above is that for all practical purposes, the weak solutions considered will behave as *global* and *classical* solutions of the respective polynomial nonlinear partial differential equations. In this way, the results in this Chapter can be seen as originating a *Polynomial Nonlinear Operational Calculus* for the respective types of nonlinear partial differential equations

Concerning *weak solutions* of *nonconservative* nonlinear partial differential equations, see Sections 5-8, it was for the *first time* in the literaature that in Rosinger [2] a rigorous treatment was given which, among others, eliminated the possibility of *stability paradoxes*. This shows the interest in the mentioned Polynomial Nonlinear Operational Calculus.

Since we only consider weak solutions, their singularities will have to be concentrated on *closed nowhere dense* subsets with *zero Lebesque measure*. Yet, as seen with the mentioned global version of the Cauchy-Kovalevskaia theorem, the second condition above, that is, of zero Lebesque measure, is only of convenience and not of necessity. The fact however is that the weak solutions considered in this Chapter are sufficiently general in order to include as rather simple particular cases many of the known types of singularities of solutions of first and second order nonlinear partial differential equations in applications.

There is however no limitation on the order of polynomial nonlinear partial differential equations which can be dealt with.

Indeed, in Sections 5-8 in the sequel, *junctions conditions* are found for rather general classes of polynomial nonlinear partial differential equations and their weak solutions. These junction conditions are rather wide ranging *nonconservative* extensions of the classical Rankine-Hugoniot shock conditions and they determine the class of *resoluble* systems of polynomial nonlinear partial differential equations, which contains as particular casaes many of the equations of physics, such as those of fluid dynamics, general relativity and magnetohydrodynamics.

For simplicity, we shall deal with the case when the domain of the independent variables is $\mathbb{R}^n$. The general case of a domain given by an arbitrary open set $\Omega \subset \mathbb{R}^n$ does not involve additional complications. In view of that, we shall freely switch to the case of arbitrary domains, whenever needed.

Finally, since we only deal with weak solutions, we shall use the index set $\Lambda = \mathbb{N}$ when constructing various chains of algebras according to the general method in Chapter 6, Section 4.

The Chapter ends with a strengthened form of the *global* Cauchy-Kovalevskaia theorem in Chapter 2, which shows the existence of *global chain generalized solutions*. That *global existence* result is also a *first* in the literature, see Rosinger [3].

In this Chapter we shall present the main concepts and results only. For the proofs and further details one can consult Rosinger [2, pp. 121-162] and Rosinger [3, pp. 349-390].

§2. SIMPLE POLYNOMIAL NONLINEAR PDEs AND RESOLUTION OF SINGULARITIES

An m-th order polynomial nonlinear partial differential operator is called *simple*, if and only if it can be written in the form

$$T(D)U(x) = \sum_{1 \leq i \leq a} L_i(D)T_iU(x), \quad x \in \mathbb{R}^n \tag{7.2.1}$$

where $L_i(D)$ are m-th order linear partial differential operators with $\mathcal{C}^\infty$-smooth coefficients, while T_i are polynomials of the form

(7.2.2) $$T_i U(x) = \sum_{1 \leq j \leq b_i} c_{ij}(x)(U(x))^j, \quad x \in \mathbb{R}^n$$

with $c_{ij} \in C^\infty(\mathbb{R}^n)$.

In the present Section and the next two, we shall deal with nonlinear partial differential equations corresponding to (7.2.1), (7.2.2), that is, having the form

(7.2.3) $$T(D)U(x) = f(x), \quad x \in \mathbb{R}^n$$

where $f \in C^\infty(\mathbb{R}^n)$ is given.

The nonlinear hyperbolic conservation laws, as well as the nonlinear second order wave equations studied in Sections 3 and 4, are obviously of the above form (7.2.3).

In general, the following large class of *quasilinear* partial differential operators

(7.2.4) $$T(D)U(x) = \sum_{\substack{p \in \mathbb{N}^n \\ |p|=m}} c_p(x) D^p U(x) + T'(D)UI(x), \quad x \in \mathbb{R}^n$$

where $c_p \in C^\infty(\mathbb{R}^n)$ and $T'(D)$ is an (m-1)-th order simple polynomial nonlinear partial differential operator, are obviously of the above form (7.2.1), (7.2.2).

A function

$$U : \mathbb{R}^n \longrightarrow \mathbb{R}$$

is called a *piece wise* C^∞-*smooth weak solution* of the simple polynomial nonlinear partial differential eqauation in (7.2.3), if and only if the following *five* conditions are satisfied:

There exists a family G of C^∞-smooth mappings $\gamma : \mathbb{R}^n \longrightarrow \mathbb{R}^{g_\gamma}$, such that the set

(7.2.5) $$\Gamma = \{x \in \mathbb{R}^n \mid \exists \quad \gamma \in G : \gamma(x) = 0 \in \mathbb{R}^{g_\gamma}\}$$

is closed, has zero Lebesque measure - therefore it is nowhere dense - and

(7.2.6) $$U \in C^\infty(\mathbb{R}^n \setminus \Gamma)$$

Further

(7.2.7) $$U^b \in L^1_{loc}(\mathbb{R}^n)$$

where $b = \max\{b_i | 1 \le i \le a\}$, with notation in (7.2.2), and moreover, the following weak solution property holds:

(7.2.8) $$\int_{\mathbb{R}^n} \Big(\sum_{1 \le i \le h} T_i U(x) L_i^*(D)\psi(x) - f(x)\psi(x)\Big)dx = 0, \quad \psi \in D(\mathbb{R}^n)$$

where $L_i^*(D)$ is the formal adjoint of $L_i(D)$.

Finally, for each $\gamma \in G$, there exists a bounded and balanced neighbourhood B_γ of $0 \in \mathbb{R}^{g_\gamma}$, such that

(7.2.9) $$\{\gamma^{-1}(B_\gamma) | \gamma \in G\} \text{ is locally finite in } \mathbb{R}^n$$

Solutions of important classes of nonlinear hyperbolic conservation laws, as well as nonlinear second-order wave equations are known to be piece wise smooth weak solutions in the above sense, see Sections 3 and 4.

A first result is presented next on the *resolution of singularities of weak solutions* for nonlinear partial differential equations. The proof can be found in Rosinger [3, pp. 353-360].

<u>Theorem 1</u>

Suppose $U : \mathbb{R}^n \to \mathbb{R}$ is a *piece wise* C^∞-*smooth weak solution* of the m-th order simple polynomial nonlinear partial differential equation (7.2.3). Then it is possible to construct regularizations (6.4.22) and algebras (6.4.34) such that

(1) $U = s + I_\ell(V,S) \in A^\ell$, $\ell \in \mathbb{N}$, where $s \in S$ does *not* depend on ℓ

(2) U satisfies (7.2.3) in the usual algebraic sense, with multiplication in A^k and the partial dedrivative operators
$D^p : A^\ell \to A^k$, $p \in \mathbb{N}^n$, $|p| \le m$ with $\ell, k \in \mathbb{N}$, $k+m \le \ell$

In view of pct. (1), s is a *chain weak solution* of (7.2.3), see Section 5, Chapter 6.

Remark 1

Theorem 1 above remains valid for arbitrary open $\Omega \subset \mathbb{R}^n$.

§3. RESOLUTION OF SINGULARITIES OF NONLINEAR SHOCK WAVES

Suppose given the nonlinear hyperbolic conservation law

$$(7.3.1) \qquad U_t(t,x) + c(U(t,x))\cdot U_x(t,x) = 0, \quad t > 0, \quad x \in \mathbb{R}$$

with the initial condition

$$(7.3.2) \qquad U(0,x) = u(x), \quad x \in \mathbb{R}$$

We shall suppose that the function

$$(7.3.3) \qquad c : \mathbb{R} \longrightarrow \mathbb{R}$$

in (7.3.1) is an *arbitrary polynomial*. Then it is obvious that (7.3.1) is a *first-order simple polynomial nonlinear* partial differential equation on $\Omega = (0,\infty) \times \mathbb{R} \subset \mathbb{R}^2$. Indeed, the left hand term in (7.3.1) can be written in the form in (7.2.1), provided that we take $a = 2$,

$$L_1(D) = D_t, \ L_2(D) = D_x, \quad T_1 U = U \ \text{ and } \ T_2 U = b(U)$$

where

$$(7.3.4) \qquad b : \mathbb{R} \longrightarrow \mathbb{R}$$

is a primitive of the function in (7.3.3), and thus again a polynomial.

It is known that under rather general conditions, Schaeffer, Golubitsky & Schaeffer, for C^∞-smooth or piece wise smooth initial data u, the equation (7.3.1) has *shock wave* solutions $U : \Omega \longrightarrow \mathbb{R}$, with the following properties.

There exists a *finite* set G of C^∞-smooth functions $\gamma : \Omega \longrightarrow \mathbb{R}$, defining C^∞-smooth *curves*

$$(7.3.5) \qquad \Gamma_\gamma = \{x \in \Omega \mid \gamma(x) = 0\}$$

which describe the propagation of the shocks, and such that

$$(7.3.6) \qquad U \in C^\infty(\Omega\backslash\Gamma), \quad \text{where } \Gamma = \bigcup_{\gamma \in G} \Gamma_\gamma$$

(7.3.7) $\quad U$ is locally bounded on Ω

(7.3.8) $$\int_{\Omega} (U(t,x)\psi_t(t,x) + b(U(t,x))\psi_x(t,x))dt\ dx = 0, \quad \psi \in \mathcal{D}(\Omega)$$

Obviously, such a solution U will be a *piece wise* $\mathcal{C}^\infty$*-smooth weak solution* of the partial differential equation in (7.3.1), in the sense of the definition in Section 2. Therefore Theorem 1 in Section 2, will yield the following result.

Theorem 2

Suppose $U : (0,\infty) \times \mathbb{R} \to \mathbb{R}$ is a *shock wave* solution of the nonlinear hyperbolic conservation law in (7.3.1) and that it satisfies the conditions (7.3.5)-(7.3.8). Then it is possible to construct $\mathcal{C}^\infty$-smooth regularizations (6.4.22) and algebras (6.4.34), such that

(1) $U = s + I_\ell(V,S) \in A^\ell, \quad \ell \in \mathbb{N}$
where $s \in \mathcal{S}$ does *not* depend on ℓ

(2) U satisfies (7.3.1) in the *usual algebraic sense*, with multiplications in A^k and the partial derivative operators $D_t, D_x : A^\ell \to A^k$, with $\ell, k \in \mathbb{N}$, $k+1 \le \ell$

In view of pct. (1), s is a *chain weak solution* of (7.3.1), see Section 5, Chapter 6.

§4. RESOLUTION OF SINGULARITIES OF KLEIN-GORDON TYPE NONLINEAR WAVES

Suppose given the Klein-Gordon type nonlinear wave equation

(7.4.1) $$U_{tt}(t,x) - U_{xx}(t,x) = T(D)U(t,x), \quad t > 0, \quad x \in \mathbb{R}$$

with the initial conditions

(7.4.2) $$U(0,x) = f(x), \quad x \in \mathbb{R}$$

(7.4.3) $$U_t(0,x) = g(x), \quad x \in \mathbb{R}$$

where $T(D)$ is a first-order $\mathcal{C}^\infty$-smooth simple polynomial nonlinear partial differential operator. Then it is obvious that (7.4.1) is a *second order* $\mathcal{C}^\infty$*-smooth simple polynomial nonlinear* partial differential equation on $\Omega = (0,\infty) \times \mathbb{R} \subset \mathbb{R}^2$, since it has the form in (7.2.4).

It is known that under general conditions, Reed [2,3], Reed & Berning, the equation (7.4.1) has local or global solutions $U : \Omega' \to \mathbb{R}$, with $\Omega' \subset \Omega$ open, which have the following properties.

There exist a *finite* number of points $x_1, \ldots, x_\sigma \in \mathbb{R}$, which originate *light cones* with the *boundaries* given by

$$\Gamma_\alpha^- = \{(t,x) \in \Omega' \,|\, x - x_\alpha + t = 0\}$$
$$\Gamma_\alpha^+ = \{(t,x) \in \Omega' \,|\, x - x_\alpha - t = 0\}$$

with $1 \leq \alpha \leq \sigma$, such that

(7.4.4) $\quad U \in \mathcal{C}^\infty(\Omega' \backslash \Gamma)$, where $\Gamma = \bigcup_{1 \leq \alpha \leq \sigma} (\Gamma_\alpha^- \cup \Gamma_\alpha^+)$

(7.4.5) $\quad U$ is locally bounded on Ω'

(7.4.6)
$$\int U(t,x)(\psi_{tt}(t,x) - \psi_{xx}(t,x))dt\,dx =$$
$$= \int_\Omega \Big(\sum_{1 \leq i \leq a} T_i U(x) L_i^*(D)\psi(t,x)\Big)dt\,dx, \quad \psi \in \mathcal{D}(\Omega')$$

where it has been assumed that $T(D)$ in (7.4.1) has the form (7.2.1), and $L_i^*(D)$ is the formal adjoint of $L_i(D)$.

Obviously, such a solution U will be a *piece wise* $\mathcal{C}^\infty$*-smooth weak solution* of (7.4.1), in the sense of the definition in Section 2. Therefore Theorem 1 in Section 2 will yield the following result, similar to that of Theorem 2 in Section 3:

Theorem 3

Suppose $U : \Omega \to \mathbb{R}$, with $\Omega \subset (0,\infty) \times \mathbb{R}$, is a solution of the Klein-Gordon nonlinear wave equation (7.4.1) and that it satisfies the conditions (7.4.4)-(7.4.6). Then it is possible to construct $\mathcal{C}^\infty$-smooth regularizations (6.4.22) and algebras (6.4.34), such that

(1) $U = s + I_\ell(V,S) \in A^\ell$, $\ell \in \mathbb{N}$
where $s \in S$ does *not* depend on ℓ

(2) U satisfies (7.4.1) in the *usual algebraic sense*, with multiplication in A^k and the partial derivative operators
$D^p : A^\ell \to A^k$, $p \in \mathbb{N}^2$, $|p| \leq 2$, with $\ell, k \in \mathbb{N}$, $k+2 \leq \ell$.

In view of pct (1), s is a *chain weak solution* of (7.4.1), see Section 5, Chapter 6.

§5. JUNCTION CONDITIONS AND RESOLUTION OF SINGULARITIES OF WEAK SOLUTIONS FOR THE EQUATIONS OF MAGNETOHYDRODYNAMICS AND GENERAL RELATIVITY

The problem of finding *junction conditions* across hypersurfaces of discontinuities for solutions of the equations of magnetohydrodynamics or general relativity is usually approached either by applying integral conditions or by introducing certain simplifying assumptions.

Such methods present obvious deficiences when compared with direct methods which would be based on the idea of obtaining the junction conditions from *weak solution conditions* formulated across the hypersurfaces of discontinuities. As the nonlinearity of the equations involved will imply the presence of products of the Heaviside function with the Dirac δ distribution and its partial derivatives, such direct methods *cannot* be implemented within the distributional framework.

In the present Section, the polynomial nonlinear operations on the singular distributions mentioned will be performed within the quotient algebras containing the distributions. The resulting method has the major advantage that among others, a clear and algebraically simple insight is obtained into the *structure of the nonlinearities* of a particularly large class of systems of polynomial nonlinear partial differential equations encountered in the study of physics. This result is presented under its general form in Section 6. The method of establishing the junction conditions presented in the sequel will at the same time yield the resolution of singularities across the hypersurfaces involved in these junction conditions.

In order to avoid rather trivial technical complications and also to make it possible at the same time to point out the essential underlying *algebraic* phenomena, only the case of polynomial nonlinearities will be dealt with, a case which obviously covers the situation in magnetohydrodynamics, as well as general relativity.

The fact that in the sequel we shall deal with *systems* and not single nonlinear partial differential equations need not cause concern. Indeed, as seen in Section 14, Chapter 1, see also Rosinger [2, pp. 32-34], the study of *sequential solutions* of such systems is obtained as an easy and direct generalization of the situation for single nonlinear partial differential equations, see Section 10, Chapter 1.

For convenience, we shall only consider the case of $\mathcal{C}^\infty$-smooth coefficients. The general case of continuous coefficients is in principle similar and is dealt with in detail in Rosinger [2, pp. 139-162].

Suppose we are given a system of polynomial nonlinear partial differential equations

(7.5.1) $$\sum_{1\le i\le h_\beta} c_{\beta i}(x) \prod_{1\le j\le k_{\beta i}} D^{p_{\beta ij}} U_{\alpha_{\beta ij}}(x) = f_\beta(x), \quad x \in \Omega$$

where $U = (U_1, \ldots, U_a) : \Omega \longrightarrow \mathbb{R}^a$ are unknown functions, while $c_{\beta i}$, $f_\beta \in C^\infty(\Omega)$ are given.

The system in (7.5.1) is called of *type* (MH), if and only if each of the associated partial differential operators

(7.5.2) $$T_\beta(D)U(x) = \sum_{1\le i\le h_\beta} c_{\beta i}(x) \prod_{1\le j\le k_{\beta i}} D^{p_{\beta ij}} U_{\alpha_{\beta ij}}, \quad x \in \Omega$$

can be written in the form

(7.5.3) $$T_\beta(D)U(x) = \sum_{1\le\rho\le r_\beta} L_{\beta\rho}(D)(U_{\alpha_{\beta\rho}}(x)P_{\beta\rho}(D)U_{\alpha_{\beta\rho}}(x)), \quad x \in \Omega$$

where $L_{\beta\rho}(D)$ are linear partial differential operators with C^∞-smooth coefficients and order $m_{\beta\rho}$, while $P_{\beta\rho}(D)$ are linear partial differential operators with C^∞-smooth coefficients and order at *most* one. Sometimes it will be convenient to write the conditions (7.5.3) under the form

(7.5.4) $$T_\beta(D)U(x) = \sum_{1\le\rho\le r_\beta} L_{\beta\rho}(D)\Big(\sum_{1\le\alpha,\alpha'\le a} U_\alpha(x)P_{\beta\rho\alpha\alpha'}(D)U_{\alpha'}(x)\Big),$$
$$x \in \Omega, \quad 1 \le \beta \le b$$

where $P_{\beta\rho\alpha\alpha'}(D)$ are of the same type with $P_{\beta\rho}(D)$.

As can easily be seen, the equations of magnetohydrodynamics as well as those of general relativity are of type (MH). The Navier-Stokes equations are also of type (MH).

If the order of the system (7.5.1) is

$$m = \max\{|p_{\beta ij}| \,\big|\, 1 \le \beta \le b,\ 1 \le i \le h_\beta,\ 1 \le j \le k_{\beta i}\}$$

then, if (7.5.1) is of type (MH), it may be assumed that

$$m = 1 + \max\{m_{\beta\rho} \mid 1 \le \beta \le b,\ 1 \le \rho \le r_\beta\}$$

Suppose now given a hypersurface $\Gamma \subset \Omega$ defined by

(7.5.5) $$\Gamma = \{x \in \Omega \mid \gamma(x) = 0\}$$

where $\gamma : \Omega \rightarrow \mathbb{R}$, $\gamma \in C^\infty$. Obviously the subset Γ in Ω is closed. We shall also suppose that

(7.5.6) Γ has zero Lebesque measure

We shall be interested in finding *weak solutions* $U = (U_1, \ldots, U_a) : \Omega \rightarrow \mathbb{R}^a$ for an (MH) type system (7.5.1) such that U is C^∞-smooth on $\Omega \backslash \Gamma$ and discontinuous across the hypersurface Γ.

It follows immediately that U will have the form

$$U(x) = U_-(x) + (U_+(x) - U_-(x)) \cdot H(x), \quad x \in \Omega \tag{7.5.7}$$

where $U_-, U_+ : \Omega \rightarrow \mathbb{R}^a, U_-, U_+ \in C^\infty$, are classical solutions for the system (7.5.1), while $H : \Omega \rightarrow \mathbb{R}$ defined by

$$H(x) = \begin{cases} 0 & \text{if} \quad \gamma(x) \leq 0 \\ 1 & \text{if} \quad \gamma(x) > 0 \end{cases} \tag{7.5.8}$$

is the Heaviside function associated with the representation of the hypersurface Γ given in (7.5.5).

The *problem* can now be formulated as follows: find the necessary and/or sufficient *junction conditions* on the system (7.5.1), the hypersurface Γ and its representation through γ in (7.5.5), as well as on the classical solutions U_- and U_+ of (7.5.1), such that U given in (7.5.7) is a weak solution for (7.5.1).

If the above problem is considered within the framework of the chains algebras containing the distributions constructed in Section 4, Chapter 6, it is easy to obtain *necessary and sufficient* junction conditions. To do so, we shall first construct C^∞-smooth regularizations of the Heaviside function (7.5.8), according to the method presented in the following lemma.

Lemma 1

Suppose given any function $\eta : \mathbb{R} \rightarrow [0,1]$, $\eta \in C^\infty$, such that

$$\begin{aligned} \eta &= 0 \quad \text{on} \quad (-\infty, -1] \\ \eta &= 1 \quad \text{on} \quad [1, +\infty] \end{aligned} \tag{7.5.9}$$

We define $s_\eta \in (C^\infty(\Omega))^{\mathbb{N}}$ by

$$s_{\eta\nu}(x) = \eta((\nu+1)\gamma(x)), \quad \nu \in \mathbb{N}, \; x \in \Omega \tag{7.5.10}$$

Then the following two relations hold

$$(7.5.11) \qquad (s_\eta)^\ell = s_{(\eta)^\ell}, \quad \ell \in \mathbb{N}\setminus\{0\}$$

$$(7.5.12) \qquad (s_\eta)^\ell \in S^\infty, \ \langle (s_\eta)^\ell, \cdot \rangle = H, \quad \ell \in \mathbb{N}\setminus\{0\}$$

Proof

The case $\ell = 1$ is obvious. If $\ell > 1$, it is easy to see that $(\eta)^\eta : \mathbb{R} \longrightarrow [0,1]$, $(\eta)^\ell \in C^\infty$ and $(\eta)^\ell$ also satisfies (7.5.9), therefore the problem is reduced to the case $\ell = 1$. □

The C^∞-smooth regularizations of H obtained above will generate C^∞-smooth regularizations

$$(7.5.13) \qquad s = (s_1, \ldots, s_a) \in ((C^\infty(\Omega))^{\mathbb{N}})^a$$

of the intended weak solution U in (7.5.7), given by the relation

$$(7.5.14) \qquad s_{\alpha\nu}(x) = (U_-)_\alpha(x) + ((U_+)_\alpha(x) - (U_-)_\alpha(x)) s_{\eta\nu}(x),$$

$$1 \leq \alpha \leq a, \quad \nu \in \mathbb{N}, \quad x \in \Omega$$

which will obviously imply

$$(7.5.15) \qquad s_\alpha \in S^\infty, \ \langle s_\alpha, \cdot \rangle = U_\alpha, \quad 1 \leq \alpha \leq a$$

At the stage the problem is to identify the conditions which will make s given in (7.5.13) a weak solution for the system (7.5.1). In this connection, the following result will be useful. Suppose given on Ω a first order, linear and homogeneous partial differential operator $P(D)$ with C^∞-smooth coefficients. For given functions $\psi_-, \psi_+, \psi'_-, \psi'_+, \in C^\infty(\Omega)$ define the regularizations $t, t' \in (C^\infty(\Omega))^{\mathbb{N}}$ as follows

$$t_\nu(x) = \psi_-(x) + (\psi_+(x) - \psi_-(x)) s_{\eta\nu}(x)$$

$$t'_\nu(x) = \psi'_-(x) + (\psi'_+(x) - \psi'_-(x)) s_{\eta\nu}(x0$$

with $\nu \in \mathbb{N}$, $x \in \Omega$, assuming that η is as in (7.5.9). Then obviously

$$t, t' \in S^\infty, \ \langle t, \cdot \rangle = \psi_- + (\psi_+ - \psi_-)H, \ \langle t', \cdot \rangle = \psi' + (\psi'_+ - \psi'_-)H$$

Proposition 1

The following relations hold

(7.5.16) $\quad tP(D)t' \in S^\infty$

(7.5.17) $\quad \langle tP(D)t', \cdot\rangle = \psi_- P(D)\psi' + (\psi_+ P(D)\psi'_+ - \psi_- P(D)\psi'_-)H + \frac{1}{2}(\psi_+ + \psi_-)(\psi'_+ - \psi'_-)P(D)H$

Proof

It is easy to see that for each $\nu \in \mathbb{N}$, the following relation holds

$$t_\nu P(D)t'_\nu = (\psi_- + (\psi_+ - \psi_-)s_{\eta\nu})P(D)(\psi'_- + (\psi'_+ - \psi'_-)s_{\eta\nu}) =$$

$$= (\psi_- + (\psi_+ - \psi_-)s_{\eta\nu})P(D)\psi'_- + (P(D)(\psi'_+ - \psi'_-)s_{\eta\nu} + (\psi'_+ - \psi'_-)P(D)s_{\eta\nu}) =$$

$$= \psi_- P(D)\psi' + (\psi_- P(D)(\psi'_- + (\psi'_+ - \psi'_-) + (\psi_+ - \psi_-)P(D)\psi'_-)s_{\eta\nu}) +$$

$$+ ((\psi_+ - \psi_-)P(D)(\psi'_+ - \psi'_-))(s_{\eta\nu})^2 + \psi_-(\psi'_+ - \psi'_-)P(D)s_{\eta\nu} +$$

$$+ (\psi_+ - \psi_-)(\psi'_+ - \psi'_-)s_{\eta\nu}P(D)s_{\eta\nu}$$

But since $P(D)$ is of first order, linear and homogeneous, we have

$$s_{\eta\nu}P(D)s_{\eta\nu} = \frac{1}{2} P(D)(s_{\eta\nu})^2$$

Now the relation (7.5.11) in Lemma 1 will imply

$$(s_{\eta\nu})^2 = s_{\xi\nu}$$

where $\xi = (\eta)^2$ also satisfies (7.5.11). This completes the proof, if (7.5.12) in Lemma 1 is taken into account. □

Corollary 1

If $P(D)$ is a linear partial differential operator on Ω with C^∞-smooth coefficients and order at most one, then the following relations hold

(7.5.18) $\quad tP(D)t' \in S^\infty$

(7.5.19)
$$<tP(D)t',\cdot> = \psi_- P(D)\psi'_- + (\psi_+ P(D)\psi'_+ - \psi_- P(D)\psi'_-) + \frac{1}{2}(\psi_+ + \psi_-)(\psi'_+ - \psi'_-)Q(D)H$$

where $Q(D)$ is the first order homogeneous part of $P(D)$.

Proof

Assume $P(D)$ has the form

$$P(D)\psi(x) = Q(D)\psi(x) + d(x)\psi(x) + e(x), \quad x \in \Omega$$

where $Q(D)$ is the first-order homogeneous part of $P(D)$, while $d,e \in C^\infty(\Omega)$. Then relation (7.5.18) follows easily from (7.5.16).

Further, for given $\nu \in \mathbb{N}$, we have

$$t_\nu d t'_\nu = (\psi_- + (\psi_+ - \psi_-)s_{\eta\nu})d(\psi'_- + (\psi'_+ - \psi'_-)s_{\eta\nu}) =$$

$$= \psi_- d\psi'_- + (\psi_- d(\psi'_+ - \psi'_-) + (\psi_+ - \psi_-)d\psi'_-)s_{\eta\nu} + (\psi_+ - \psi'_-)d(\psi'_+ - \psi'_-)(s_{\eta\nu})^2$$

therefore, in view of Lemma 1

$$<tdt',\cdot> = \psi_- d\psi'_- + (\psi_+ d\psi'_+ - \psi_- d\psi'_-)H$$

Finally, it is easy to see that

$$<te,\cdot> = \psi_- e + (\psi_+ e - \psi_- e)H$$

The last two relations together with (7.5.17) will obviously yield (7.5.19). □

The result on *junction conditions* for discontinuous solutions of systems of partial differential equations of type (MH) will be presented in Theorem 4. First we need the following result.

Propositions 2

Suppose $U_-, U_+ : \Omega \to \mathbb{R}^a$ are two C^∞-smooth solutions of the m-th order polynomial nonlinear system of type (MH) in (7.5.1). Then for any C^∞-smooth regularization s given in (7.5.13), the following relations hold for every $1 \le \beta \le b$

(7.5.20) $\quad T_\beta(D)s \in S^\infty$

(7.5.21)
$$\langle T_\beta(D)s,\cdot\rangle =$$
$$= \sum_{1\le\rho\le r_\beta} L_{\beta\rho}(D)((\sum_{1\le\alpha,\alpha'\le a}(U_+)_\alpha P_{\beta\rho\alpha\alpha'}(D)(U_+)_{\alpha'})H) -$$
$$- \sum_{1\le\rho\le r_\beta} L_{\beta\rho}(D)((\sum_{1\le\alpha,\alpha'\le a}(U_-)_\alpha P_{\beta\rho\alpha\alpha'}(D)(U_-)_{\alpha'})H +$$
$$+ \frac{1}{2}\sum_{1\le\rho\le r_\beta} L_{\beta\rho}(D)((\sum_{1\le\alpha,\alpha'\le a}((U_+)_\alpha + (U_-)_\alpha)((U_+)_{\alpha'} -$$
$$- (U_-)_{\alpha'}))Q_{\beta\rho\alpha\alpha'}(D)H)$$

where $Q_{\beta\rho\alpha\alpha'}(D)$ is the first order homogeneous part of $P_{\beta\rho\alpha\alpha'}(D)$.

<u>Proof</u>

In view of (7.5.4) and (7.5.13), it follows that
$$T_\beta(D)s = \sum_{1\le\rho\le r_\beta} L_{\beta\rho}(D)(\sum_{1\le\alpha,\alpha'\le a} s_\alpha P_{\beta\rho\alpha\alpha'}(D)s_{\alpha'})$$

But (7.5.18) implies the relations
$$s_\alpha P_{\beta\rho\alpha\alpha'}(D)s_{\alpha'} \in S^\infty$$

therefore the linearity of $L_{\beta\rho}(D)$ will give
$$T_\beta(D)s \in S^\infty$$

as well as
$$\langle T_\beta(D)s,\cdot\rangle = \sum_{1\le\rho\le r_\beta} L_{\beta\rho}(D)(\sum_{1\le\alpha,\alpha'\le a} \langle s_\alpha P_{\beta\rho\alpha\alpha'}(D)s_{\alpha'},\cdot\rangle)$$

Now (7.5.19) will give
$$\langle s_\alpha P_{\beta\rho\alpha\alpha'}(D)s_{\alpha'},\cdot\rangle = (U_-)_\alpha P_{\beta\rho\alpha\alpha'}(D)(U_-)_{\alpha'} + ((U_+)_\alpha P_{\beta\rho\alpha\alpha'}(D)(U_+)_{\alpha'} -$$
$$- (U_-)_\alpha P_{\beta\rho\alpha\alpha'}(D)(U_-)_{\alpha'})H +$$
$$+ \frac{1}{2}((U_+)_\alpha + (U_-)_\alpha)((U_+)_{\alpha'} - (U_-)_{\alpha'})Q_{\beta\rho\alpha\alpha'}(D)H$$

where $Q_{\beta\rho\alpha\alpha'}(D)$ is the first order homogeneous part of $P_{\beta\rho\alpha\alpha'}(D)$. It follows that for $1 \leq \beta \leq b$, the following relation holds

$$<T_\beta(D)s,\cdot> = T_\beta(D)U_- + \sum_{1\leq\rho\leq r_\beta} L_{\beta\rho}(D)((\sum_{1\leq\alpha,\alpha'\leq a} (U_+)_\alpha P_{\beta\rho\alpha\alpha'}(D)(U_+)_{\alpha'})H) -$$

$$- \sum_{1\leq\rho\leq r_\beta} L_{\beta\rho}(D)((\sum_{1\leq\alpha,\alpha'\leq a} (U_-)_\alpha P_{\beta\rho\alpha\alpha'}(D)(U_-)_{\alpha'})H) +$$

$$+ \frac{1}{2} \sum_{1\leq\rho\leq r_\beta} L_{\beta\rho}(D)((\sum_{1\leq\alpha,\alpha'\leq a} ((U_+)_\alpha + (U_-)_\alpha)(U_+)_{\alpha'} -$$

$$- (U_-)_{\alpha'}))Q_{\beta\rho\alpha\alpha'}(D)H)$$

Since U_- was supposed to be a classical solution of the system (7.5.1), we have

$$T_\beta(D)U_- = 0 \quad \text{on} \quad \Omega,$$

which completes the proof □

Before presenting the result in Theorem 4, we need the following definition.

Given two functions $U_-, U_+ : \Omega \longrightarrow \mathbb{R}^a$, $U_-, U_+ \in C^\infty$, we define $U : \Omega \longrightarrow \mathbb{R}^a$ by

$$U(x) = U_-(x) + (U_+(x) - U_-(x))H(x), \quad x \in \Omega$$

and use the notation

$$I = \{\alpha | 1 \leq \alpha \leq a, \ U_\alpha \notin C^\infty(\Omega)\}$$

Then the functions U_-, U_+ will be called *independent* on Γ, if and only if for any $\lambda_\alpha \in \mathbb{R}$, with $\alpha \in I$, the following implication is valid:

$$(\sum_{\alpha\in I} \lambda_\alpha U_\alpha \in C^\infty(\Omega)) \Rightarrow (\lambda_\alpha = 0, \quad \alpha \in I)$$

If $a = 1$, then U_-, U_+ are trivially independent on Γ, which is why the above condition was not demanded in Theorem 1 in Section 2. In terms of the system in (7.5.1), the case $a = 1$ corresponds to the situation when one unknown function $U : \Omega \longrightarrow \mathbb{R}$ has to satisfy b partial differential equations.

Obviously, if $U \in C^\infty$ then U_-, U_+ are independent on Γ.

Moreover, $U \in C^\infty$ if and only if

$$D^p(U_-)_\alpha(x) = D^p(U_+)_\alpha(x), \quad 1 \leq \alpha \leq a,\ x \in \Gamma,\ p \in \mathbb{N}^n$$

Theorem 4

Suppose $U_-, U_+ : \Omega \to \mathbb{R}^a$ are two C^∞-smooth solutions of the m-th order polynomial nonlinear system of (MH) in (7.5.1) and suppose given a C^∞-smooth hypersurface (7.5.5). Then the function

$$U(x) = U_-(x) + (U_+(x) - U_-(x))H(x), \quad x \in \Omega \tag{7.5.22}$$

where H is the Heaviside function (7.5.8) associated with the hypersurface (7.5.5), is a *weak solution* of the system (7.5.1), if and only if the following *junction conditions* are satisfied for each $1 \leq \beta \leq b$

$$\begin{aligned}
&\sum_{1\leq\rho\leq r_\beta} L_{\beta\rho}(D)\Big(\Big(\sum_{1\leq\alpha,\alpha'\leq a} (U_+)_\alpha P_{\beta\rho\alpha\alpha'}(D)(U_+)_{\alpha'} - \\
&- \sum_{1\leq\alpha,\alpha'\leq a} (U_-)_\alpha P_{\beta\rho\alpha\alpha'}(D)(U_-)_{\alpha'} H\Big) + \\
&+ \frac{1}{2}\sum_{1\leq\rho\leq r_\beta} L_{\beta\rho}(D)\Big(\Big(\sum_{1\leq\alpha,\alpha'\leq a} ((U_+)_\alpha + (U_-)_\alpha) \\
&((U_+)_{\alpha'} - (U_-)_{\alpha'})\Big) Q_{\beta\rho\alpha\alpha'}(D)H = f_\beta
\end{aligned} \tag{7.5.23}$$

where $Q_{\beta\rho\alpha\alpha'}(D)$ is the first order homogeneous part of $P_{\beta\rho\alpha\alpha'}(D)$.

If (7.5.23) is satisfied and the functions U_-, U_+ are independent on Γ, it is possible to construct C^∞-smooth regularizations (6.4.22) and algebras (6.4.34), such that

(7.5.24) $U = s_\alpha + I_\ell(V,S) \in A^\ell, \quad 1 \leq \alpha \leq a, \quad \ell \in \mathbb{N}$

where $s_\alpha \in S$, with $1 \leq \alpha \leq a$, do *not* depend on ℓ

(7.5.25) U satisfies each of the equations of the system (7.5.1) in the *usual algebraic sense*, with multiplication in A^k and the partial derivative operators $D^p : A^\ell \to A^k$, $p \in \mathbb{N}^n$, $|p| \leq m$, with $\ell, k \in \mathbb{N}$, $k+m \leq \ell$

Proof

If the junction conditions (7.5.23) hold, then in view of Proposition 2, the function (7.5.22) will be a weak solution of the system (7.5.1)

Conversely, assume U in (7.5.22) is a weak solution of the system (7.5.1). The operations in the definition of U can be performed with $D'(\Omega)$, however the same does not hold for the operations on U performed by $T_\beta(D)$, with $1 \leq \beta \leq b$, as these operations will involve products $H \cdot P_{\beta\rho\alpha\alpha'}(D)H$. Nevertheless, according to (7.5.20) the C^∞-smooth regularization s of U constructed in (7.5.14) has the property that $T_\beta(D)_s$ is weakly convergent for every $1 \leq \beta \leq b$. Therefore the assumption that U is a weak solution of the system (7.5.1) implies that

$$\langle T_\beta(D)s, \cdot \rangle = f_\beta, \quad 1 \leq \beta \leq b$$

This, in view of the relations (7.5.21), completes the proof of the converse.

For the rest of the proof, see Rosinger [3, pp. 373-377].

§6. RESOLUBLE SYSTEMS OF POLYNOMIAL NONLINEAR PARTIAL DIFFERENTIAL EQUATIONS

The necessary and sufficient junction conditions across hypersurfaces of discontinuities of weak solutions for systems of type (MH) and the resolution of the corresponding singularities presented in the previous Section are extended here to a large class of systems of nonlinear partial differential equations which contains many of the equations modelling various physical phenomena.

For convenience, we shall only deal with the case of C^∞-smooth coefficients. The general case of continuous coefficients can be treated in a similar way, see Rosinger [2,pp. 152-162].

Definition 1

The system of polynomial nonlinear partial differential equations in (7.5.1) is called *resoluble*, if and only if each of the associated partial differential operators in (7.5.2) can be written in the form

$$T_\beta(D)U(x) = \sum_{1 \leq \rho \leq r_\beta} (T_{\beta\rho}(D)(\psi(x),\chi(x))) \cdot D^{p_{\beta\rho}}(\omega(x))^{\ell_{\beta\rho}}, \quad x \in \Omega,\ 1 \leq \beta \leq b \tag{7.6.1}$$

whenever

(7.6.2) $$U(x) = \psi(x) + \chi(x)\cdot\omega(x), \quad x \in \Omega$$

where $\psi,\chi : \Omega \longrightarrow \mathbb{R}^a$, $\omega : \Omega \longrightarrow \mathbb{R}$, $\psi,\chi,\omega \in C^\infty$, $p_{\beta\rho} \in \mathbb{N}^n$, $\ell_{\beta\rho} \in \mathbb{N}$ and $T_{\beta\rho}$ are $m'_{\beta\rho}$-th order polynomial nonlinear partial differential operators in ψ and χ.

The pair (m',m''), where

(7.6.3) $$m' = \max\{m'_{\beta\rho} \mid 1 \leq \beta \leq b,\ 1 \leq \rho \leq r_\beta\}$$
$$m'' = \max\{|p_{\beta\rho}| \,\big|\, 1 \leq \beta \leq b,\ 1 \leq \rho \leq r_\beta\}$$

is called the *split order* of the resoluble system (7.5.1).

Proposition 3

A system of type (MH) is resoluble.

Proof

Assume the partial differential operators in (7.5.2) corresponding to the system (7.5.1) are of form (7.5.3). Let us take $U : \Omega \longrightarrow \mathbb{R}^a$ given in (7.6.2). Then (7.5.3) or equivalently (7.5.4) yields

$$T_\beta(D)U = \sum_{1\leq\rho\leq r_\beta} L_{\beta\rho}(D)\Big(\sum_{1\leq\alpha,\alpha'\leq a} (\psi_\alpha+\chi_\alpha\cdot\omega)P_{\beta\rho\alpha\alpha'}(D)(\psi_{\alpha'}+\chi_{\alpha'}\cdot\omega)\Big)$$

But

$$(\psi_\alpha + \chi_\alpha\cdot\omega)P_{\beta\rho\alpha\alpha'}(D)(\psi_{\alpha'} + \chi_{\alpha'}\cdot\omega) =$$

$$= (\psi_\alpha + \chi_\alpha\cdot\omega)(P_{\beta\rho\alpha\alpha'}(D)\psi_{\alpha'} + (P_{\beta\rho\alpha\alpha'}(D)\chi_{\alpha'})\cdot\omega + \chi_{\alpha'}\cdot Q_{\beta\rho\alpha\alpha'}(D)\omega)$$

where $Q_{\beta\rho\alpha\alpha'}(D)$ is the first order homogeneous part of $P_{\beta\rho\alpha\alpha'}(D)$, therefore

$$\omega\cdot Q_{\beta\rho\alpha\alpha'}(D)\omega = \frac{1}{2}\, Q_{\beta\rho\alpha\alpha'}(D)(\omega)^2$$

Now the relation (7.6.1) follows easily □

The characteristic behaviour of the *resoluble* systems of polynomial nonlinear partial differential equations with respect to weak solutions with discontinuities across hypersurfaces is presented now.

Theorem 5

Suppose the m-th order polynomial nonlinear system of partial differential equations in (7.5.1) is resoluble.

Given $U_-, U_+ : \Omega \to \mathbb{R}^a$, two C^∞-smooth solutions of (7.5.1), and a C^∞-smooth hypersurface (7.5.5), define the function $U : \Omega \to \mathbb{R}^a$ by

(7.6.4) $$U_-(x) = U_-(x) + (U_+(x) - U_-(x))H(x), \quad x \in \Omega$$

where H is the Heaviside function (7.5.8) associated with the hypersurface (7.5.5).

Then U is a *weak solution of* (7.5.1), if and only if *the following junction conditions* are satisfied for each $1 \leq \beta \leq b$

(7.6.5) $$\sum_{1 \leq \rho \leq r_\beta} (T_{\beta\rho}(D)(U_-, U_+ - U_-)(\cdot D^{p_{\beta\rho}} H = f_\beta$$

In that case, if the functions U_-, U_+ are independent on Γ, it is possible to construct C^∞-smooth regularizations (6.4.22) and algebras (6.4.34) such that

(7.6.6)
$$U_\alpha = s_\alpha + I_\ell(V,S) \in A^\ell, \quad 1 \leq \alpha \leq a, \quad \ell \in \mathbb{N}$$
where $s_\alpha \in S$, with $1 \leq \alpha \leq a$, do *not* depend on ℓ

(7.6.7)
U satisfies each of the equations of the system (7.5.1) in the *usual algebraic sense*, with the multiplication in A^k and the partial derivative
$D^p : A^\ell \to A^k$, $p \in \mathbb{N}^n$, $|p| \leq m$, with $\ell, k \in \mathbb{N}$, $k+m \leq \ell$

Proof

For U given in (7.6.4) let us consider s given in (7.5.14). Then, for each $1 \leq \beta \leq b$, the following relations hold

(7.6.8) $$T_\beta(D)s \in S^\infty$$

(7.6.9) $$\langle T_\beta(D)s, \cdot \rangle = \sum_{1 \leq \rho \leq r_\beta} (T_{\beta\rho}(D)(U_-, U_+ - U_-)) \cdot D^{p_{\beta\rho}} H$$

Indeed, in view of (7.6.1) we have for $1 \leq \beta \leq b$ and $\nu \in \mathbb{N}$, the relations

$$(7.6.10) \qquad T_\beta(D)s_\nu = \sum_{1\leq\rho\leq r} (T_{\beta\rho}(D)(U_-,U_+-U_-))\cdot D^{p_{\beta\rho}}(s_{\eta\nu})^{\ell_{\beta\rho}}$$

But, in view of (7.5.11), we have

$$(s_{\eta\nu})^{\ell_{\beta\rho}} = s_{\xi\nu}, \quad \text{where} \quad \xi = (\eta)^{\ell_{\beta\rho}}$$

Moreover

$$T_{\beta\rho}(D)(U_-,U_+-U_-) \in C^\infty(\Omega)$$

therefore the products $T_{\beta\rho}(D)(U_-,U_+-U)\cdot D^{p_{\beta\rho}}H$ in (7.6.20) are well defined in $D'(\Omega)$.

In this way (7.6.18) and (7.6.19) will follow easily from (7.5.12).

Assume now that the junction conditons (7.6.15) hold. Then in view of (7.6.18) and (7.6.19), U will obviously be a weak solution of (7.5.1).

Conversely, if U is a weak solution of (7.5.1) then (7.6.18) and (7.6.19) will obviously imply the junction conditions (7.6.15).

The second part of the proof is similar to that given for Theorem 4 in Section 5.

§7. COMPUTATION OF THE JUNCTION CONDITIONS

The *junction conditions* (7.6.15) contain as a *particular* case the junction conditions (7.5.23), and at the same time have a more *compact* form, therefore we shall present here only the way they can be computed explicitely.

We shall assume the notations and conditions in Theorem 5, Section 6.

The basic relations used in the sequel are

$$(7.7.1) \qquad D^pH = (D^p\gamma)\cdot\delta, \quad p \in \mathbb{N}^n, \quad |p| = 1$$

and

$$(7.7.2) \qquad D^p(D^\ell\delta) = (D^p\gamma)\cdot(D^{\ell+1}\delta), \quad p \in \mathbb{N}^n, \ |p| = 1, \ \ell \in \mathbb{N}$$

as well as

(7.7.3) $$\gamma\cdot\delta = 0$$

(7.7.4) $$\gamma\cdot D^{\ell+1}\delta + (\ell+1)D^{\ell}\delta = 0, \quad \ell \in \mathbb{N}$$

where δ is the Dirac distribution concentrated on the hypersurface Γ in (7.5.5), which is represented by the mapping $\gamma : \Omega \longrightarrow \mathbb{R}$, $\gamma \in C^{\infty}$. It is well known that the relations (7.7.1) and (7.7.2) are valid if for instance

$$\operatorname{grad} \gamma(x) \neq 0, \quad x \in \Gamma$$

It is easy to see that the relations (7.7.1) and (7.7.2) will yield

(7.7.5) $$D^p H = \sum_{0\leq\ell\leq|p|-1} (K_{p\ell}(D)\gamma)\cdot D^{\ell}\delta, \quad p \in \mathbb{N}^n$$

where $K_{p\ell}(D)$ are polynomial nonlinear partial differential operators of order $\ell + 1$, which can be obtained from the recurrent relations

(7.7.6) $$K_{po}(D)\gamma = D^p\gamma, \quad p \in \mathbb{N}^n, \ |p| = 1$$

(7.7.7) $$\begin{aligned} &K_{p+q\ell}(D)\gamma = D^q(K_{p\ell}(D)\gamma) + (K_{p\ell-1}(D)\gamma)D^q\gamma, \\ &p,q \in \mathbb{N}^n, |q| = 1, \ \ell \in \mathbb{N}, \ \ell \leq |p| \end{aligned}$$

Substituing now the relations (7.7.6), (7.7.7) into the junction conditions (7.6.15) and rearranging the terms according to the increasing order of the derivatives of the Dirac δ distribution, we have

(7.7.8) $$\begin{aligned} &(G_{\beta 0}(D)(U_-,U_+-U_-,\gamma))\cdot H + \\ &+ \sum_{1\leq\ell\leq m''} (G_{\beta\ell}(D)(U_-,U_+-U_-,\gamma))\cdot D^{\ell-1}\delta = f_\beta, \quad 1 \leq \beta \leq b \end{aligned}$$

where $G_{\beta\ell}(D)$ are polynomial nonlinear partial differential operators in U_-, $U_+ - U_-$ and γ. With the help of the relations (7.7.3), (7.7.4), it is possible to eliminate in (7.7.8) the Dirac δ distribution and its lower derivatives and finally obtain relations of the form

(7.7.9) $$\begin{aligned} &(T'_\beta(D)(U_-,U_+-U_-,\gamma))\cdot H + \\ &+ (T''_\beta(D))(U_-,U_+-U_-,\gamma))\cdot D^{m''-1}\delta = f_\beta, \quad 1 \leq \beta \leq b \end{aligned}$$

with $T'_\beta(D)$, $T''_\beta(D)$ polynomial nonlinear partial differential operators in U_-, $U_+ - U_-$ and γ, relations which can be seen as the *explicit form* of the *junction conditions* in (7.6.15). Indeed, as U_- and U_+ were assumed to be known solutions of the resoluble system (7.5.1), the relations (7.7.9) will give condtions on γ in the form of polynomial nonlinear partial differential equations, i.e., conditions on the possible *hypersurfaces of discontinuities* of *weak solutions* of the resoluble system (7.5.1.)

§8. EXAMPLES OF RESOLUBLE SYSTEMS OF POLYNOMIAL NONLINEAR PARTIAL DIFFERENTIAL EQUATIONS

The aim of this section is to explicitate in a particular case of low order the conditions (7.6.1)-(7.6.3) defining the resoluble systems of partial differential equations.

Given the system of polynomial nonlinear partial differential equations in (7.5.1), its *order of nonlinearity* is by definition

$$\pi = \max\{k_{\beta i} \mid 1 \leq \beta \leq b,\ 1 \leq i \leq h_\beta\} \tag{7.8.1}$$

Obviously, the system of partial differential equations in (7.5.1) is nonlinear, if and only if $\pi \geq 2$.

The *order* of the system of partial differential equations in (7.5.1) is called the integer

$$m = \max\{|p_{\beta ij}| \,\big|\, 1 \leq \beta \leq b,\ 1 \leq i \leq h_\beta,\ 1 \leq j \leq k_{\beta i}\} \tag{7.8.2}$$

We consider the simplest nontrivial systems of polynomial nonlinear partial differential equations in (7.5.1) which correspond to

$$\pi = 2, \quad m = 1 \tag{7.8.3}$$

and which contain the nonlinearity exhibited by the Navier-Stokes equations.

Obviously, the associated partial differential operators in (7.5.2) will have the form

(7.8.4)
$$\begin{aligned} T_\beta(D)U(x) = & \sum_{1\le\alpha\le a} c'_{\beta\alpha}(x)U_\alpha(x) + \sum_{\substack{1\le\alpha\le a\\ 1\le i\le n}} c''_{\beta\alpha i}(x)D_iU_\alpha(x) + \\ & + \sum_{1\le\alpha,\alpha'\le a} c'''_{\beta\alpha\alpha'}(x)U_\alpha(x)U_{\alpha'}(x) + \\ & + \sum_{\substack{1\le\alpha,\alpha'\le a\\ 1\le i\le n}} c^{1v}_{\beta\alpha\alpha' i}(x)U_\alpha(x)D_iU_{\alpha'}(x) + \\ & + \sum_{\substack{1\le\alpha,\alpha'\le a\\ 1\le i,j\le n}} c^{v}_{\beta\alpha\alpha' ij}(x)D_iU_\alpha(x)D_jU_{\alpha'}(x), \quad x\in\Omega,\ 1\le\beta\le b \end{aligned}$$

where

(7.8.5)
$$D_i = D^{p_i}, \quad \text{with} \quad p_i = (\overbrace{0,\ldots,0,1}^{i}, 0,\ldots),\ 1 \le i \le n$$

Now, a direct and easy computation will show that the partial differential operators in (7.8.4) satisfy the condition (7.6.1), if and only if the associated partial differential operators

(7.8.6)
$$T'_\beta(D) = \sum_{\substack{1\le\alpha,\alpha'\le a\\ 1\le i,j\le n}} c^{v}_{\beta\alpha\alpha' ij}(x)D_iU_\alpha(x)D_jU_{\alpha'}(x), \quad x \in \Omega,\ 1 \le \beta \le b$$

also satisfy that condition.

Substituting (7.6.2) into (7.8.6), it follows that all the resulting terms in (7.8.6) can be written in the form required in (7.6.1), with the exception of the terms in

(7.8.7)
$$\sum_{1\le i,j\le n} \Big(\sum_{1\le\alpha,\alpha'\le a} c^{v}_{\beta\alpha\alpha' ij}(x)(\chi_\alpha(x)\chi_{\alpha'}(x))D_i\omega(x)D_j\omega(x)\Big), \quad x \in \Omega,\ 1 \le \beta \le b$$

Therefore, the system (7.5.1) will be resoluble, if and only if

(7.8.8)
$$c^{v}_{\beta\alpha\alpha' ij}(x) = -c^{v}_{\beta\alpha\alpha' ji}(x),\ x \in \Omega,\ 1\le\beta\le b,\ 1\le\alpha,\alpha'\le a,\ 1\le i,\ j\le n$$

Obviously, the condition (7.8.8) is satisfied by the Navier-Stokes equations.

More complicated examples of resoluble systems of polynomial nonlinear partial differential equations can be found in Rosinger [2, pp. 157-162].

§9. GLOBAL VERSION OF THE CAUCHY-KOVALEVSKAIA THEOREM IN CHAINS OF ALGEBRAS OF GENERALIZED FUNCTIONS

We recall that the classical Cauchy-Kovalevskaia theorem has a wealth of outstanding features which during the last four or five decades, with the emergence of theories of generalized solutions, has somewhat been overlooked.

Let us review some of these features.

First perhaps is the fact that the mentioned theorem is by far the most general nonlinear method in obtaining existence of solutions.

The depth of insight this theorem offers is illustrated by the 'non hard' character of its proof, which only uses classical - in fact rather elementary and straightforward - analysis, in which the only price paid is in some computation involving majorants of power series in several complex variables, a computation which is known to boil down to summing a geometric series. Indeed, that proof gives a clear understanding of the fact that the result in the Cauchy-Kovalevskaia theorem is locally the best possible in the given conditions. This fact is further confirmed by the Holmgren uniqueness result.

In this way we obtain one of the best possible local existence, uniqueness and regularity results concerning classical solutions of very general nonlinear partial differential equations.

The limitations of the result in the Cauchy-Kovalevskaia theorem are as follows.

It only deals with noncharacteristic initial value problems. The fact however that the initial values have to be given on a noncharacteristic hypersurface is not so much a limitation but rather a necessity, as simple counter examples can illustrate it. The limitation is rather in the fact that it does not apply to mixed, that is, initial and boundary value problems.

A second limitation is that both the equation and initial values have to be analytic, as well as the noncharacteristic hypersurface involved. This may present inconvenience in the case of linear or nonlinear partial differential equations

$$(7.9.1) \qquad T(D)U(x) = f(x), \quad x \in \Omega \subset \mathbb{R}^n$$

where $T(D)$ is analytic, but f or the associated initial values are not. However, in many - if not most - of such situations which present interest, one can assume that all the functions involved are piece wise analytic. And then, by breaking up the initial domain Ω in smaller subdomains, one can get back to the analyticity conditions required in the Cauchy-Kovalevskaia theorem. The rest is a job in patching up the analytic pieces of solutions obtained, which can be done for instance, in the way mentioned in Chapter 2, or the sequel.

Finally, the third limitation is that we only obtain local solutions, that is, in a neighbourhood of the noncharacteristic hypersurface on which the initial values are specified. Here the feeling that the local character of the resulting solutions is a limitation rather seems to come from the conservative estimates used in the proof of the Cauchy-Kovalevskaia theorem in order to obtain the convergence of the power series giving the solution. Indeed, these estimates will yield domains of convergence - and thus existence of solutions - which can in fact be much smaller than the respective maximal domains may actually be. But the other fact is equally obvious: in many - if not most - of the cases of analytic equations and noncharacteristic analytic initial values, we cannot expect the existence of global classical solutions. A good example for that is offered by the shock wave equations. Therefore the legitimate objection should rather be that the proof of the Cauchy-Kovalevskaia theorem is not geared to give information on the maximal domain of existence of solutions.

There is also sometime the objection that the proof of the Cauchy-Kovalevskaia theorem does not give indications about the dependence of solutions on the initial values. This however may be included into the third limitation above. Indeed, we already have a difficulty in establishing the dependence of the maximal size of the domain of existence of solutions on the initial values. So that, perhaps owing to the general and elementary nature of the proof, we cannot expect to obtain more.

The above suggest the possibility that a wide range of global solutions of interest for nonlinear partial differential equations with initial and/or boundary value problems may belong to the following class: they are analytic on each of the pair wise disjoint subdomains $\Omega_i \subset \Omega$ of a suitable countable family

$$(\Omega_i \mid i \in I), \ \Omega_i \cap \Omega_j = \phi \ \text{ for } \ i,j \in I, \ \ i \neq j \tag{7.9.2}$$

such that the closed set

$$\Gamma = \Omega \setminus \bigcup_{i \in I} \Omega_i \ \text{ is nowhere dense in } \ \Omega \tag{7.9.3}$$

On each Ω_i, with $i \in I$, the local solution may be given by the Cauchy-Kovaleskaia theorem, subjected to the Holmgren uniqueness. The existence of such kind of solutions is quaranteed by Chapter 2, where it is shown that one can further assume satisfied the condition

$$\text{mes } \ \Gamma = 0 \tag{7.9.4}$$

However, wishing to allow greater generality, we shall not assume this latter condition, since the method of resolution of singularities we use is powerful enough to work without it.

Such solutions, while unique and analytic locally, that is, on each Ω_i, with $i \in I$, could have globally a great flexibility, owing to the possibilities allowed by the two conditions in (7.9.2) and (7.9.3), which are obviously sufficiently large in order to include for instance Cantor-set type of Γ or fractals, which have lately been associated with the study of turbulence and chaos.

It is in this way, among others, that the patching up of local solutions into a global one can be of interest.

This problem was approached and given one of the possible solutions in Chapter 2, where further comments on the relevance of this patching up method are presented..

Here we give a stronger version of that globalized Cauchy-Kovalevskaia theorem, which in Chapter 2 only constructed an $A^{\ell} \rightarrow \bar{A}$ sequential solution.

Indeed, we shall this time construct a sequence s of C^{∞}-smooth functions which is a *chain generalized solution* in suitable chains of algebras of generalized function, see Section 5, Chapter 6.

We can start with a general *analytic* nonlinear partial differential equation as in (2.5.1)

$$(7.9.5) \qquad D_t^m U(t,y) = G(t,y,\ldots,D_t^p D_y^q U(t,y),\ldots), \quad (t,y) \in \Omega$$

where $\Omega \subset \mathbb{R}^n$ is nonvoid, open, $t \in \mathbb{R}$, $y \in \mathbb{R}^{n-1}$, $m \geq 1$, $p \in \mathbb{N}$, $p < m$, $q \in \mathbb{N}^{n-1}$, $p + |q| \leq m$ and G is analytic.

As a consequence of the mentioned result in Theorem 2, Section 6, Chapter 2, there exist $\Gamma \subset \Omega$ with

$$(7.9.6) \qquad \Gamma \text{ closed, nowhere dense in } \Omega$$

$$(7.9.7) \qquad \text{mes } \Gamma = 0$$

and $U : \Omega \setminus \Gamma \rightarrow \mathbb{C}$ an *analytic* solution of (7.9.5) on $\Omega \setminus \Gamma$.

Now using the construction in (2.5.5) - (2.5.9) we obtain the sequence of C^{∞}-smooth functions

$$(7.9.8) \qquad s \in (C^{\infty}(\Omega))^{\mathbb{N}}$$

which satisfies (2.5.14) with

$$(7.9.9) \qquad T(D) = D_t^m - G(t,y,\ldots,D_t^p D_y^q,\ldots) \quad \text{on } \Omega$$

Our construction of chains of algebras of generalized functions in which a global version of the Cauchy-Kovalevskaia theorem holds, will be based on Corollary 1 in Section 3, Chapter 6, according to which

(7.9.10) $I = I_{nd} \cap (C^\infty(\Omega))^{\mathbb{N}}$

is a C^∞-*smooth regular* ideal in $(C^\infty(\Omega))^{\mathbb{N}}$. Let us take

(7.9.11) V

a vector subspace in $I \cap V^\infty$. Then, according to Theorem 1 in Section 2, Chapter 6, there exist C^∞-*smooth regularizations*

(7.9.12) $(V, v \oplus S, I, (C^\infty(\Omega))^{\mathbb{N}})$

of

(7.9.13) $D'(\Omega) = S^\infty/V^\infty$

for a suitable choice of the vector subspaces $S \subset S^\infty$.

Let us take any derivative invariant subalgebra $A \subset (C^\infty(\Omega))^{\mathbb{N}}$, such that

(7.9.14) $s \in A$

(7.9.15) $I + (V \oplus S) \subset A$

In view of (7.9.12) and (7.9.15) it follows easily that

(7.9.16) $(V, V \oplus S, I, A)$ is a C^∞-smooth regularization

hence (2.2.6) and (7.9.10) yield

(7.9.17) I is a derivative invariant, C^∞-smooth regular ideal in A

Now we can *modify* (6.4.25)-(6.4.29) as follows.

First, let us denote by

(7.9.18) I_{w_S}

the ideal in $(C^\infty(\Omega))^{\mathbb{N}}$ generated by $\{D^p w_S \mid p \in \mathbb{N}^n\}$. Then (2.2.6) and (7.9.10) yield

(7.9.19) $I_{w_S} \subset I$

Now, instead of (6.4.26), we define for $\ell \in \mathbb{N}$

(7.9.20) $\qquad A_\ell(V,S)$

as being the derivative invariant subalgebra generated by $I_{w_s} + V_\ell + S + s$ in A. Further, instead of (6.4.27), we define for $\ell \in \mathbb{N}$

(7.9.21) $\qquad I_\ell(V,S)$

as being the ideal generated by $I_{w_s} + V_\ell$ in $A_\ell(V,S)$.

It is easy to see that similar to Theorem 6 in Section 4, Chapter 6, we obtain for $\ell \in \mathbb{N}$, that

(7.9.22) $\qquad (V_\ell, S_\ell, I_\ell, A_\ell)$ is a C^∞-smooth regularization

while in view (7.9.18)-(7.9.21), it follows that for $\ell \in \mathbb{N}$, we have

(7.9.23) $\qquad w_s \in I_\ell, \quad s \in A_\ell$

(7.9.24) $\qquad I_{w_s} \subset I_\ell \subset I$

Furthermore, Theorems 6 and 7 in Section 4, Chapter 6, will hold for the C^∞-smooth regularizations (7.9.22) and the corresponding quotient algebras of generalized functions

(7.9.25) $\qquad A^\ell = A_\ell(V,S)/I_\ell(V,S), \quad \ell \in \mathbb{N}$

which form the chain of algebras of generalized functions

(7.9.26) $\qquad D'(\Omega) \subset A^\infty \rightarrow \ldots \rightarrow A^\ell \rightarrow .. A^0, \quad \ell \in \mathbb{N}$

Now we come to the following *global* version of the Cauchy-Kovalevskaia theorem, which for convenience is only formulated in the *polynomial nonlinear* case.

Theorem 6

Suppose given the analytic and polynomial nonlinear partial differential equation

(7.9.27) $$D_t^m U(t,y) = G(t,y,\ldots,D_t^p D_y^q U(t,y),\ldots),\ (t,y) \in \Omega$$

where $\Omega \subset \mathbb{R}^n$ is open, $0 \in \Omega$, $t \in \mathbb{R}$, $y \in \mathbb{R}^{n-1}$, $p \in \mathbb{N}$, $p < m$, $q \in \mathbb{N}^{n-1}$, $p + |q| \leq m$. Further, suppose given the analytic initial values

(7.9.28) $$D_t^p U(0,y) = g_p(y),\quad p \in \mathbb{N},\ p < m,\ y \in V$$

where $V \subset \mathbb{R}^{n-1}$ is open.

Then, one can construct sequences of functions

(7.9.29) $$s \in (C^\infty(\Omega))^{\mathbb{N}}$$

and C^∞-smooth regularizations (7.9.16) with the corresponding algebras (7.9.25), such that

(7.9.30) $$U_\ell = s + I_\ell(V,S) \in A^\ell,\quad \ell \in \mathbb{N},\quad \ell \geq m$$

satisfy the equation (7.9.27) in the *usual algebraic sense*, with multiplication in A^k and partial derivatives $D^p : A^\ell \to A^k$, $p \in \mathbb{N}^n$, $|p| \leq m$, $\ell,k \in \mathbb{N}$, $k+m \leq \ell$

Morevover, $U_\ell \in A^\ell$, with $\ell \in \mathbb{N}$, are usual analytic functions on an open set containing $\{0\} \times V$, and they satisfy (7.9.28).

It follows that s is a *chain weak solution* of (7.9.27), see Section 5, Chapter 6.

Proof

It follows from (7.9.23) and the extended versions of Theorem 6 and 7 in Section 4, Chapter 6, corresponding to the C^∞-smooth regularizations in (7.9.22).

CHAPTER 8

THE PARTICULAR CASE OF COLOMBEAU'S ALGEBRAS

§1. SMOOTH APPROXIMATIONS AND REPRESENTATIONS

Recently, a particularly efficient, yet simple and elementary *nonlinear theory* of generalized functions has been introduced by J.F. Colombeau.

This nonlinear theory is based on the construction of certain differential algebras $\mathcal{G}(\mathbb{R}^n)$ of generalized functions, algebras which happen to be particular cases of the construction in Chapter 6, see for details Section 3 in the sequel.

The special interest in Colombeau's algebras $\mathcal{G}(\mathbb{R}^n)$ comes from their rather natural and central role within the general nonlinear theory constructed in Chapter 6, see for details Appendices 1 and 3, at the end of this Chapter.

Although results such as in Chapters 2, 3 and 4 could so far not been possible to reproduce in Colombeau's algebras, nevertheless, the natural and central role of these algebras allow the proof of the existence in Colombeau's algebras of generalized solutions for large classes of earlier unsolved or unsolvable linear or nonlinear partial differential equations, some of the latter having for instance a basic role in quantum field interaction theory. A first systematic account of this theory was presented in Colombeau [1], where the method employed was based on the De Silva differential in locally convex spaces. Soon after, in Colombeau [2, 4], a completely *elementary* presentation of the mentioned nonlinear theory of generalized functions has been achieved. In view of its rather surprising ease, which makes it readily accessible to wide groups of mathematicians, plysicists, engineers, etc., we shall only deal with that latter version. It is particularly interesting to note that the version in Colombeau [2, 4], although needs some rudiments of the linear theory of distributions, is in fact *much simpler* than any nontrivial presentation of that latter linear theory known up until now, presentation which would go at least as far as the solution of *variable* coefficient linear partial differential equations.

In fact, Colombeau's theory appears to lead to one of the ultimate possible simplification in the study of linear and nonlinear partial differential equations. Indeed, it only uses two basic tools: elementary calculus and topology in Euclidean spaces, as well as quotient structures in commutative rings of smooth functions. The effect is the reduction of most of the mathematics involved to usual partial derivatives and multiple integrals, respectively to chasing arrows in diagrams involving rings and ideals of smooth functions. What turns all that into a surprisingly *efficient* method for *solving* linear and nonlinear partial differential equations is an *asymptotic interpretation*, based on the specific structure of the *index set* - see (8.1.13) - which is used in the definition of Colombeau's generalized functions in (8.1.18)-(8.1.21). See also Appendices 2 and 4 at the end of this Chapter. That asymptotic interpretation proves to be much more efficient than *topological properties* on various rather sophisticated spaces of functions. In this way, we are presented with one of the deeper insight into the workings involved in connection with the solution of linear and nonlinear partial differential equations.

In this Chapter, we shall only present a short account of Colombeau's nonlinear theory. For further details one can consult Colombeau [1, 2], Biagioni [2], and Rosinger [3, pp. 49-192].

Before we present Colombeau's method, let us in essence recapitulate the *classical* and *linear distributional* framework used in the study of linear or nonlinear partial differential equations, and for simplicity, let us assume that the domain of the independent variables is $\Omega = \mathbb{R}^n$. Then, the spaces of possible solutions are

(8.1.1) $$\mathcal{A}n \subset \mathcal{C}^\infty \subset \ldots \subset \mathcal{C}^0 \subset \mathcal{L}^1_{loc} \subset \mathcal{D}'$$

where by $\mathcal{A}n$ we denoted the analytic functions.

From the point of view of nonlinear operations, in particular unrestricted multiplication, the above spaces, except for $\mathcal{L}^1_{loc}$ and $\mathcal{D}'$, are particularly suitable, since they are associative and commutative algebras with unit element, moreover, they are closed under a wide range of nonlinear operations of appropriate smoothness.

From the point of view of *partial derivability*, only $\mathcal{A}n$, $\mathcal{C}^\infty$ and $\mathcal{D}'$ allow indefinite iterations of such operations.

It follows that in (8.1.1), $\mathcal{A}n$ and $\mathcal{C}^\infty$ alone are the suitable kind of *differential algebras* for a convenient study of *nonlinear* partial differential equations.

Unfortunately however, as seen in Chapter 5, $\mathcal{A}n$ and $\mathcal{C}^\infty$ are not sufficiently large for the above purpose, and in fact, we would need some *differential algebras* A, sufficiently *large* in order to contain the $\mathcal{D}'$ distributions, i.e.,

(8.1.2) $$\mathcal{D}' \subset A$$

It is important to note that in view of Section 5, Chapter 5, extensions of the space $\mathcal{D}'$ of distributions are needed even for the solution of *linear* variable coefficient partial differential equations.

And then, let us present the way Colombeau is constructing an extension such as in (8.1.2).

The basic idea is very simple: as is well known, Schwartz [1], every distribution can be approximated by $\mathcal{C}^\infty$-smooth functions, in particular according to the following 'convolution with a δ-sequence' procedure, see also Mikusinski [1].

Suppose given an arbitrary, fixed distribution $T \in \mathcal{D}'$.

For any $\psi \in \mathcal{D}$ with

(8.1.3) $$\int_{\mathbb{R}^n} \psi(x)dx = 1$$

and $\epsilon \in (0,\infty)$, let us define $\psi_\epsilon \in \mathcal{D}$ by

(8.1.4) $$\psi_\epsilon(x) = \psi(x/\epsilon)/\epsilon^n, \quad x \in \mathbb{R}^n$$

Then, using the convolution of distributions, we define

(8.1.5) $$f_\epsilon(x) = (T*\psi_\epsilon)(x), \quad x \in \mathbb{R}^n, \ \epsilon \in (0,\infty)$$

and obtain that

(8.1.6) $$f_\epsilon \in \mathcal{C}^\infty, \quad \epsilon \in (0,\infty)$$

But $\psi_\epsilon \longrightarrow \delta$ in $\mathcal{D}'$, when $\epsilon \longrightarrow 0$, hence, in view of a basic property convolution we obtain the following *smooth approximation* property

(8.1.7) $$f_\epsilon \longrightarrow T \text{ in } \mathcal{D}', \text{ when } \epsilon \longrightarrow 0$$

In this way, to each distribution $T \in \mathcal{D}'$, one can associate sequences of functions $(f_\epsilon | \epsilon \in (0,\infty)) \in (\mathcal{C}^\infty)^{(0,\infty)}$ which converge to T in $\mathcal{D}'$. In other words, we obtain

(8.1.8) $$\mathcal{D}' \text{ '}\subset\text{' } (\mathcal{C}^\infty)^{(0,\infty)}$$

where the so called inclusion '$\subset$' - which is not yet a proper inclusion - means the above *multivalent* association

(8.1.9) $$T \longmapsto (f_\epsilon | \epsilon \in (0,\infty))$$

The *importance* of (8.1.8) is that $(\mathcal{C}^\infty)^{(0,\infty)}$ is obviously a *differential algebra* with the term-wise operations on the sequences of functions. Therefore (8.1.8) is *nearly* one of the desired extension (8.1.2) of the distributions. Indeed, all we need is to manage to take away the quotation marks in (8.1.8), that is, to replace (8.1.9) by a convenient *univalent* mapping.

In Colombeau's method, this is done by a particular choice of an index set Λ, of a subalgebra $\mathcal{A}$ in $(\mathcal{C}^\infty)^\Lambda$ and finally of an ideal $\mathcal{I}$ in $\mathcal{A}$, which together give us the following extension, i.e., embedding

$$(8.1.10) \qquad \mathcal{D}' \subset A = \mathcal{A}/\mathcal{I}$$

of the distributions $\mathcal{D}'$ into the *differential algebra* $A = \mathcal{A}/\mathcal{I}$. In this way, we obtain, among others, a *representation* of distributions as classes of sequences of *smooth* functions. One of the special features of Colombeau's method is the choice of the above *index* set Λ, to which we turn now.

In order to define the index set Λ, we proceed as follows. For $m \in \mathbb{N}_+ = \mathbb{N}\setminus\{0\}$, we denote

$$(8.1.11) \qquad \Phi_m = \left\{ \phi \in \mathcal{D}(\mathbb{R}^n) \;\middle|\; \begin{array}{l} *) \ \int_{\mathbb{R}^n} \phi(x)dx = 1 \\ **) \ \forall \ p \in \mathbb{N}^n, \ 1 \leq |p| \leq m : \\ \qquad \int_{\mathbb{R}^n} x^p \phi(x)dx = 0 \end{array} \right\}$$

The avove condition *) is of course the same with (8.1.3) and is needed in order to obtain 'δ-sequences' by the method in (8.1.4). The condition **) wwill be needed in connection with the embedding in the diagram (8.2.27).

From (8.1.11) it follows obviously that we have the inclusions

$$(8.1.12) \qquad \Phi_1 \supset \ldots \supset \Phi_m \supset \ldots$$

We shall take as our basic *index* set

$$(8.1.13) \qquad \Lambda = \Phi = \Phi_1$$

The nontriviality of Φ results from

Lemma 1

$$(8.1.14) \qquad \Phi_m \neq \phi, \quad m \in \mathbb{N}_+$$

$$(8.1.15) \qquad \bigcap_{m \in \mathbb{N}_+} \Phi_m = \phi \qquad \square$$

It should be noted that, while property (8.1.14) is of course essential for Colombeau's method, property (8.1.15) will not be used in the sequel and it was only given as a general information connected with (8.1.12).

As a last remark on the index set Φ, we note that, for $m \in \mathbb{N}_+$ and with the definition in (8.1.4), we have

(8.1.16) $\phi \in \Phi_m \Rightarrow (\phi_\epsilon \in \Phi_m, \ \epsilon > 0)$

italics elitesequences which is the framework of Colombeau's method is given by

(8.1.17) $\mathcal{E}[\mathbb{R}^n] = (\mathcal{C}^\infty(\mathbb{R}^n))^\Phi$

which is obviously an *associative*, *commutative differential algebra* with the term-wise operations on the respective sequences of functions.

Let us now define the following subalgebra $\mathcal{A}$ in $\mathcal{E}[\mathbb{R}^n]$, which consists of all $f \in \mathcal{E}[\mathbb{R}^n]$ such that

(8.1.18)
$$\begin{aligned}
&\forall \ K \subset \mathbb{R}^n \text{ compact}, \ p \in \mathbb{N}^n : \\
&\exists \ m \in \mathbb{N}_+ : \\
&\forall \ \phi \in \Phi_m : \\
&\exists \ \eta, \ c < 0 : \\
&\forall \ x \in K, \ \ \epsilon \in (0,\eta) : \\
&\qquad |D^p f(\phi_\epsilon, x)| \leq \frac{c}{\epsilon^m}
\end{aligned}$$

where D^p is the usual partial derivative of order $p \in \mathbb{N}^n$, of the $\mathcal{C}^\infty$-smooth function

$$\mathbb{R}^n \ni x \longmapsto f(\phi_\epsilon, x) \in \mathbb{C}$$

Further, let us define the following *ideal* $\mathcal{I}$ in $\mathcal{A}$, given by all $f \in \mathcal{A}$ such that

(8.1.19)
$$\begin{aligned}
&\forall \ K \subset \mathbb{R}^n \text{ compact}, \ p \in \mathbb{N}^n : \\
&\exists \ \ell \in \mathbb{N}_+, \ \beta \in B : \\
&\forall \ m \in \mathbb{N}_+, \ m \geq \ell, \ \phi \in \phi_m : \\
&\exists \ \eta, \ c < 0 : \\
&\forall \ x \in K, \ \ \epsilon \in (0,\eta) : \\
&\qquad |D^p f(\phi_\epsilon, x)| \leq c\epsilon^{\beta(m)-\ell}
\end{aligned}$$

where we denoted

(8.1.20)
$$B = \left\{ \beta : \mathbb{N}_+ \longrightarrow (0,\infty) \ \middle| \ \begin{array}{ll} *) & \beta \text{ increasing} \\ **) & \lim\limits_{m\to\infty} \beta(m) = \infty \end{array} \right\}$$

Finally, the *algebra of generalized functions* of Colombeau is given by

(8.1.21) $\mathcal{G}(\mathbb{R}^n) = \mathcal{A}/\mathcal{I}$

An easy direct check shows that $\mathcal{A}$ is indeed a subalgebra in $\mathcal{E}[\mathbb{R}^n]$ and $\mathcal{I}$ is an ideal in $\mathcal{A}$, therefore, $\mathcal{G}(\mathbb{R}^n)$ is an *associative*, *commutative algebra*. Further, we obviously have

(8.1.22) $D^p\mathcal{I} \subset \mathcal{I},\ D^p\mathcal{A} \subset \mathcal{A},\ p \in \mathbb{N}^n$

therefore, we can define the *partial derivative operators*

(8.1.23) $D^p : \mathcal{G}(\mathbb{R}^n) \longrightarrow \mathcal{G}(\mathbb{R}^n), \quad p \in \mathbb{N}^n$

by

(8.1.24) $D^p(f+\mathcal{I}) = D^pf + \mathcal{I}, \quad p \in \mathbb{N}^n, \quad f \in \mathcal{A}$

Obviously, the above partial derivative operators on $\mathcal{G}(\mathbb{R}^n)$ are *linear* and satisfy the Leibnitz rule of *product derivative*.

There are various rather natural *motivations* for the particular, somewhat complicated and unexpected way the algebra $\mathcal{A}$ and ideal $\mathcal{I}$ were defined above. These motivations become apparent as Colombeau's theory and its applications are presented. A rather direct, functional analytic motivation for the *necessary structure* of $\mathcal{A}$ and $\mathcal{I}$ is given in Appendix 1, at the end of this Chapter. Here we should only like to point out a certain analogy with the quotient space respresentation of distributions in (5.A1.24). Indeed, similar to $\mathcal{V}^\infty$, the ideal $\mathcal{I}$ has to model a certain 'convergence to zero' property, which in (8.1.19) is one of the weakest possible for an ideal, thus subalgebra, being only polynomial in ϵ. This makes $\mathcal{I}$ in a way the largest possible subalgebra, in particular ideal, which is convenient from the point of view of the *stability* of generalized solutions. But as $\mathcal{I}$ has to be an ideal in $\mathcal{A}$, this sets a constraint on the possible growth of the elements in $\mathcal{A}$. From here, we obtain the growth condition (8.1.18) which is polynomial in $1/\epsilon$. The role played by the other parameters, such as K, p, m, η, etc., in (8.1.18) and (8.1.19), will become clear in the sequel, see also the mentioned Appendix 1.

Before we go further, it is useful to note the *nontriviality* of $\mathcal{A}$ and $\mathcal{I}$, which follows among others from the fact that

(8.1.25) $\mathcal{A} \subsetneqq \mathcal{E}[\mathbb{R}^n]$ and $\mathcal{I}$ is *not* an ideal in $\mathcal{E}[\mathbb{R}^n]$

Indeed, let us define $f : \Phi \times \mathbb{R}^n \longrightarrow \mathbb{C}$ by

$$f(\phi,x) = e^{1/d(\phi)}, \quad \phi \in \Phi,\ x \in \mathbb{R}^n$$

where

$$d(\phi) = \sup\{\|x-y\| \mid x,y \in \operatorname{supp} \phi\}, \quad \phi \in \Phi$$

Further, let us define $g : \Phi \times \mathbb{R}^n \to \mathbb{C}$ by

$$g(\phi,x) = 1/f(\phi,x), \quad \phi \in \Phi, \ x \in \mathbb{R}^n$$

Since $d(\phi_\epsilon) = \epsilon d(\phi)$, with $\phi \in \Phi$, $\epsilon > 0$, it follows easily that

$$f \in \mathcal{E}[\mathbb{R}^n] \setminus \mathcal{A}, \quad g \in \mathcal{I}, \ f \cdot g = 1 \notin \mathcal{I}$$

hence (8.1.25).

Finally, we should note that the above definitions of $\mathcal{A}$ and $\mathcal{I}$ may suggest the following question: Isn't it that $\mathcal{A}$ is precisely the set of Cauchy sequences in a certain topology on $\mathcal{C}^\infty(\mathbb{R}^n)$, and furthermore, $\mathcal{I}$ is precisely the set of sequences convergent to zero in the same topology? In which case $\mathcal{G}(\mathbb{R}^n)$ given in (8.1.21) would obviously be the completion of $\mathcal{C}^\infty(\mathbb{R}^n)$ in the mentioned topology?

As seen in Appendix 2 at the end of this Chapter, the answer is *negative*, insofar as $\mathcal{A}$ and $\mathcal{I}$ are not the Cauchy and convergent to zero sequences in *any* topology on $\mathcal{C}^\infty(\mathbb{R}^n)$. In particular, the quotient structure which defines $\mathcal{G}(\mathbb{R}^n)$ in (8.1.21) is not a usual topological construction of a completion of $\mathcal{C}^\infty(\mathbb{R}^n)$.

§2. PROPERTIES OF THE DIFFERENTIAL ALGEBRA $\mathcal{G}(\mathbb{R}^n)$

Of course, our first interest is to prove the embeddings

(8.2.1) $$\mathcal{C}^\infty \subset \ldots \subset \mathcal{C}^0 \subset \mathcal{D}' \subset \mathcal{G}(\mathbb{R}^n)$$

and establish the way the algebraic and/or differential operations on $\mathcal{C}^\infty$, $\mathcal{C}^0$ and $\mathcal{D}'$ extend to $\mathcal{G}(\mathbb{R}^n)$.

In this respect, it will be more convenient to move from particular to general, as in this way we shall have the opportunity to become more familiar with the differential algebra $\mathcal{G}(\mathbb{R}^n)$.

Therefore, let us start with the easiest embedding

(8.2.2) $$\mathcal{C}^\infty \subset \mathcal{G}(\mathbb{R}^n)$$

which is defined by the mapping

(8.2.3) $$\mathcal{C}^\infty \ni f \longmapsto \tilde{f} + \mathcal{I} \in \mathcal{G}(\mathbb{R}^n)$$

where

(8.2.4) $$\tilde{f}(\phi,x) = f(x), \quad \phi \in \Phi,\ x \in \mathbb{R}^n$$

In order to show that the mapping (8.2.3) is well defined, we have to prove that

(8.2.5) $$\mathcal{C}^\infty \ni f \longmapsto \tilde{f} \in \mathcal{A}$$

which follows easily from (8.1.18), by noticing that

(8.2.6) $$D^p\tilde{f}(\phi_\epsilon,x) = D^pf(x), \quad p \in \mathbb{R}^n,\ \phi \in \Phi,\ \epsilon > 0,\ x \in \mathbb{R}^n$$

Further, we note that the mapping (8.2.3) is injective, as in view of (8.1.19) and (8.2.6), we obtain

(8.2.7) $$(f \in \mathcal{C}^\infty,\ \tilde{f} \in \mathcal{I}) \Rightarrow f = 0$$

In view of (8.2.4) and (8.2.5), it is obvious that the mapping (8.2.3) is an algebra homomorphism. Finally, in view of (8.1.23), (8.1.24), (8.2.3) and (8.2.4), it follows that the partial derivative operators D^p, with $p \in \mathbb{N}^n$, on $\mathcal{G}(\mathbb{R}^n)$, will coincide with the usual partial derivatives of functions, when restricted to $\mathcal{C}^\infty(\mathbb{R}^n)$.

We can resume the above in the following:

Theorem 1

The embedding $\mathcal{C}^\infty(\mathbb{R}^n) \subset \mathcal{G}(\mathbb{R}^n)$ defined in (8.2.3) is an embedding of *differential algebras*, and the function $1 \in \mathcal{C}^\infty(\mathbb{R}^n)$ is the unit in the algebra $\mathcal{G}(\mathbb{R}^n)$. □

Let us now proceed further and define the embedding

(8.2.8) $$\mathcal{C}^0 \subset \mathcal{G}(\mathbb{R}^n)$$

by mapping

(8.2.9) $$\mathcal{C}^0 \ni f \longmapsto \bar{f} + \mathcal{I} \in \mathcal{G}(\mathbb{R}^n)$$

where

$$(8.2.10) \qquad \bar{f}(\phi,x) = \int_{\mathbb{R}^n} f(x+y)\phi(y)dy = \int_{\mathbb{R}^n} f(y)\phi(y-x)dy, \quad \phi \in \Phi,\ x \in \mathbb{R}^n$$

As above, in order to prove that (8.2.9) is well defined, we have to show that

$$(8.2.11) \qquad \mathcal{C}^0 \ni f \longmapsto \bar{f} \in \mathcal{A}$$

First we note that in view of (8.2.10), we have $\bar{f} \in \mathcal{E}[\mathbb{R}^n]$, since $\phi \in \mathcal{D}(\mathbb{R}^n)$. Further obviously

$$(8.2.12) \quad D^p\bar{f}(\phi_\epsilon,x) = \frac{(-1)^{|p|}}{\epsilon^{n+|p|}} \int_{\mathbb{R}^n} f(y)D^p\phi(\tfrac{y-x}{\epsilon}), \; p \in \mathbb{N}^n,\ \phi \in \Phi,\ \epsilon > 0,\ x \in \mathbb{R}^n$$

hence (8.2.11) follows easily from (8.1.18). Now, from (8.1.19) and (8.2.12), it follows easily that

$$(8.2.13) \qquad (f \in \mathcal{C}^0,\ \bar{f} \in \mathcal{I}) \Rightarrow f = 0$$

hence the mapping (8.2.9) is injective.

In this way we obtain:

Theorem 2

The embedding $\mathcal{C}^0(\mathbb{R}^n) \subset \mathcal{G}(\mathbb{R}^n)$ defined in (8.2.9) is an embedding of *vector spaces.* □

Before we study further properties of the above embedding, it is useful to define the embedding

$$(8.2.14) \qquad \mathcal{D}' \subset \mathcal{G}(\mathbb{R}^n)$$

by the mapping

$$(8.2.15) \qquad \mathcal{D}' \ni T \longmapsto f_T + \mathcal{I} \in \mathcal{G}(\mathbb{R}^n)$$

where we define

$$(8.2.16) \qquad f_T(\phi,x) = T_y(\phi(y-x)), \quad \phi \in \Phi,\ x \in \mathbb{R}^n$$

Obviously (8.2.15) is an extension of (8.2.9), as it reduces to the latter when the distribution T is generated by a continuous function f. The fact that $f_T \in \mathcal{E}[\mathbb{R}^n]$, with $T \in \mathcal{D}'$, follows easily from basic results concerning the convolution of distributions, Schwartz [1]. In the same way one obtains

(8.2.17) $$D^p f_T(\phi_\epsilon, x) = T_y(D^p \phi_\epsilon(y-x)),\ p \in \mathbb{N}^n,\quad \phi \in \Phi,\ \epsilon > 0,\quad x \in \mathbb{R}^n$$

and then (8.1.18) yields

(8.2.18) $$\mathcal{D}' \ni T \longmapsto f_T \in \mathcal{A}$$

thus the mapping (8.2.15) is well defined.

From (8.2.12) and (8.1.19) also follows that

(8.2.19) $$(T \in \mathcal{D}',\ f_T \in \mathcal{I}) \Rightarrow T = 0$$

which means that the mapping (8.2.15) is injective.

Finally, in view of (8.1.23), (8.2.15), and (5.4.3), it follows that the partial derivative operators D^p, with $p \in \mathbb{N}^n$, on $\mathcal{G}(\mathbb{R}^n)$, will coincide with the distributional partial derivatives, when restricted to $\mathcal{D}'(\mathbb{R}^n)$. In particular, D^p on $\mathcal{G}(\mathbb{R}^n)$, coincides with the usual partial derivative D^p of smooth functions, when restricted to $\mathcal{C}^\ell(\mathbb{R}^n)$, with $\ell \in \mathbb{N}$, $\ell \geq |p|$.

We can conclude as follows:

Theorem 3

The embedding $\mathcal{D}(\mathbb{R}^n) \subset \mathcal{G}(\mathbb{R}^n)$ defined in (8.2.15) is an embedding of *vector spaces* which *extends* the distributional partial derivatives. □

Remark 1

As seen in Example 1 below, the vector space embedding $\mathcal{C}^0(\mathbb{R}^n) \subset \mathcal{G}(\mathbb{R}^n)$ in (8.2.9) is *not* an embedding of algebras, i.e., the multiplication in $\mathcal{G}(\mathbb{R}^n)$, when restricted to $\mathcal{C}^0(\mathbb{R}^n)$, *does not always* coincide with the usual multiplication of continuous functions.

It is *particularly important* to note that many of the effects of the above *deficiency* concerning the multiplication in $\mathcal{G}(\mathbb{R}^n)$ of the usual continuous functions will be *eliminated* with the introduction of the *coupled calculus* presented in Section 5. This *coupled calculus* is a specific and essential feature of Colombeau's method and it is one way to overcome the conflict between insufficient smoothness, multiplication and differentiation presented in Chapter 1. However, as seen in (1.1.15), the use of *one single* differential algebra in (8.1.23) may lead to *undesirable basic limitations*, which so far, can be overcome only by using chains of algebras, such as those in Chapter 6, or going even beyond that framework, as is done in Chapter 4.

Now, let us consider several examples which help to clarify the above embedding properties. For simplicity, we consider the one dimensional case, when $n = 1$.

Example 1

In connection with Remark 1, let us take $f_1, f_2 \in \mathcal{C}^0(\mathbb{R}^n)$ defined by

$$(8.2.20) \quad f_1(x) = x_- = \begin{cases} x & \text{if } x \leq 0 \\ 0 & \text{if } x \geq 0 \end{cases}, \quad f_2(x) = x_+ = \begin{cases} 0 & \text{if } x \leq 0 \\ x & \text{if } x \geq 0 \end{cases}, \quad x \in \mathbb{R}$$

Then, with the product in $\mathcal{C}^0(\mathbb{R}^n)$, we have

$$(8.2.21) \qquad f_1 \cdot f_2 = 0$$

However, in $\mathcal{G}(\mathbb{R})$ we shall have

$$(8.2.22) \qquad (\bar{f}_1 + \mathcal{I}) \cdot (\bar{f}_2 + \mathcal{I}) = g + \mathcal{I} \in \mathcal{G}(\mathbb{R})$$

where, in view of (8.1.10), we have

$$(8.2.23) \qquad g(\phi_\epsilon, 0) = \epsilon^2 \left(\int_{-\infty}^{0} y\phi(y)dy \right) \left(\int_{0}^{\infty} y\phi(y)dy \right), \quad \epsilon > 0$$

and then, in view of (8.1.19), it follows that

$$(8.2.24) \qquad g \notin \mathcal{I}$$

since we have

$$(8.2.25) \qquad \begin{array}{l} \forall \ \phi \in \mathbb{N}_+ : \\ \exists \ \phi \in \Phi_m : \\ \displaystyle\int_{-\infty}^{0} x\phi(x)dx, \ \int_{0}^{\infty} x\phi(x)dx \neq 0 \end{array}$$

Obviously (8.2.22) and (8.2.24) yield

$$(8.2.26) \qquad (\bar{f} + \mathcal{I}) \cdot (\bar{f} + \mathcal{I}) \neq 0 \in \mathcal{G}(\mathbb{R})$$

In this way, the product of the continuous functions x_- and x_+ is zero in $\mathcal{C}^0(\mathbb{R})$, but it is no longer zero in $\mathcal{G}(\mathbb{R})$.

Returning now to the above general embedding results, one can note that, while the embedding $\mathcal{C}^0 \subset \mathcal{G}$ is a particular case of the embedding $\mathcal{D}' \subset \mathcal{G}$, the embedding $\mathcal{C}^\infty \subset \mathcal{G}$ does *not* at first *seem* to be a particular case of these latter two ones. Indeed, both (8.2.10) and (8.2.16) are the same kind of convolution formulas, while (8.2.4) is obviously not. However, this difference is only apparent, since we have the following *commutative diagram*

(8.2.27)
$$\begin{array}{ccc} \mathcal{C}^\infty(\mathbb{R}^n) \ni f & \longmapsto & \tilde{f}+\mathcal{I} \in \mathcal{G}(\mathbb{R}^n) \\ \subset \Big\downarrow & & \Big\updownarrow \text{id} \\ \mathcal{C}^0(\mathbb{R}^n) \ni f & \longmapsto & \bar{f}+\mathcal{I} \in \mathcal{G}(\mathbb{R}^n) \end{array}$$

Indeed, one can prove, see Rosinger [3], that

(8.2.28) $$f \in \mathcal{C}^\infty(\mathbb{R}^n) \Rightarrow \bar{f} - \tilde{f} \in \mathcal{I}$$

§3. COLOMBEAU'S ALGEBRA $\mathcal{G}(\mathbb{R}^n)$ AS A COLLAPSED CASE OF CHAINS OF ALGEBRAS

We show that Colombeau's algebra of generalized functions, see (8.1.21)

(8.3.1) $$\mathcal{G}(\mathbb{R}^n) = \mathcal{A}/\mathcal{I}$$

is of the type (6.1.24), and in particular, $\mathcal{I}$ is a $\mathcal{C}^\infty$-smooth regular ideal in the sense of Definition 2 in Section 2, Chapter 6.

In view of (8.1.18) and (8.1.19) we shall take, see (8.1.13)

(8.3.2) $$\Lambda = \Phi$$

and $\ell = \infty$. That places us within the framework in Section 2, Chapter 6, since $\mathcal{A}$ is a subalgebra in $(\mathcal{C}^\infty)^\Lambda$ and $\mathcal{I}$ is an ideal in $\mathcal{A}$. Furthermore, in view of (8.2.2)-(8.2.7), condition (6.1.27) is obviously satisfied.

Within this Section, it will be convenient to consider the elements of $(\mathcal{C}^\infty)^\Lambda$ as given by functions

$$f : \Phi \times \mathbb{R}^n \to \mathbb{C}$$

such that $f(\phi, \cdot) \in \mathcal{C}^\infty(\mathbb{R}^n)$, for $\phi \in \Phi$.

Now, let us define

$$(8.3.3) \qquad \mathcal{S}^\infty_\Lambda$$

as the set of all $f \in (\mathcal{C}^\infty)^\Lambda$ such that

$$(8.3.4) \qquad \begin{array}{l} \exists \ T \in \mathcal{D}'(\mathbb{R}^n) : \\ \forall \ \Omega \subset \mathbb{R}^n \text{ open, bounded} : \\ \exists \ m \in \mathbb{N}_+ : \\ \forall \ \phi \in \Phi_m : \\ \quad \lim\limits_{\epsilon \downarrow 0} f(\phi_\epsilon, \cdot)\Big|_\Omega = T\Big|_\Omega \end{array}$$

where the limit is taken in the sense of the weak topology of $\mathcal{D}'(\Omega)$. Obviously, the conditions (6.1.2)-(6.1.5) will be satisfied. Therefore, with $\mathcal{S}^\infty_\Lambda$ in (8.3.3), the kernel

$$(8.3.5) \qquad \mathcal{V}^\infty_\Lambda$$

of the linear surjection (6.1.3) will satisfy (6.1.6) and (6.1.7).

With $\mathcal{V}^\infty_\Lambda$, $\mathcal{S}^\infty_\Lambda$, $\mathcal{A}$ and $\mathcal{I}$ given as above, it only remains to define $\mathcal{S}$ and $\mathcal{V}$ in order to be able to enter within the framework of Section 2, Chapter 6. And then, we take

$$(8.3.6) \qquad \mathcal{S} = \left\{ f \in \mathcal{A} \;\middle|\; \begin{array}{l} \exists \ T \in \mathcal{D}'(\mathbb{R}^n) : \\ T = f + \mathcal{I} \in \mathcal{G}(\mathbb{R}^n) \end{array} \right\}$$

In view of (8.2.14)-(8.2.19), it follows that

$$(8.3.7) \qquad \mathcal{U}_{\mathcal{C}^\infty, \Lambda} \subset \mathcal{S}$$

and the mapping

$$(8.3.8) \qquad \mathcal{S} \ni f \longmapsto T \in \mathcal{D}'(\mathbb{R}^n)$$

is a *linear surjection* which satisfies (6.1.5). Finally, let

$$(8.3.9) \qquad \mathcal{V}$$

be the kernel of (8.3.8).

<u>Proposition 1</u>

The following relations hold

(8.3.10) $\mathcal{I} \cap \mathcal{S} = \mathcal{V}$

(8.3.11) $\mathcal{I} \subset \mathcal{V}_\Lambda^\infty$

(8.3.12) $\mathcal{S} \subset \mathcal{S}_\Lambda^\infty$

(8.3.13) $\mathcal{V}_\Lambda^\infty \cap \mathcal{S} = \mathcal{V}$

(8.3.14) $\mathcal{I} \cap \mathcal{S}_\Lambda^\infty \subset \mathcal{V}_\Lambda^\infty$

<u>Proof</u>

(8.3.10).
The inclusion $\subset$ result as follows. Let $f \in \mathcal{I} \cap \mathcal{S}$. Then $f \in \mathcal{I}$ yields

$$T = f + \mathcal{I} = 0 \in \mathcal{G}(\mathbb{R}^n)$$

hence (8.3.8), (8.3.9) yield $f \in \mathcal{V}$. Conversely, if $f \in \mathcal{V}$ then (8.3.8), (8.3.9) yield $T = 0 \in \mathcal{D}'(\mathbb{R}^n)$, hence $T = 0 \in \mathcal{G}(\mathbb{R}^n)$. But $\mathcal{V} \subset \mathcal{S} \subset \mathcal{A}$, thus $f \in \mathcal{A}$. Now $f + \mathcal{I} = T = 0 \in \mathcal{G}(\mathbb{R}^n)$ yields $f \in \mathcal{I}$.

(8.3.11).
Take $f \in \mathcal{I}$, then (8.1.19), (8.3.4) and (8.3.5) yield $f \in \mathcal{V}_\Lambda^\infty$.

(8.3.12).
Take $f \in \mathcal{S}$, then (8.3.6) implies

(8.3.15) $f \in \mathcal{A},\ T = f + \mathcal{I} \in \mathcal{D}'(\mathbb{R}^n) \subset \mathcal{G}(\mathbb{R}^n)$

Assume given $\Omega \subset \mathbb{R}^n$ open and $\psi \in \mathcal{D}(\Omega)$. Then, as seen in Rosinger [3, p. 302],

(8.3.16)
$$\begin{aligned} &\exists\ m \in \mathbb{N}_+ : \\ &\forall\ \phi \in \Phi_m : \\ &\quad \lim_{\epsilon \downarrow 0} g(\phi_\epsilon) = T(\psi) \end{aligned}$$

where

(8.3.17) $$g(\phi) = \int_{\mathbb{R}^n} \psi(x) f(\phi, x) dx, \quad \phi \in \Phi(\mathbb{R}^n)$$

And (8.3.16), (8.3.17) obviously imply that $f \in \mathcal{S}^{\infty}_{\Lambda}$.

(8.3.13).
We note that for $f \in \mathcal{S}$, the relations (8.3.15), (8.3.16) provide for $T \in \mathcal{D}'(\mathbb{R}^n)$, such that

(8.3.18)
$$\begin{aligned} &\forall \quad \Omega \subset \mathbb{R}^n \text{ open} : \\ &\exists \quad m \in \mathbb{N}_+ : \\ &\lim_{\epsilon \downarrow 0} f(\phi_\epsilon, \cdot)\Big|_\Omega = T\Big|_\Omega \end{aligned}$$

where the limit is taken in the sense of the weak topology on $\mathcal{D}'(\Omega)$. If $f \in \mathcal{V}$ then (8.3.9) implies that $T = 0 \in \mathcal{D}'(\mathbb{R}^n)$. Hence (8.3.18) and (8.3.5) yield $f \in \mathcal{V}^{\infty}_{\Lambda}$. Thus, we obtain the inclusion

$$\mathcal{V} \subset \mathcal{V}^{\infty}_{\Lambda} \cap \mathcal{S}$$

Conversely, if $f \in \mathcal{V}^{\infty}_{\Lambda}$ then (8.3.5) gives $T = 0 \in \mathcal{D}'(\mathbb{R}^n)$, hence (8.3.9) yields $f \in \mathcal{V}$.

(8.3.14).
It follows from (8.3.11) □

Before going further, we should note that

(8.3.19) $$\mathcal{V}^{\infty}_{\Lambda} \not\subset \mathcal{A}$$

hence

(8.3.20) $$\mathcal{S}^{\infty}_{\Lambda} \not\subset \mathcal{A}$$

Indeed, let us take $\alpha, \gamma : (0,\infty) \to (0,\infty)$ such that

(8.3.21) $$\lim_{\epsilon \downarrow 0} \alpha(\epsilon)/e^{1/\epsilon} = \lim_{\epsilon \downarrow 0} \gamma^n(\epsilon)/\alpha(\epsilon) = \infty$$

and define $f_{\alpha,\gamma} : \Phi \times \mathbb{R}^n \to \mathbb{C}$ by

(8.3.22) $$f_{\alpha,\gamma}(\phi,x) = \alpha(d(\phi))\phi(\gamma(d(\phi))x), \quad \phi \in \Phi, \ x \in \mathbb{R}^n$$

where $d(\phi)$ is given in Section 1. Then, in view of (8.3.21), it follows easily that

(8.3.23) $$f_{\alpha,\gamma} \in \mathcal{V}_\Lambda^\infty$$

However

(8.3.24) $$f_{\alpha,\gamma} \notin \mathcal{A}$$

since, for $p \in \mathbb{N}^n$, $\epsilon > 0$, $x \in \mathbb{R}^n$, we have

$$D^p f_{\alpha,\gamma}(\phi_\epsilon,x) = \alpha(\epsilon d(\phi))(\gamma(d(\phi)))^{|p|} D^p\phi(\gamma(\epsilon d(\phi))x/\epsilon)/\epsilon^{n+|p|}$$

hence (8.1.18) and (8.3.24) will obviously imply (8.3.24).

A useful consequence of Proposition 1 is the following:

Corollary 1

We have

(8.3.25) $$\underset{\mathcal{I}\cap\mathcal{S}_\Lambda^\infty}{\operatorname{codim}} \mathcal{I} \cap \mathcal{V}_\Lambda^\infty = 0$$

therefore

(8.3.26) $\mathcal{I}$ is a small ideal in $\mathcal{A}$

Proof

The relation (8.3.25) follows directly from (8.3.14). Now (8.3.26) is obvious, in view of (6.2.15). □

Now we can show that Colombeau's differential algebra of generalized functions $\mathcal{G}(\mathbb{R}^n) = \mathcal{A}/\mathcal{I}$, see (8.3.1), is a *particular case* of the quotient algebras of generalized functions constructed in (6.1.24), Section 1, Chapter. Indeed, we obtain the following result.

Theorem 4

Given Colombeau's differential algebra of generalized functions

(8.3.27) $$\mathcal{G}(\mathbb{R}^n) = \mathcal{A}/\mathcal{I}$$

then, with the notations in (8.3.3), (8.3.5), (8.3.6) and (8.3.9), we have that

(8.3.28) $\qquad (\mathcal{V},\mathcal{S},\mathcal{I},\mathcal{A})$ is a $\mathcal{C}^\infty$-smooth regularization

of the representation of distributions

(8.3.29) $\qquad \mathcal{D}'(\mathbb{R}^n) = \mathcal{S}_\Lambda^\infty/\mathcal{V}_\Lambda^\infty$

therefore

(8.3.30) $\qquad \mathcal{I}$ is a $\mathcal{C}^\infty$-smooth regular ideal in $\mathcal{A}$

Furthermore, we have the inclusion diagram

(8.3.31)
$$\begin{array}{ccccc}
\mathcal{I} & \longrightarrow & \mathcal{A} & \longrightarrow & (\mathcal{C}^\infty)^\Lambda \\
\uparrow & & \uparrow & & \\
\mathcal{V} & \longrightarrow & \mathcal{S} & \longleftarrow & \mathcal{U}_{\mathcal{C}^\infty,\Lambda} \\
\downarrow & & \downarrow & & \\
\mathcal{V}_\Lambda^\infty & \longrightarrow & \mathcal{S}_\Lambda^\infty & &
\end{array}
\qquad (\mathcal{I} \longrightarrow \mathcal{V}_\Lambda^\infty)$$

which satisfies the relations

(8.3.32) $\qquad \mathcal{I} \cap \mathcal{S} = \mathcal{V}$

(8.3.33) $\qquad \mathcal{V}_\Lambda^\infty \cap \mathcal{S} = \mathcal{V}$

(8.3.34) $\qquad \mathcal{V}_\Lambda^\infty + \mathcal{S} = \mathcal{S}_\Lambda^\infty$

Proof

First we prove the inclusions in (8.3.31). The inclusions

$$\mathcal{I} \longrightarrow \mathcal{A} \longrightarrow (\mathcal{C}^\infty)^\Lambda$$

follow from (8.1.21). The inclusions

$$\mathcal{V} \longrightarrow \mathcal{S} \longleftarrow \mathcal{U}_{\mathcal{C}^\infty,\Lambda}, \quad \mathcal{S} \longrightarrow \mathcal{A}$$

follow from (8.3.6), (8.3.7) and (8.3.9). The inclusion

$$\mathcal{V}_\Lambda^\infty \longrightarrow \mathcal{S}_\Lambda^\infty$$

follows from (8.3.3) and (8.3.5). The inclusions

$$\mathcal{V} \rightarrow \mathcal{V}^{\infty}_{\Lambda}, \quad \mathcal{S} \rightarrow \mathcal{S}^{\infty}_{\Lambda}$$

follow from (8.3.13) and (8.3.12) respectively. Finally, the inclusions

$$\mathcal{V} \rightarrow \mathcal{I} \rightarrow \mathcal{V}^{\infty}_{\Lambda}$$

follow from (8.3.10) and (8.3.11).

Now (8.3.32) and (8.3.33) are the same with (8.3.10) and (8.3.13) respectively.

Concerning (8.3.34), the inclusion $\subset$ follows from (8.3.31). The converse inclusions $\supset$ is obtained easily. Indeed, $\mathcal{S}^{\infty}_{\Lambda}$ defined in (8.3.3) satisfies (6.1.3), that is, with the notation in (8.3.4), the mapping

$$\mathcal{S}^{\infty}_{\Lambda} \ni f \longmapsto T \in \mathcal{D}'(\mathbb{R}^n)$$

is a linear surjection. Now, it suffices to take into account (8.3.8) and (8.3.5), and the proof of (8.3.34) is completed.

The relations (8.3.31)-(8.3.34) will obviously yield (8.3.28) and (8.3.30) □

Remark 2

The weaker version of the above property (8.3.30) according to which

$\mathcal{I}$ is a regular ideal in $\mathcal{A}$

can obtained in a direct way from (8.3.26) and Theorem 2 in Section 2, Chapter 6.

Remark 3

Concerning the fact that Colombeau's algebra $\mathcal{G}(\mathbb{R}^n)$ is a *collapsed* case of the chains of algebras (6.4.7), we note the following.

In view of the results in this Section, it is easy to see that in the case of Colombeau's algebra of generalized functions

(8.3.35) $$\mathcal{G}(\mathbb{R}^n) = \mathcal{A}/\mathcal{I}$$

the conditions required in Theorem 8 in Section 4, Chapter 6, are satisfied. In this way, the results in (8.2.14) and (8.2.2) are particular cases of Theorem 8 and pct. (2) in Theorem 6 in Section 4, Chapter 6.

Indeed, we can take (6.4.52) as given by (8.3.28). Further we can replace $\mathcal{I}_\ell$ and $\mathcal{A}_\ell$ in (6.4.58) with $\mathcal{I}$ and $\mathcal{A}$ respectively and still obtain $\mathcal{C}^\infty$-smooth regularizations

(8.3.36) $$(\mathcal{V}_\ell,\mathcal{S}_\ell,\mathcal{I}_\ell,\mathcal{A}_\ell) = (\mathcal{V},\mathcal{S},\mathcal{I},\mathcal{A}), \quad \ell \in \mathbb{N}$$

In this way, we shall obtain

(8.3.37) $$\mathcal{G}(\mathbb{R}^n) = A^\infty = A^\ell, \quad \ell \in \mathbb{N}$$

and the results in Theorems 5, 6, 7 and 8 in Section 4, Chapter 6, will still hold.

§4. INTEGRALS OF GENERALIZED FUNCTIONS

As mentioned in Section 2, the inconvenience of the fact that the embedding $\mathcal{C}^0(\mathbb{R}^n) \subset \mathcal{G}(\mathbb{R}^n)$ is only an embedding of vector spaces and not of algebras, will to a good extent be overcome by a *coupled caluclus* introduced in Section 5.

One of the prerequisites of this *coupled calculus* is the notion of the *value* of a generalized function $F \in \mathcal{G}(\mathbb{R}^n)$ at a point $x \in \mathbb{R}^n$.

In order to define the *value* notion, we need an *extenstion* of the complex numbers.

It is interesting to note that the way this extension is made, recalls certain basic constructions in Nonstandard Analysis, Schmieden & Laughwitz, Stroyan & Luxemburg, etc.

Indeed, suppose given a generalized function

(8.4.1) $$F = f + \mathcal{I} \in \mathcal{G}(\mathbb{R}^n) = \mathcal{A}/\mathcal{I}, \quad f \in \mathcal{A}$$

then

(8.4.2) $$\Phi \times \mathbb{R}^n \ni (\phi,x) \longmapsto f(\phi,x) \in \mathbb{C}$$

Therefore, for given, fixed $x \in \mathbb{R}^n$, it is natural to define the value $F(x)$ of F at x, as follows: let us *fix* x in (8.4.2) and thus obtain the mapping

(8.4.3) $$\Phi \ni \phi \longmapsto h(\phi) = f(\phi,x) \in \mathbb{C}$$

and define $F(x)$ as generated in a suitable way by h in (8.4.3). In this respect, we note that the mapping $f \longmapsto F$ in (8.4.1) is of course not injective, hence, we can expect the same with the mapping $h \longmapsto F(x)$. It follows that just as in (8.4.1), all we have to do is to factor out the noninjectivity of $h \longmapsto F(x)$, by using a suitable quotient structure, similar to (8.1.21).

Then, let us denote

(8.4.4) $$\mathcal{E}_0 = \mathbb{C}^{\Phi}$$

which is obviously and associative and commutative algebra. Further, let us denote by $\mathcal{A}_0$ the set of all $h \in \mathcal{E}_0$, such that

(8.4.5) $$\begin{aligned} &\exists \ m \in \mathbb{N}_+ : \\ &\forall \ \phi \in \Phi_m : \\ &\exists \ \eta, c > 0 : \\ &\forall \ \epsilon \in (0,\eta) : \\ &\quad |h(\phi_\epsilon)| \leq \frac{c}{\epsilon^m} \end{aligned}$$

It is easy to see that $\mathcal{A}_0$ is a subalgebra in $\mathcal{E}_0$. Finally, let us denote by $\mathcal{I}_0$ the set of all $h \in \mathcal{A}_0$, such that

(8.4.6) $$\begin{aligned} &\exists \ \ell \in \mathbb{N}_+, \ \beta \in B : \\ &\forall \ m \in \mathbb{N}_+, \ m \geq \ell, \ \phi \in \Phi_m : \\ &\exists \ \eta, c > 0 : \\ &\forall \ \epsilon \in (0,\eta) \ ; \\ &\quad |h(\phi_\epsilon)| \leq c\epsilon^{\beta(m)-\ell} \end{aligned}$$

It follows that $\mathcal{I}_0$ is an ideal in $\mathcal{A}_0$, and similar to (8.1.25), we have

(8.4.7) $$\mathcal{A}_0 \subsetneqq \mathcal{E}_0 \text{ and } \mathcal{I}_0 \text{ is } \textit{not} \text{ an ideal in } \mathcal{E}_0$$

Now, the associative and cummutative algebra $\mathfrak{C}$ of *generalized complex numbers* is defined by

(8.4.8) $$\mathfrak{C} = \mathcal{A}_0/\mathcal{I}_0$$

Obviously $\mathfrak{C}$ may depend on the dimension n of the underlying Euclidean space $\mathbb{R}^n$. In other words, we may have *various* sets of generalized complex numbers, corresponding to different values of n. However, that will not be of interest in the sequel.

Let us define the embedding

(8.4.9) $$\mathbb{C} \subset \mathfrak{C}$$

by the mapping

$$(8.4.10) \qquad \mathbb{C} \ni z \longmapsto \tilde{z} + \mathcal{I}_0 \in \bar{\mathbb{C}} = \mathcal{A}_0/\mathcal{I}_0$$

where

$$(8.4.11) \qquad \tilde{z}(\phi) = z, \quad \phi \in \Phi$$

hence (8.4.10) is obviously injective. In this way, we obtain:

<u>Proposition 2</u>

The embedding $\mathbb{C} \subset \bar{\mathbb{C}}$ defined in (8.4.10) is an embedding of *algebras*. □

An essential operation, in fact binary relation, needed in the sequel, is the *association* of a usual complex number $z \in \mathbb{C}$ with *some* of the generalized complex numbers $\bar{z} \in \bar{\mathbb{C}}$, which is defined as follows. The usual complex number $z \in \mathbb{C}$ is said to be *associated* with the generalized complex number $\bar{z} \in \bar{\mathbb{C}}$, in which case we shall denote $\bar{z} \longmapsto z$, if there is a representation $\bar{z} = h + \mathcal{I} \in \bar{\mathbb{C}} = \mathcal{A}_0/\mathcal{I}_0$, such that

$$(8.4.12) \qquad \begin{array}{l} \exists \quad m \in \mathbb{N}_+ : \\ \forall \quad \phi \in \phi_m : \\ \quad \lim\limits_{\epsilon \downarrow 0} h(\phi_\epsilon) = z \end{array}$$

It is important to note that *not* every $\bar{z} \in \bar{\mathbb{C}}$ has an associated $z \in \mathbb{C}$.

In view of the above, let us denote

$$(8.4.13) \qquad \bar{\mathbb{C}}_0 = \left\{ \bar{z} \in \;\middle|\; \begin{array}{l} \exists \quad z \in \mathbb{C} : \\ \bar{z} \longmapsto z \end{array} \right\}$$

Given a generalized function

$$(8.4.14) \qquad F = f + \mathcal{I} \in \mathcal{G}(\mathbb{R}^n) = \mathcal{A}/\mathcal{I}, \quad f \in \mathcal{A}$$

and $x \in \mathbb{R}^n$, the *value* $F(x)$ of F at x, is by definition the *generalized complex number*

$$(8.4.15) \qquad F(x) = f_x + \mathcal{I}_0 \in \bar{\mathbb{C}} = \mathcal{A}_0/\mathcal{I}_0$$

where

(8.6.16) $$f_x(\phi) = f(\phi,x), \quad \phi \in \Phi$$

Now, we present the property of the embedding $\mathcal{C}^0(\mathbb{R}^n) \subset \mathcal{G}(\mathbb{R}^n)$, which is one of the essential features of the *coupled calculus* in Colombeau's method:

<u>Theorem 5</u>

Suppose given a generalized function which corresponds to a *continuous* function, i.e., $F = f \in \mathcal{C}^0(\mathbb{R}^n) \subset \mathcal{G}(\mathbb{R}^n)$. If $x \in \mathbb{R}^n$, then

(8.4.17) $$F(x) \in \mathbb{C}_0 \quad \text{and} \quad F(x) \longmapsto f(x)$$

that is, the generalized complex number $F(x)$ which is the value of the generalized function F at x, has as *associated* usual complex number $f(x)$, which is the usual value of the *continuous* function f at x.

In other words, we have the following *commutative diagram*

(8.4.18) $$\begin{array}{ccc} \mathcal{C}^0(\mathbb{R}^n)\times\mathbb{R}^n \ni (f,x) & \xrightarrow{\quad (1) \quad} & f(x) \in \mathbb{C} \\ (2)\downarrow & & \uparrow(4) \\ \mathcal{G}(\mathbb{R}^n)\times\mathbb{R}^n \ni (F,x) & \xrightarrow[\quad (3) \quad]{} & F(x) \in \mathbb{C}_0 \end{array}$$

where (1) is the usual computation of the value of f at x, (2) is defined by (8.2.9), (3) is defined by (8.4.15), and (4) is the relation of association $\longmapsto$, defined in (8.4.12).

A further prerequisite of the *coupled calculus* is the notion of *integral* of a generalized function, which is defined as follows. Suppose given a generalized function

(8.4.19) $$F = f + \mathcal{I} \in \mathcal{G}(\mathbb{R}^n) = \mathcal{A}/\mathcal{I}, \quad f \in \mathcal{A}$$

and $K \subset \mathbb{R}^n$ compact. Then we define the *integral* of F on K, as the generalized complex number

(8.4.20) $$\int_K F(x)dx = h + \mathcal{I}_0 \in \mathbb{C} = \mathcal{A}_0/\mathcal{I}_0$$

where

(8.4.21) $$h(\phi) = \int_K f(\phi,x)dx, \quad \phi \in \Phi$$

It is easy to see that this is a correct definition. Indeed, $f(\phi,\cdot) \in \mathcal{C}^\infty(\mathbb{R}^n)$, with $\phi \in \Phi$, since $f \in \mathcal{A}$. Further, (8.4.21) yields

$$(8.4.22) \qquad h(\phi_\epsilon) = \int_K f(\phi_\epsilon, x)dx, \quad \phi \in \Phi, \quad \epsilon > 0$$

hence in view of (8.1.18) and (8.4.5), we obtasin $h \in \mathcal{A}_0$. Finally, (8.4.20) does not depend on f in (8.4.19), since (8.1.19) and (8.4.22) will obviously yield $f \in \mathcal{I} \Rightarrow h \in \mathcal{I}_0$.

In the particular case of a generalized function which corresponds to a $\mathcal{C}^\infty$-smooth functions, i.e., $F = f \in \mathcal{C}^\infty(\mathbb{R}^n) \subset \mathcal{G}(\mathbb{R}^n)$, the above notion of integral *coincides* with the usual one. Indeed, in view of (8.2.3), we have

$$(6.4.23) \qquad F = \tilde{f} + \mathcal{I} \in \mathcal{G}(\mathbb{R}^n)$$

with

$$(8.4.24) \qquad \tilde{f}(\phi, x) = f(x), \quad \phi \in \Phi, \quad x \in \mathbb{R}^n$$

and then, (8.4.20) and (8.4.21) yield

$$(8.4.25) \qquad \int_K F(x)dx = \tilde{z} + \mathcal{I}_0 \in \tilde{\mathbb{C}}$$

with, see (8.4.10), the relation

$$(8.4.26) \qquad \tilde{z} = \int_K f(x)dx, \quad \phi \in \Phi$$

thus, in view of the embedding (8.4.10), we can write

$$(8.4.27) \qquad \int_K F(x)dx = \int_K f(x)dx \in \mathbb{C} \subset \tilde{\mathbb{C}}$$

With the help of the relation of association $\longmapsto$, the above generalizes to *continuous functions*. Indeed, we have:

Theorem 6

Suppose given a generalized function which corresponds to a continuous function, i.e. $F = f \in \mathcal{C}^0(\mathbb{R}^n) \subset \mathcal{G}(\mathbb{R}^n)$. Then for every $K \subset \mathbb{R}^n$ compact, we have

(8.4.28) $$\int_K F(x)dx \in \overline{\mathbb{C}}_0 \quad \text{and} \quad \int_K F(x)dx \longmapsto \int_K f(x)dx$$

i.e., the integral over K of the generalized function F is a generalized complex number which has as *associated* usual complex number the usual integral of f over K.

It follows that we have the *commutative diagram*

(8.4.29) $$\begin{array}{ccc} \mathcal{C}^0(\mathbb{R}^n) \ni f & \xrightarrow{(1)} & \int_K f(x)dx \in \mathbb{C} \\ (2)\downarrow & & \uparrow (4) \\ \mathcal{G}(\mathbb{R}^n) \ni F & \xrightarrow[(3)]{} & \int_K F(x)dx \in \overline{\mathbb{C}}_0 \end{array}$$

where (1) is the usual integral of f over K, (2) is the embedding (8.2.9), (3) is the integral (8.4.20) of the generalized function F over K, and (4) is the association $\int_K F(x)dx \longmapsto \int_K f(x)dx$ defined in (8.4.12).

Remark 4

The following two properties will be useful in the sequel. Suppose given $F \in \mathcal{G}(\mathbb{R}^n)$, $\psi \in \mathcal{D}(\mathbb{R}^n)$ amd $K \subset \mathbb{R}^n$ compact, with $\text{supp}\,\psi \subset K$. Then, with the product $\psi \cdot F$ defined in $\mathcal{G}(\mathbb{R}^n)$, the integral of $\psi \cdot F$ over K, as defined in (8.4.20), does not depend on K. Therefore, we shall denote

(8.4.30) $$\int_{\mathbb{R}^n} (\psi \cdot F)(x)dx = \int_K (\psi \cdot F)(x)dx \in \overline{\mathbb{C}}$$

Suppose given $T, S \in \mathcal{D}'(\mathbb{R}^n)$, $\psi \in \mathcal{D}(\mathbb{R}^n)$ and $\Omega \subset \mathbb{R}^n$ open, such that $\text{supp}\,\psi \subset \Omega$. Then

(8.4.31) $$T\big|_\Omega = S\big|_\Omega \Rightarrow \int_{\mathbb{R}^n} (\psi \cdot T)(x)dx = \int_{\mathbb{R}^n} (\psi \cdot S)(x)dx$$

where the products $\psi \cdot T$ and $\psi \cdot S$ are computed in $\mathcal{G}(\mathbb{R}^n)$. Indeed, (8.4.31) follows easily from (8.2.16) and (8.4.20).

An extension of (8.4.27) which is essential in the *coupled calculus* defined in the next Section, is presented in:

Theorem 7

Suppose given $\psi \in \mathcal{D}(\mathbb{R}^n)$ and $T \in \mathcal{D}'(\mathbb{R}^n)$, then, with the product in $\mathcal{G}(\mathbb{R}^n)$, we have

$$(8.4.32) \qquad \int_{\mathbb{R}^n} (\psi \cdot T)(x)dx = T(\psi) \in \mathbb{C} \subset \overline{\mathbb{C}}$$

§5. COUPLED CALCULUS IN $\mathcal{G}(\mathbb{R}^N)$

As follows from Chapter 1, *no single* differential algebra is fully suited to handle in a sufficiently general way the *conflicting interplay* between arbitrary multiplication and indefinite derivability or partial derivability of generalized functions. In view of that, it follows that *additional structures* may be required on such differential algebras.

This of course applies to $\mathcal{G}(\mathbb{R}^n)$ as well. And then, Colombeau's method defines an additional structure by *two* special *equivalence relations* on $\mathcal{G}(\mathbb{R}^n)$, which together with the usual equality, arbitrary multiplication and indefinite partial derivation on $\mathcal{G}(\mathbb{R}^n)$, can be seen as a *coupled calculus*

A *motivation* for the way this coupled calculus is defined, is presented first.

One of the basic features of the *linear* theory of distributions, Schwartz [1], is the following. Given a fixed *distribution* $T \in \mathcal{D}'(\mathbb{R}^n)$, then, for every *test* function $\psi \in \mathcal{D}(\mathbb{R}^n)$, one can define the *integral* of the *product* $\psi \cdot T \in \mathcal{D}'(\mathbb{R}^n)$, by

$$(8.5.1) \qquad \int_{\mathbb{R}^n} (\psi \cdot T)(x)dx = T(\psi) \in \mathbb{C}$$

Indeed, in the particular case when $T = f \in \mathcal{L}^1_{loc}(\mathbb{R}^n) \subset \mathcal{D}'(\mathbb{R}^n)$ we shall have $\psi \cdot T = \psi \cdot f \in \mathcal{L}^1_{loc}(\mathbb{R}^n) \subset \mathcal{D}'(\mathbb{R}^n)$, hence (8.5.1) holds in the usual sense of (5.4.2).

It follows that the distributions $T \in \mathcal{D}'(\mathbb{R}^n)$ can be *characterized* by the *integrals* (8.5.1) of their distributional *products* $\psi \cdot T$ with arbitrary *test* functions $\psi \in \mathcal{D}(\mathbb{R}^n)$. Indeed, when ψ in (8.5.1) ranges over all of $\mathcal{D}(\mathbb{R}^n)$, the corresponding numbers

$$\int_{\mathbb{R}^n} (\psi \cdot T)(x)dx = T(\psi) \in \mathbb{C}$$

offer a *local characterization* of the fixed distribution T. And then, through the converse of the distribution localization principle mentioned in (5.A1.7)-(5.A1.9), we obtain an overall, *global characterization* of the fixed distribuiton T.

It should be noted that within the linear theory of distributions, no point value $T(x)$, at $x \in \mathbb{R}^n$, is associated with arbitrary distributions $T \in \mathcal{D}'(\mathbb{R}^n)$, therefore the above local characterization in (8.5.1) does indeed play a special role. In particular, we obtain

$$(8.5.2) \qquad T = 0 \in \mathcal{D}'(\mathbb{R}^n) \Longleftrightarrow \left(\int_{\mathbb{R}^n} (\psi \cdot T)(x)dx = 0, \psi \in \mathcal{D}(\mathbb{R}^n) \right)$$

In various applications of the generalized functions in $\mathcal{G}(\mathbb{R}^n)$ - for instance, the solution of linear and nonlinear partial differential equations, as presented in the sequel - it proves to be particularly useful to *extend* the above properties (8.5.1) and (8.5.2) from $\mathcal{D}'(\mathbb{R}^n)$ to $\mathcal{G}(\mathbb{R}^n)$. In fact, this is the *essence* of the *coupled calculus* in the method of Colombeau.

Let us now present the three concepts involved in the mentioned extension. Then, we shall present their basic properties, which will elucidate their role, as well as the way Colombeau's *coupled caluclus* operates. It is *particularly important* to point out that, although the next three definitions and related basic properties, as well as those in the previous Sections 3-5, may at first seem somewhat unusual and involved, Colombeau's coupled calculus is in fact *by far the simplest* way known so far in the literature in order to overcome the constraints inherent in any *nonlinear* theory of generalized functions, mentioned in Chapter 1. However, in view of Chapter 6, that simplicity of approach may as well happen the impinge on its effectiveness. The extent to which that may indeed be the case is illustrated by the fact that the results in Chapters 2-4 and 7, could not so far been possible to reproduce within Colombeau's algebra $\mathcal{G}(\mathbb{R}^n)$.

It may appear that a convenient extension of (8.5.2) would be given by the following definition. A generalized function $F \in \mathcal{G}(\mathbb{R}^n)$ is called *test null*, denoted $F \sim 0$, if for every $\psi \in \mathcal{D}(\mathbb{R}^n)$ we have

$$\int_{\mathbb{R}^n} (\psi \cdot F)(x)dx = 0, \psi \in \mathbb{T} \tag{8.5.3}$$

where the product is computed in $\mathcal{G}(\mathbb{R}^n)$ and the integral is in the sense of (8.4.30).

Two generalized functions $F_1, F_2 \in \mathcal{G}(\mathbb{R}^n)$ are called *test equal*, denoted $F_1 \sim F_2$, if $F_2 - F_1$ is test null, i.e., $F_2 - F_1 \sim 0$.

Remark 5

Obviously, $\sim$ is an equivalence relation on $\mathcal{G}(\mathbb{R}^n)$.

However, as seen below in Theorem 7, the equivalence relation $\sim$ is too restrictive, therefore, we need a more general equivalence relation defined as follows.

A distribution $T \in \mathcal{D}'(\mathbb{R}^n)$ is said to be *associated* with a generalized function $F \in \mathcal{G}(\mathbb{R}^n)$, in which case we denote $F \Vdash T$, if, for every $\psi \in \mathcal{D}(\mathbb{R}^n)$, we have

$$\int_{\mathbb{R}^n} (\psi \cdot F)(x)dx \longmapsto \int_{\mathbb{R}^n} (\psi \cdot T)(x)dx \tag{8.5.4}$$

where both products are computed in $\mathcal{G}(\mathbb{R}^n)$, while the integrals are taken in the sense of (8.4.30).

Finally, two generalized functions $F_1, F_2 \in \mathcal{G}(\mathbb{R}^n)$ are said to be *associated*, denoted $F_1 \approx F_2$, if $F_2 - F_1$ has $0 \in \mathcal{D}'(\mathbb{R}^n)$ as associated distribution, i.e., $F_2 - F_1 \Vdash 0$.

Remark 6

Obviously, $\approx$ is an *equivalence* relation on $\mathcal{G}(\mathbb{R}^n)$.

As suggested by (8.5.4), the binary relation $\Vdash \subset \mathcal{G}(\mathbb{R}^n) \times \mathcal{D}'(\mathbb{R}^n)$ is neither reflexive, nor symmetric.

We shall now present the basic properties connected with the above definitions. The properties will, among others, settle the relation between the usual function, respectively distribution multiplications

$$\mathcal{C}^0(\mathbb{R}^n) \times \mathcal{C}^0(\mathbb{R}^n) \ni (f_1, f_2) \longmapsto f_1 \cdot f_2 \in \mathcal{C}^0(\mathbb{R}^n) \tag{8.5.5}$$

(8.5.6) $$\mathcal{C}^\infty(\mathbb{R}^n) \times \mathcal{D}'(\mathbb{R}^n) \ni (\chi,T) \longmapsto \chi T \in \mathcal{D}'(\mathbb{R}^n)$$

and their corresponding versions in $\mathcal{G}(\mathbb{R}^n)$.

First, we note that in view of (8.5.3) and (8.5.4), we have for $F_1, F_2 \in \mathcal{G}(\mathbb{R}^n)$ the relation

(8.5.7) $$F_1 \sim F_2 \Rightarrow F_1 \approx F_2$$

Similarly, if $F \in \mathcal{G}(\mathbb{R}^n)$ and $T \in \mathcal{D}'(\mathbb{R}^n)$, then

(8.5.8) $$F \Vdash T \Longleftrightarrow F \approx T$$

Further, in view of (8.4.32), we have for every distribution $T \in \mathcal{D}'(\mathbb{R}^n)$

(8.5.9) $$T \sim 0 \Rightarrow T = 0$$

(8.5.10) $$T \approx 0 \Rightarrow T = 0$$

in other words, the equivalence relations $\sim$ and $\approx$ defined on $\mathcal{G}(\mathbb{R}^n)$, coincide with the usual equality $=$ when restricted to $\mathcal{D}'(\mathbb{R}^n)$.

Now, we can present the relation between the classical product of continuous functions in (8.5.5) and their product in $\mathcal{G}(\mathbb{R}^n)$.

<u>Theorem 8</u>

Suppose given two continuous functions $f_1, f_2 \in \mathcal{C}^0(\mathbb{R}^n)$. Their usual product $f_1 \cdot f_2 \in \mathcal{C}^0(\mathbb{R}^n)$ being a continuous function, is also a distribution, i.e., $f_1, f_2 \in \mathcal{D}'(\mathbb{R}^n)$. On the other hand, we can consider the generalized functions $F_1, F_2 \in \mathcal{G}(\mathbb{R}^n)$ which correspond to f_1 and f_2 respectively, according to the embedding (8.2.8).

Then, the product $F_1 \cdot F_2 \in \mathcal{G}(\mathbb{R}^n)$ of these generalized functions computed in $\mathcal{G}(\mathbb{R}^n)$ is such that

(8.5.11) $$F_1 \cdot F_2 \vdash f_1 \cdot f_2, \; F_1 \cdot F_2 \approx f_1 \cdot f_2$$

i.e. the distribution $f_1 \cdot f_2$ is *associated* with the generalized function $F_1 \cdot F_2$.

In other words, we have the *commutative diagram*

(8.5.12)
$$\begin{array}{ccc} \mathcal{C}^0(\mathbb{R}^n)\times\mathcal{C}^0(\mathbb{R}^n)\ni(f_1,f_2) & \xmapsto{\quad(1)\quad} & f_1\cdot f_2\in\mathcal{C}^0(\mathbb{R}^n)\subset\mathcal{D}'(\mathbb{R}^n) \\ (2)\big\downarrow & & \big\uparrow(4) \\ \mathcal{G}(\mathbb{R}^n)\times\mathcal{G}(\mathbb{R}^n)\ni(F_1,F_2) & \xmapsto[\quad(3)\quad]{} & F_1\cdot F_2\in\mathcal{G}(\mathbb{R}^n) \end{array}$$

where (1) is the usual product of continuous functions, (2) is defined by (8.2.9), (3) is the product of generalized functions in $\mathcal{G}(\mathbb{R}^n)$, and (4) is the relation of association $\Vdash$ defined in (8.5.4), or the equivalence relation $\approx$ □

It is *particularly important* to note that the result in Theorem 7 concerning the relation between products of continuous functions effectuated in $\mathcal{C}^0(\mathbb{R}^n)$ and $\mathcal{G}(\mathbb{R}^n)$ does not hold if in (8.5.11) we replace $\approx$ by $\sim$. In fact, this is one of the main reasons why Colombeau's *coupled calculus* uses the *weaker equivalence* relation $\approx$.

Concerning the relation between the distributional product in (8.5.6) and is version in $\mathcal{G}(\mathbb{R}^n)$, we have:

Theorem 9

Suppose given $\chi \in \mathcal{C}^\infty(\mathbb{R}^n)$ and $T \in \mathcal{D}'(\mathbb{R}^n)$ and let us denote by $S = \chi\cdot T \in \mathcal{D}'(\mathbb{R}^n)$ their distributional product. On the other hand, considering their product as generalized functions in $\mathcal{G}(\mathbb{R}^n)$, we have $F = \chi_o T \in \mathcal{G}(\mathbb{R}^n)$, where for the sake of clarity within the present theorem, we denoted by $\cdot$ the multiplication in $\mathcal{D}'(\mathbb{R}^n)$, see (8.5.6), and by 0 the multiplication in $\mathcal{G}(\mathbb{R}^n)$.

Then

(8.5.13) $\qquad F \sim S$

i.e., the distributional product $S = \chi\cdot T$ is *associated* with the generalized function product $F = \chi_o T$.

In other words, we have the *commutative diagram*

(8.5.14)
$$\begin{array}{ccc} \mathcal{C}^\infty(\mathbb{R}^n)\times\mathcal{D}'(\mathbb{R}^n)\ni(\chi,T) & \xmapsto{\quad(1)\quad} & S=\chi\cdot T\in\mathcal{D}'(\mathbb{R}^n) \\ (2)\big\downarrow & & \big\uparrow(4) \\ \mathcal{G}(\mathbb{R}^n)\times\mathcal{G}(\mathbb{R}^n)\ni(\chi,T) & \xmapsto[\quad(3)\quad]{} & F_1\cdot F_2\in\mathcal{G}(\mathbb{R}^n) \end{array}$$

where (1) is the distributional product (8.5.6), (2) is defined by (8.2.2) and (8.2.14), (3) is the product of generalized functions in $\mathcal{G}(\mathbb{R}^n)$, and (4) is the equivalence relation $\sim$, defined in (8.5.3). □

An important property of the equivalence relations $\sim$ and $\approx$ is their compatibility with the partial derivatives in $\mathcal{G}(\mathbb{R}^n)$. Indeed we have:

Theorem 10

If $F_1, F_2 \in \mathcal{G}(\mathbb{R}^n)$ and $p \in \mathbb{N}^n$, then

(8.5.15) $$F_1 \sim F_2 \Rightarrow D^p F_1 \sim D^p F_2$$

(8.5.16) $$F_1 \approx F_2 \Rightarrow D^p F_1 \approx D^p F_2$$ □

We can recapitulate by noting the following components of Colombeau's *coupled calculus*:

(1) $\mathcal{G}(\mathbb{R}^n)$ is an associative and commutative algebra, with arbitrary partial derivative operators

(8.5.17) $$D^p : \mathcal{G}(\mathbb{R}^n) \longrightarrow \mathcal{G}(\mathbb{R}^n), \quad p \in \mathbb{N}^n$$

which are linear mapping and satisfy the Leibnitz rule of product derivatives.

(2) The vector space embedding

(8.5.18) $$\mathcal{D}'(\mathbb{R}^n) \subset \mathcal{G}(\mathbb{R}^n)$$

is such that the partial derivative operators (8.5.17) coincide with the usual distributional partial derivatives, when restricted to $\mathcal{D}'(\mathbb{R}^n)$. In particular, D^p in (8.5.17) coincides with the usual partial derivative D^p of smooth functions, when restricted to $\mathcal{C}^\ell(\mathbb{R}^n)$, with $\ell \in \mathbb{N}$, $\ell \geq |p|$.

(3) The particular case of (8.5.18) given by

(8.5.19) $$\mathcal{C}^\infty(\mathbb{R}^n) \subset \mathcal{G}(\mathbb{R}^n)$$

is an embedding of differential algebras.

(4) The particular vector space embedding

(8.5.20) $\quad \mathcal{C}^0(\mathbb{R}^n) \subset \mathcal{G}(\mathbb{R}^n)$

defined by (8.5.18) is *not* an embedding of algebras. This fact is unavoidable, owing to the conflict between insufficient smoothness, multiplication and differentiation, in particular, owing to the so called Schwartz impossibility and other related results, see Chapter 1.

Colombeau's *coupled calculus* aims, amongh others, to overcome the difficulty in (4) above.

This is done in the following way.

An equivalence relation $\sim$ is defined on $\mathcal{G}(\mathbb{R}^n)$, i.e., for arbitrary $F, G \in \mathcal{G}(\mathbb{R}^n)$, by the relation

(8.5.21) $$F \sim G \iff \int_{\mathbb{R}^n} (\psi \cdot (F-G))(x)dx = 0, \quad \psi \in \mathcal{D}(\mathbb{R}^n)$$

This equivalence relation $\sim$ is compatible with the vector space structure of $\mathcal{G}(\mathbb{R}^n)$ and the partial derivative operators (8.5.17) on $\mathcal{G}(\mathbb{R}^n)$. Moreover, if $T, S \in \mathcal{D}'(\mathbb{R}^n)$ then

(8.5.22) $$T \sim S \iff T = S$$

The interest in the equivalence relation $\sim$ comes from the following property: if $\chi \in \mathcal{C}^\infty(\mathbb{R}^n)$, $T \in \mathcal{D}'(\mathbb{R}^n)$ then

(8.5.23) $$\chi \circ T \sim \chi \cdot T$$

where 0 and $\cdot$ denote the multiplications in $\mathcal{G}(\mathbb{R}^n)$ and $\mathcal{D}'(\mathbb{R}^n)$ respectively.

However, in order to handle the difficulty in (4) above, the equivalence relation $\sim$ is too strong. Therefore, a weaker equivalence relation $\approx$ is defined on $\mathcal{G}(\mathbb{R}^n)$, i.e. for arbitrary $F, G \in \mathcal{G}(\mathbb{R}^n)$, in the following way

(8.5.24) $$F \approx G \iff \int_{\mathbb{R}^n} (\psi \cdot (F-G))(x)dx \longmapsto 0, \psi \in \mathcal{D}(\mathbb{R}^n)$$

This equivalence relation $\approx$ is again compatible with the vector space structure of $\mathcal{G}(\mathbb{R}^n)$, as well as the partial derivative operators (8.5.17) on $\mathcal{G}(\mathbb{R}^n)$. Further, if $T, S \in \mathcal{D}'(\mathbb{R}^n)$ then again

(8.5.25) $$T \approx S \iff T = S$$

The *essential* property of the equivalence relation $\approx$ which settles the issue connected with (4) above, is the following. If $f,g \in \mathcal{C}^0(\mathbb{R}^n)$ and $F,G \in \mathcal{G}(\mathbb{R}^n)$ are the generalized functions which corresponds to f and g respectively, then

$$F \cdot G \approx f \cdot g \tag{8.5.26}$$

where the multiplications in the left and right hand terms are in $\mathcal{G}(\mathbb{R}^n)$ and $\mathcal{C}^0(\mathbb{R}^n)$ respectively.

In this way, Colombeau's *coupled calculus* on the differential algebra $\mathcal{G}(\mathbb{R}^n)$ means in fact the *additional* consideration of the *two equivalence relations* on $\mathcal{G}(\mathbb{R}^n)$ given by $\sim$ and $\approx$.

One can obviously ask whether it would be convenient to factor $\mathcal{G}(\mathbb{R}^n)$ by $\sim$ and/or $\approx$, and in view of (8.5.22), (8.5.25), obtain the following embeddings of vector spaces

$$\mathcal{D}'(\mathbb{R}^n) \subset \mathcal{G}(\mathbb{R}^n)/_{\approx} \subset \mathcal{G}(\mathbb{R}^n)/_{\sim} \tag{8.5.27}$$

and hence do away with the additional complication brought about by Colombeau's coupled calculus.

However, it is easy to see that $\mathcal{G}(\mathbb{R}^n)/_{\approx}$ and $\mathcal{G}(\mathbb{R}^n)/_{\sim}$ are *not* algebras, Rosinger [3].

Nevertheless, the equivalnce relation $\sim$, and especially $\approx$, prove to be particularly useful in the study of *generalized solutions* of *nonlinear* partial differential equations. Indeed, let us take for instance, the shock wave equation

$$U_t + U_x \cdot U = 0, \quad t > 0, \quad x \in \mathbb{R} \tag{8.5.28}$$

Even in case $U \in \mathcal{C}^1((0,\infty)\times\mathbb{R}) \setminus \mathcal{C}^\infty((0,\infty)\times\mathbb{R})$ is a classical solution of (8.5.28), it is likely that U will *not* be a generalized solution of (8.5.28) when considered with the multiplication in $\mathcal{G}((0,\infty)\times\mathbb{R})$, owing to the difficulty in (4) above. But, in view of (8.5.26), U will obviously satisfy

$$U_t + U_x \cdot U \approx 0 \tag{8.5.29}$$

with the multiplication in $\mathcal{G}((0,\infty)\times\mathbb{R})$. It follows that in order to find the classical, weak, generalized, etc., solutions of the usual nonlinear partial differential equation (8.5.28), we have *to solve the equivalence relation* in (8.5.30), within $\mathcal{G}((0,\infty)\times\mathbb{R})$. Details are presented in Rosinger [3].

§6. GENERALIZED SOLUTIONS OF NONLINEAR WAVE EQUATIONS IN QUANTUM FIELD INTERACTION

For simplicity we shall only consider the scalar valued case of nonlinear wave equations.

The class of equations are of the form

$$(8.6.1) \qquad (\frac{\partial^2}{\partial t^2} - \Delta)U = F(U) \quad \text{in} \quad \mathcal{G}(\mathbb{R}^4)$$

where $F : \mathbb{R} \longrightarrow \mathbb{R}$ are $\mathcal{C}^\infty$-smooth functions, such that

$$(8.6.2) \qquad F(0) = 0$$

and

$$(8.6.3) \qquad \sup_{x \in \mathbb{R}} |D^p F| < \infty, \quad p \in \mathbb{N}_+$$

It follows that F need not be bounded. For instance, we can have

$$(8.6.4) \qquad F(u) = au + b \sin u, \quad u \in \mathbb{R}$$

with given $a,b \in \mathbb{R}$, in which case (8.6.1) is a version of the Sine-Gordon equation.

Before going further, we have to note that since F is only defined on $\mathbb{R}$, it can be applied but to *real valued* generalized functions U which are defined as follows.

The generalized function $G \in \mathcal{G}(\mathbb{R}^n)$ is called real valued, if and only if there exists a representation

$$(8.6.5) \qquad G = g + \mathcal{I} \in \mathcal{G}(\mathbb{R}^n), \quad g \in \mathcal{A}$$

such that $g(\phi,x)$ is real vlued for all real valued $\phi \in \Phi$ and all $x \in \mathbb{R}^n$.

In view of the above we have $F(G) \in \mathcal{G}(\mathbb{R}^n)$, for every real valued $G \in \mathcal{G}(\mathbb{R}^n)$.

The nonlinear wave equation (8.6.1) will be considered with the Cauchy initial value problem

$$(8.6.6) \qquad U\Big|_{t=0} = u_0, \quad \frac{\partial U}{\partial t}\Big|_{t=0} = u_1$$

where $u_0, u_1 \in \mathcal{G}(\mathbb{R}^3)$ are arbitrary, given real valued generalized functions.

Within the above rather general framework, we have the following *existence result*.

Theorem 11 (Colombeau [2,4])

The initial value problem

$$(8.6.7) \qquad (\frac{\partial^2}{\partial t^2} - \Delta)U(t,x) = F(U(t,x)), \quad (t,x) \in \mathbb{R}^4$$

$$(8.6.8) \qquad U(0,x) = u_0(x), \quad x \in \mathbb{R}^3$$

$$(8.6.9) \qquad \frac{\partial}{\partial t} U(0,x) = u_1(x), \quad x \in \mathbb{R}^3$$

has real valued generalized function solutions $U \in \mathcal{G}(\mathbb{R}^4)$, for every pair of real valued generalized functions $u_0, u_1 \in \mathcal{G}(\mathbb{R}^3)$.

Theorem 12 (Colombeau [2,4])

The solution $U \in \mathcal{G}(\mathbb{R}^4)$ of the problem (8.6.7)-(8.6.9) in Theorem 10 is unique. □

The above uniqueness result shows that inspite of the generality of the framework in which the nonlinear wave equation and initial value problem (8.6.7)-(8.6.9) are considred the solution method is rather *focussed*, as it delivers a unique solution within that general framework.

In order to see the appropriateness of that focussing it is useful to compare the unique generalized solutions with the unique classical solutions whenever the latter exist. In this respect, we shall consider the following *two* cases.

Case 1 The classical $\mathcal{C}^\infty$-smooth situation when $u_0, u_1 \in \mathcal{C}^\infty(\mathbb{R}^3)$ and we have a unique classical solution $V \in \mathcal{C}^\infty(\mathbb{R}^4)$. Then, in view of the embedding (8.2.2), it is obvious that

$$(8.6.10) \qquad U = V \quad \text{in } \mathcal{G}(\mathbb{R}^4)$$

where $U \in \mathcal{G}^\infty(\mathbb{R}^4)$ is the unique generalized solution.

Case 2 When $u_0 \in \mathcal{C}^3(\mathbb{R}^3)$, $u_1 \in \mathcal{C}^2(\mathbb{R}^2)$ and we have a unique classical solution $V \in C^2(\mathbb{R}^4)$, see Colombeau [2, p. 220]. In that case we have the following coherence result.

Theorem 13 (Colombeau [2, 4])

Suppose $V \in \mathcal{C}^2(\mathbb{R}^4)$ and $U \in \mathcal{G}(\mathbb{R}^4)$ are the unique classical and generalized solutions of the nonlinear wave equation and initial value problem (8.6.7)-(8.6.9), corresponding to

$$u_0 \in \mathcal{C}^3(\mathbb{R}^3),\ u_1 \in \mathcal{C}^2(\mathbb{R}^3) \tag{8.6.11}$$

Then $U \in \mathcal{D}'(\mathbb{R}^4)$ and

$$U \approx V \quad \text{in} \quad \mathcal{G}(\mathbb{R}^4) \tag{8.6.12}$$

and for every $t_0 \in \mathbb{R}$, we have $V\big|_{t=t_0} \in \mathcal{D}'(\mathbb{R}^3)$ as well as

$$U\big|_{t=t_0} \approx V\big|_{t=t_0} \quad \text{in} \quad \mathcal{G}(\mathbb{R}^3). \tag{8.6.13}$$

§7. GENERALIZED SOLUTIONS FOR LINEAR PARTIAL DIFFERENTIAL EQUATIONS.

As mentioned in Chapter 5, the early history of the linear theory of the Schwartz distributions had known quite a number of momentous events, both for the better and for the worse.

One of the first major successes was the proof of the existence of an elementary solution for every linear constant coefficient partial differential equation, which was obtained in the early fifties by Ehrenpreis, and independently, Malgrange. Soon after, in 1954, came the famous and improperly understood, so called impossibility result in Schwartz [2].

Another, rather anecdotic event, is mentioned in Treves [4], who in 1955 was given the theses problem to prove that every linear partial differential equation with $\mathcal{C}^\infty$-smooth coefficients not vanishing identically at a point, has a distribution solution in a neighbourhood of that point. The particularly instructive aspect involved is that the thesis director who suggested the above thesis problem was at the time, and for quite a while after, one of the leading analysts. That can only show the fact that around i955, there was hardly any understanding of the problems involved in the local distributional solvability of linear partial differential equations with $\mathcal{C}^\infty$-smooth coefficients, see Treves [4].

As mentioned in Chapter 5, a very simple and clear *negative* answer to the above thesis problem was soon given by Lewy in 1957, who showed that the following quite simple linear partial differential equation

$$(8.7.1) \qquad \frac{\partial}{\partial x_1} U + \frac{\partial}{\partial x_2} U - 2i(x_1+x_2) \frac{\partial}{\partial x_3} U = f, \quad x = (x_1, x_2, x_3) \in \mathbb{R}^3$$

cannot have distribution solutions in any neighbourhood of any point $x \in \mathbb{R}^3$, if $f \in \mathcal{C}^\infty(\mathbb{R}^3)$ belongs to a rather large class.

The solvability of linear partial differential equations with $\mathcal{C}^\infty$-smooth coefficients failed to be achieved even later when, the Schwartz $\mathcal{D}'$ distributions were extended by other *linear* spaces of generalized functions, such as the hyperfunctions, Sato et. al. That failure was proved for instance in 1967, in Shapira.

As mentioned in Chapter 5, a sufficiently general characterization of solvability, and thus of unsolvability, for linear partial differential equations with $\mathcal{C}^\infty$-smooth coefficients has not yet been obtained within the framework of Schwartz's linear theory of distributions. And that inspite of several quite far reaching partial results which make use of rather hard tools from linear functional analysis as well as complex functions of several variables.

In view of the above it is the more remarkable that *for the first time* ever in the study of various generalized functions, Colombeau's nonlinear theory does yield *local* generalized solutions for practically arbitrary *systems* of linear partial differential equations. Furthermore, under certain natural growth conditions on coefficients, one can also obtain *global* generalized solutions for large classes of *systems* of linear partial differential equations, systems which contain as particular cases most of the so far unsolvable linear, $\mathcal{C}^\infty$-smooth coefficient partial differential equations, see Colombeau [3,4].

Without going into the full details - which can be found in Rosinger [3] and Colombeau [3,4] - we shall present the main results and a few illustrations.

The systems of linear partial differential equations whose generalized solutions will be obtained within Colombeau's nonlinear theory, contain as particular cases systems of the form

$$(8.7.2) \qquad D_t U_i(t,x) = \sum_{\substack{1 \leq j \leq \ell \\ p \in P_{ij}}} a_{ijp}(t,x) D_x^p U_j(t,x) + b_i(t,x), \quad 1 \leq i \leq \ell, \quad (t,x) \in \mathbb{R}^{n+1}$$

with the initial value problem

$$(8.7.3) \qquad U_i(0,x) = u_i(x), \quad 1 \leq i \leq \ell, \; x \in \mathbb{R}^n$$

where $P_{ij} \subset \mathbb{N}^n$ are finite, while a_{ijp}, b_i and u_i are $\mathcal{C}^\infty$-smooth. It is well known, Treves [2], that under very general conditions, arbitrary systems of linear partial differential equations with $\mathcal{C}^\infty$-smooth coefficients and initial value problems can be written in the equivalent form (8.7.2), (8.7.3).

Our aim is to find generalized functions

$$(8.7.4) \qquad U_1, \ldots, U_\ell \in \mathcal{G}(\mathbb{R}^{n+1})$$

which in a suitable sense, are solutions of (8.7.2) and (8.7.3), or of even more general systems, see (8.7.19).

A basic remark which conditions much of the way such generalized solutions are and can be found is the following.

The system (8.7.2), (8.7.3) *cannot* in general have solutions (8.7.4), if D_t, D_x^p and the respective equality relation $=$ are considered in $\mathcal{G}(\mathbb{R}^{n+1})$, in the usual way defined in this Chapter. The argument for that is rather classical - see for details Colombeau [3] - and it is based on a contradiction between Holmgren type uniqueness and general nonuniqueness results in the $\mathcal{C}^\infty$-smooth coefficient case, see also Treves [2] and Colombeau [4].

Furthermore, if we replace the equality relation $=$ by the equivalence relation $\sim$ or $\approx$ in $\mathcal{G}(\mathbb{R}^{n+1})$, the system (8.7.3), (8.7.4) will still fail to have solutions (8.7.4), see again Colombeau [3].

A way our from this impasse is to replace the partial derivatives D_x^p in $\mathcal{G}(\mathbb{R}^{n+1})$ by the *more smooth* partial derivatives ${}_hD_x^p$ defined next. The effect of such a replacement is very simple, yet crucial. Indeed, as seen earlier in this Chapter the partial derivatives D_x^p in $\mathcal{G}$ coincide with the classical partial derivative when restricted to $\mathcal{C}^\infty$-smooth functions. Therefore, they are *not* sufficiently smooth in the following well known sense that an $\mathcal{L}^\infty$-type bound on a $\mathcal{C}^\infty$-smooth function does not imply any $\mathcal{L}^\infty$-type bound on its classical derivatives. Contrary to that situation, the more smooth partial derivatives ${}_hD_x^p$ on $\mathcal{G}(\mathbb{R}^{n+1})$, defined next, will have the property (8.7.15).

Suppose given a function

$$(8.7.5)\qquad h : (0,\infty) \to (0,\infty) \quad \text{with} \quad \lim_{\epsilon\downarrow 0} h(\epsilon) = 0$$

such as for instance

$$(8.7.6)\qquad h(\epsilon) = \left| \begin{array}{l} 1/\ell n(1/\epsilon) \quad \text{if} \quad \epsilon \in (0,1) \\ \text{arbitrary otherwise} \end{array} \right.$$

Such a function h satisfying (8.7.5) will be called a *derivation rate*.

Before defining the smooth derivatives, we note the following property

$$(8.7.7)\qquad \begin{array}{l} \forall \quad \phi \in \Phi(\mathbb{R}^n) : \\ \exists \quad \psi \in \Phi(\mathbb{R}^n),\ \epsilon > 0 : \\ \quad *) \quad \text{diam supp } \psi = 1 \\ \quad **) \quad \phi = \psi_\epsilon \end{array}$$

and for given ϕ, one obtains ψ and ϵ in a *unique* way. In view of that, we shall in the sequel often exchange ϕ with ψ_ϵ, according to condition **) in (8.7.7).

Given $1 \leq i \leq n$, we define the h-*partial derivative*

$$(8.7.8)\qquad {}_hD_{x_i} : \mathcal{G}(\mathbb{R}^n) \to \mathcal{G}(\mathbb{R}^n)$$

as follows: if

$$(8.7.9)\qquad F = f + \mathcal{I} \in \mathcal{G}(\mathbb{R}^n),\quad f \in \mathcal{A}$$

then

$$(8.7.10)\qquad {}_hD_{x_i} F = g + \mathcal{I} \in \mathcal{G}(\mathbb{R}^n),\quad g \in \mathcal{A}$$

where

$$(8.7.11)\qquad g(\phi,x) = ((D_{x_i} f(\psi_\epsilon,\cdot))*\psi_{h(\epsilon)})(x),\quad \phi \in \Phi(\mathbb{R}^n),\ x \in \mathbb{R}^n$$

and $*$ is the usual convolution of functions. It is easy to see that the above definition is correct, since indeed $g \in \mathcal{A}$, while ${}_hD_{x_i} F$ does not depend on $f \in \mathcal{A}$ in (8.7.9).

Obviously (8.7.11) yields

$$g(\phi,x) = \int_{\mathbb{R}^n} D_{x_i} f(\psi_\epsilon, x-h(\epsilon)y)\psi(y)dy = \tag{8.7.12}$$

$$= \frac{1}{h(\epsilon)} \int_{\mathbb{R}^n} f(\psi_\epsilon, x-h(\epsilon)y) D_{y_i}\psi(y)dy, \quad \phi \in \Phi(\mathbb{R}^n),\ x \in \mathbb{R}^n$$

Given now $p = (p_1, \ldots, p_n) \in \mathbb{N}^n$, with $|p| \geq 1$, we can define the h-*partial derivative*

$$(8.7.13) \qquad {}_hD^p : \mathcal{G}(\mathbb{R}^n) \longrightarrow \mathcal{G}(\mathbb{R}^n)$$

as the iteration

$$(8.7.14) \qquad {}_hD^p = ({}_hD_{x_1})^{p_1} \ldots ({}_hD_{x_n})^{p_n}$$

of the h-partial derivatives in (8.7.8). It is easy to see that the above definition is correct, since the h-partial derivatives (8.7.8) are commutative. Obviously, the h-partial derivatives (8.7.8), and thus (8.7.13), are linear mappings.

The essential *smoothing* property of the h-partial derivatives defined above is apparent in (8.7.12). Indeed, given a compact $K \in \mathbb{R}^n$, we obviously have for $\phi \in \Phi(\mathbb{R}^n)$ the estimate

$$(8.7.15) \qquad \sup_{x \in K} |g(\phi,x)| \leq \frac{c}{h(\epsilon)} \sup_{x \in K'} |f(\psi_\epsilon, x)|$$

where $K' = K + h(\epsilon)\mathrm{supp}\,\psi \subset \mathbb{R}^n$ is compact and

$$c = \int_{\mathrm{supp}\,\psi} |D_{y_i}\psi(y)|dy < \infty$$

In this way, an $\mathcal{L}^\infty$-type bound for $g(\phi,\cdot)$ on K can be obtained in terms of an $\mathcal{L}^\infty$-type bound for $f(\psi_\epsilon,\cdot)$ on a bounded neighbourhood of K. Therefore, a representative g of the h-partial derivative ${}_hD_{x_i}F$ is locally bounded by a representative f of F. It should be noted that in view of (8.7.7), we obviously have ψ, c, ϵ and hence $h(\epsilon)$ in (8.7.15), depending only on ϕ and not on K. Finally, the justification of the definition (8.7.11) is in (8.7.12), which leads to (8.7.15). And all that is based on the classical property of convolution to commute with partial derivatives.

Now, we present several basic properties which show the *coherence* between the h-partial derivatives and the usual partial derivatives, when both are applied to important classes of generalized functions in $\mathcal{G}(\mathbb{R}^n)$.

Theorem 14

If $T \in \mathcal{D}'(\mathbb{R}^n)$ and $p \in \mathbb{N}^n$, then

(8.7.16) $\quad {}_hD^pT \approx D^pT \quad \text{in} \quad \mathcal{G}(\mathbb{R}^n)$

Theorem 15

For $\chi \in \mathcal{C}^\infty(\mathbb{R}^n)$ and $T \in \mathcal{D}'(\mathbb{R}^n)$ let us denote by $\chi T \in \mathcal{D}'(\mathbb{R}^n)$ and $\chi \cdot T \in \mathcal{G}(\mathbb{R}^n)$ the classical product in $\mathcal{D}'$, respectively the product in $\mathcal{G}$. Then $\chi T \sim \chi \cdot T$, see (8.5.13). Furthermore, for $p \in \mathbb{N}^n$ we have

(8.7.17) $\quad {}_hD^p(\chi T) \approx {}_hD^p(\chi \cdot T) \approx D^p(\chi T) \approx D^p(\chi \cdot T) \quad \text{in} \quad \mathcal{G}(\mathbb{R}^n)$

Corollary 2

For $\chi \in \mathcal{C}^\infty(\mathbb{R}^n)$ and $T \in \mathcal{D}'(\mathbb{R}^n)$, the h-partial derivatives ${}_hD_{x_i}$, with $1 \leq i \leq n$, satisfy the following version of the Leibnitz rule of product derivatives

(8.7.18) $\quad {}_hD_{x_i}(\chi \cdot T) \approx (D_{x_i}\chi) \cdot T + \chi \cdot (D_{x_i}T) \quad \text{in} \quad \mathcal{G}(\mathbb{R}^n)$ □

The basic result which uses h-partial derivatives and leads to the existence of generalized solutions for systems containing those in (8.7.2), (8.7.3) is presented now. It suffices to formulate it for one single linear partial differential operator of the type

(8.7.19) $\quad \sum_{p \in P} a_p(x) D^p U(x), \quad x \in \mathbb{R}^n$

where $P \subset \mathbb{N}^n$ is finite, while $a_p \in \mathcal{C}^\infty(\mathbb{R}^n)$, with $p \in P$.

Theorem 16

If $U \in \mathcal{D}'(\mathbb{R}^n)$ then the following *three* conditions are equivalent:

With the classical multiplication in $\mathcal{D}'(\mathbb{R}^n)$ we have

(8.7.20) $$\sum_{p \in P} a_p D^p U = b \in \mathcal{D}'(\mathbb{R}^n)$$

For any family $(h_p | p \in P)$ of derivation rates, we have with the multiplication in $\mathcal{G}(\mathbb{R}^n)$

(8.7.21) $$\sum_{p \in P} a_{p\, h_p} D^p U \approx b \in \mathcal{D}'(\mathbb{R}^n)$$

There exists a family $(h_p | \in P)$ of derivation rates, such that with the multiplication in $\mathcal{G}(\mathbb{R}^n)$ we have

(8.7.22) $$\sum_{p \in P} a_{p\, h_p} D^p U \approx b \in \mathcal{D}'(\mathbb{R}^n)$$ □

In view of Theorem 15 above, we are naturally led to the following:

Definition 1

Suppose given the linear partial differential equation

(8.7.23) $$\sum_{p \in P} a_p D^p U = b$$

where $P \subset \mathbb{N}^n$ is finite, $a_p \in \mathcal{C}^\infty(\mathbb{R}^n)$, with $p \in \mathbb{N}$, and $b \in \mathcal{D}'(\mathbb{R}^n)$.

A generalized function $U \in \mathcal{G}(\mathbb{R}^n)$ is called a *Colombeau weak solution* of (8.7.23), if and only if there exists a family $(h_p | p \in P)$ of derivation rates such that

(8.7.24) $$\sum_{p \in P} a_{p\, h_p} D^p U \approx b \quad \text{in} \quad \mathcal{G}(\mathbb{R}^n)$$

Indeed with this definition we obtain:

Corollary 3

A distribution $U \in \mathcal{D}'(\mathbb{R}^n)$ is a Colombeau weak solution of (8.7.23), if and only if it satisfies that equation in the sense of the classical operations in $\mathcal{D}'(\mathbb{R}^n)$. □

The above definition and corollary extend in an obvious way to linear systems such as (8.7.2).

As is known for instance from the mentioned example of Lewy, linear partial differential equations (8.7.23) do *not* in general have distribution solutions.

However, as seen next, systems (8.7.2), (8.7.3) have Colombeau weak solutions under rather general conditions. These solutions are global on $\mathbb{R}^{n+1}$, that is in t and x.

Theorem 17

Suppose that each of the coefficients a_{ijp} and b_i in (8.7.2) satisfies the condition written generically for c

(8.7.25) $$\sup_{\substack{t \in I \\ x \in \mathbb{R}^n}} |D^q_{t,x} c(t,x)| < \infty$$

for every bounded $I \subset \mathbb{R}$ and $q \in \mathbb{N}^{n+1}$.

Further, let us suppose that each of the initial values u_i in (8.7.3) satisfies the condition written generically for u

(8.7.26) $$\sup_{x \in \mathbb{R}^n} |D^p_x u(x)| \leq \infty$$

for every $p \in \mathbb{N}^n$.

Then there exists a Colombeau weak solution

(8.7.27) $$U_1, \ldots, U_\ell \in \mathcal{G}(\mathbb{R}^{n+1})$$

for the system (8.7.2), which also satisfies the initial value conditions (8.7.3). □

As an easy consequence we obtain the following *local* existence result which does not require any boundedness conditions on coefficients or initial values.

Corollary 4

In every strip

(8.7.28) $\Delta = \mathbb{R} \times \{x \in \mathbb{R}^n \mid |x| < L\} \subset \mathbb{R}^{n+1}$

with $L > 0$, there exists a Colombeau weak solution

(8.7.29) $U_1, \ldots, U_\ell \in \mathcal{G}(\Lambda)$

which satisfies (8.7.2) in Δ, and also satisfies the initial value conditions (8.7.3) for $x \in \mathbb{R}^n$, $|x| < L$ □

The power of the *local* existence result in Corollary 4 above can easily be seen, as it yields Colombeau weak solutions in *every strip* of type (8.7.28), for various distributionally unsolvable linear partial differential equations with $\mathcal{C}^\infty$-smooth coefficients.

For instance, Lewy's equation (8.7.1) can be written in the equivalent form

(8.7.30) $D_t U = -iD_{x_1} U + 2i(t+ix_1)D_{x_2} U + f, \; t \in \mathbb{R}, \; x = (x_1, x_2) \in \mathbb{R}^2$

which will have Colombeau weak solutions in every strip $\Delta = \mathbb{R} \times \{x \in \mathbb{R}^2 \mid |x| < L\} \subset \mathbb{R}^3$, $L > 0$.

Similarly, Grushin's equation can be equivalently written as

(8.7.31) $D_t U = -itD_x U + f, \quad t \in \mathbb{R}, \quad x \in \mathbb{R}$

hence it will have Colombeau weak solutions in every strip $\Delta = \mathbb{R} \times [-L,L] \subset \mathbb{R}^2$, with $L > 0$.

The same applies to the initial value problem for the Cauchy-Riemann equation

(8.7.32) $D_t = -iD_x U, \quad t \in \mathbb{R}, \; x \in \mathbb{R}$

(8.7.33) $U(0,x) = u_0(x), \quad x \in \mathbb{R}$

which, as is well known, cannot have distribution solutions even locally if $u_0 \in \mathcal{C}^\infty(\mathbb{R})$ is not anlytic, since the only distribution solutions of (8.7.32) are analytic in $z = t + ix$.

It should be noted that Lewy's equation (8.7.30) does not satisfy the boundedness conditions (8.7.25), owing to the coefficient $2i(t + ix_1)$. Hence, with the methods in this Section, we cannot obtain for it global Colombeau weak solutions.

On the other hand, Grushin's equation (8.7.31) satisfies (8.7.25), if and only if f satisfies that condition, in which case we have *global* Colombeau weak solutions for it.

The Cauchy-Riemann equation (8.7.32) obviously satisfies (8.7.25). Thus, if u_0 satisfies (8.7.26), then (8.7.32), (8.7.33) will have *global* Colombeau weak solutions.

Concerning the *coherence* between the Colombeau weak solutions obtained by the method in the proof of Theorem 16 and known classical or distributional solutions, a series of examples of familiar linear partial differential equations are studied in Colombeau [3,4]. Here we mention the following coherence results.

If (8.7.2) is constant coefficient hyperbolic, the Colombeau weak solutions coincide with the classical ones.

If (8.7.2), (8.7.3) is analytic, the Colombeau weak solutions coincide with the classical analytic ones.

Similar results hold for classes of parabolic or elliptic equations.

As mentioned, see also Treves [2], one *cannot* expect uniqueness results in Theorem 16 or Corollary 4, since systems (8.7.2), (8.7.3) can even have nonunique $\mathcal{C}^\infty$-smooth solutions.

The method of proof for Theorem 16 can be extended to systems (8.7.2), (8.7.3) with more general coefficients and intial values, which are no longer $\mathcal{C}^\infty$-smooth, but can be distributions or even generalized functions.

In fact, the method of proof for Theorem 16 is nonlinear, thus it can yield Colombeau weak solutions for *nonlinear systems* of partial differential equations, see Colombeau [3,6].

APPENDIX 1

THE NATURAL CHARACTER OF COLOMBEAU'S DIFFERENTIAL ALGEBRA

It is interesting to note that the definitions of the algebra $\mathcal{A}$ and ideal $\mathcal{I}$ given in (8.1.18) and (8.1.19) respectively, have a rather *natural* character, inspite of what at first sight may appear to be an ad-hoc one.

Indeed, let us recall a few well known properties from the linear theory of distributions, Rudin, within the framework of a given Euclidean space $\mathbb{R}^n$.

Let us define the mappping

(8.A1.1) $\mathcal{D}' \ni T \longmapsto L_T : \mathcal{D} \to \mathcal{C}^\infty$

by

(8.A1.2) $L_T(\phi) = T*\phi, \quad \phi \in \mathcal{D}$

Obviously, L_T is a well defined linear mapping of $\mathcal{D}$ into $\mathcal{C}^\infty$, which has the following three properties

(8.A1.3) L_T is continuous with the usual topologies on $\mathcal{D}$ and $\mathcal{C}^\infty$

(8.A1.4) L_T commutes with every translation $\tau_x : \mathbb{R}^n \to \mathbb{R}^n$, with $x \in \mathbb{R}^n$

(8.A1.5) L_T commutes with every partial derivative D^p, with $p \in \mathbb{N}^n$

These properties are in fact well known properties of the convolution * of distributions.

The *nontrivial* fact is given by the following two *converse* properties.

Suppose given a linear mapping

(8.A1.6) $L : \mathcal{D} \to \mathcal{C}^0$

which is continuous in the usual topologies on $\mathcal{D}$ and $\mathcal{C}^0$ If L satisfies (8.A1.4) then there exists $T \in \mathcal{D}'$ such that $L = L_T$.

Similarly, suppose given a linear mapping

(8.A1.7) $L : \mathcal{D} \to \mathcal{C}^\infty$

which is continuous in the usual topologies on $\mathcal{D}$ and $\mathcal{C}^\infty$. If L satisfies (8.A1.5) then there exists $T \in \mathcal{D}'$ such that $L = L_T$.

In short, the convolution $*$ of distributions is the *unique* bilinear form which commutes with translations or partial derivatives.

Let us now recall the arguments in Section 1 on smooth approximations and representations which were expressed in (8.1.8) by the so called inclusion

$$\mathcal{D}' \text{ 'c' } (\mathcal{C}^\infty)^{(0,\infty)} \tag{8.A1.8}$$

It was further argued that all what was to be done was to replace it by a proper inclusion or embedding, as for instance in (8.1.10). In view of that, we are obviously interested in suitable mappings

$$\mathcal{D}' \ni T \longmapsto t \in (\mathcal{C}^\infty)^{(0,\infty)} \tag{8.A1.9}$$

For convenience, let us *simplify* the issue, by only considering the following restriction of (8.A1.9)

$$\mathcal{D} \ni T \longmapsto t \in (\mathcal{C}^\infty)^{(0,\infty)} \tag{8.A1.10}$$

which is equivalent with

$$\mathcal{D} \ni T \longmapsto t_\epsilon \in \mathcal{C}^\infty, \quad \epsilon \in (0,\infty) \tag{8.A1.11}$$

Now, if we request that for each $\epsilon \in (0,\infty)$, the mappings in (8.A1.11) are *continuous* with the usual topologies on $\mathcal{D}$ and $\mathcal{C}^\infty$, and that they *commute with all partial derivatives*, then according to the above, there exist $T_\epsilon \in \mathcal{D}'$, with $\epsilon \in (0,\infty)$, such that

$$t_\epsilon = T_\epsilon * T, \quad \epsilon \in (0,\infty), \quad T \in \mathcal{D} \tag{8.A1.12}$$

But in view of the argument in Section 1, it is natural to require that $t_\epsilon \longrightarrow T$ in $\mathcal{D}'$, when $\epsilon \longrightarrow 0$. This means in view of (8.A1.12), that we can further assume the property

$$T_\epsilon \longrightarrow \delta \text{ in } \mathcal{D}', \text{ when } \epsilon \longrightarrow 0 \tag{8.A1.13}$$

Therefore we can in fact assume that

$$T_\epsilon \in \mathcal{D}, \quad \epsilon \in (0,\infty) \tag{8.A1.14}$$

in which case (8.A1.12) can be extended to $T \in \mathcal{D}'$, and that answers the initial question of the mappings (8.A1.9).

Recapitulating the above, we are led to a general form for (8.A1.9), given by mappings

(8.A1.15) $\mathcal{D}' \ni T \longmapsto t \in (\mathcal{C}^\infty)^\Phi$

with

(8.A1.16) $\Phi \subset \mathcal{D}$

(8.A1.17) $t(\phi) = T*\phi, \quad \phi \in \Phi$

since in the previous argument we could take $\Phi = \{T_\epsilon | \epsilon \in (0,\infty)\}$.

The essential points so far are in the condition (8.A1.16) on the *index set* Φ, and in the presence of the *convolution* $*$ of distributions in (8.A1.17).

Here it should be noted that, as it follows from (8.2.10), in Colombeau's theory the convolution in (8.A1.17) is replaced by the following one

(8.A1.18) $t(\phi) = T*\check{\phi}, \quad \phi \in \Phi$

where $\check{\phi}(x) = \phi(-x)$, for $\phi \in \mathcal{D}$, $x \in \mathbb{R}^n$.

Now, we are in the position to obtain the needed insight into the *necessary structure* of the sets Φ, $\mathcal{A}$ and $\mathcal{I}$, which are fundamental in Colombeau's theory.

Let us proceed first with Φ. In view of (8.A1.13) and (8.A1.14), it is natural to ask condition *) in (8.1.11), as well as the following one

(8.A1.19) $\Phi \ni \phi \longmapsto \phi_\epsilon \in \Phi, \quad \epsilon \in (0,\infty)$

where the notation in (8.1.4) was used. Condition **) in (8.1.11) is required for the diagrm (8.2.27), as mentioned in Section 1.

Turning to $\mathcal{A}$, it is now obvious that it has to be a partial derivative invariant subalgebra in $(\mathcal{C}^\infty)^\Phi$ which contains at least all the mappings

(8.A1.20) $\bar{f}(\phi,x) = \int_{\mathbb{R}^n} f(x+y)\phi(y)dy, \quad f \in \mathcal{C}^0, \ \phi \in \Phi, \quad x \in \mathbb{R}^n$

But

(8.A1.21) $D^p\bar{f}(\phi_\epsilon,x) = \dfrac{(-1)^{|p|}}{\epsilon^{|p|}} \int_{\mathbb{R}^n} f(x+\epsilon y)D^p\phi(y)dy$

for $f \in \mathcal{C}^0$, $p \in \mathbb{N}^n$, $\phi \in \Phi$, $\epsilon \in (0,\infty)$ and $x \in \mathbb{R}^n$. Hence the elements of $\mathcal{A}$ can exhibit a *polynomial growth* in $1/\epsilon$, depending on $p \in \mathbb{N}^n$ and $\phi \in \Phi$.

A usual way to measure such a growth is to restrict the above $\mathcal{C}^\infty$-smooth functions $\bar{f}(\phi,\cdot)$ and $D^p\bar{f}(\phi,\cdot)$ to compact subsets $K \subset \mathbb{R}^n$. That being done, the condition in the definition (8.1.18) of $\mathcal{A}$ will follow now in a natural way.

Concerning $\mathcal{I}$, it obviously has to be a partial derivative invariant ideal in $\mathcal{A}$, subject to the additional conditon in (8.2.28). That latter conditions means

(8.A1.22) $$D^p g \in \mathcal{I}, \quad p \in \mathbb{N}^n$$

where

(8.A1.23) $$g(\phi,x) = \int_{\mathbb{R}^n} (f(x+y)-f(x))\phi(y)dy, \quad \phi \in \Phi, \quad x \in \mathbb{R}^n$$

for $f \in \mathcal{C}^\infty(\mathbb{R}^n)$. In particular, in the one dimensional case, when $n = 1$, we have for given $m \in \mathbb{N}_+$

(8.A1.24) $$D^p g(\phi_\epsilon,x) = \sum_{1\leq q\leq m} \frac{\epsilon^q}{q!} D^{p+q}f(x) \int_{\mathbb{R}} (\tfrac{y}{\epsilon})^q \phi(\tfrac{y}{\epsilon})\frac{dy}{\epsilon} + \\ + \frac{1}{(m+1)!} \int_{\mathbb{R}} y^{m+1} D^{p+m+1} f(x+\theta y)\phi(\tfrac{y}{\epsilon})\frac{dy}{\epsilon}$$

with $\phi \in \Phi$, $x \in \mathbb{R}$, $\epsilon > 0$, $p \in \mathbb{N}$ and suitably chosen $\theta \in (0,1)$. Hence, if $\phi \in \Phi_m$ then condition **) in (8.1.11) implies that all the integrals under the above sum will vanish, and we remain with

(8.A1.25) $$D^p g(\phi_\epsilon,x) = \frac{\epsilon^{m+1}}{(m+1)!} \int_{\mathbb{R}} z^{m+1} D^{p+m+1} f(x+\epsilon\theta z)\phi(z)dz$$

In this way the elements of $\mathcal{I}$ can behave as *polynomials* in ϵ, depending on $m \in \mathbb{N}_+$, $\phi \in \Phi_m$ and $p \in \mathbb{N}^n$.

Then, an argument similar with the one used above for $\mathcal{A}$, will lead us to the definition of $\mathcal{I}$ given in (8.1.19).

We can shortly recapitulate as follows. If we want to have (8.A1.11) under any form, such as for instance

$$(8.A1.26) \qquad \mathcal{D} \ni T \xmapsto{\alpha_\phi} t_\phi \in \mathcal{C}^\infty, \ \phi \in \Phi$$

with Φ an infinite index set, and if we want that for every $\phi \in \Phi$ we have

$$(8.A1.27) \qquad \alpha_\phi : \mathcal{D} \to \mathcal{C}^\infty \text{ linear, continuous}$$

$$(8.A1.28) \qquad \begin{array}{ccc} \mathcal{D} \ni T & \xmapsto{\alpha_\phi} & t_\phi \in \mathcal{C}^\infty \\ D^p \downarrow & & \downarrow \\ \mathcal{D} \ni D^p T & \xmapsto[\alpha_\phi]{} & D^p t_\phi \in \mathcal{C}^\infty \end{array} \quad \text{commutes for } p \in \mathbb{N}^n$$

then there exist $T_\phi \in \mathcal{D}'$, with $\phi \in \Phi$, such that

$$(8.A1.29) \qquad t_\phi = T_\phi * T, \ \phi \in \Phi, \ T \in \mathcal{D}$$

and the rest of the above argument will follow.

In particular, differential algebras containing the $\mathcal{D}'$ distribution and which are based on sequential smooth approximations as in (8.A1.26), yet are different from Colombeau's algebras, will fail to satisfy at least one of the conditions (8.A1.27) or (8.A1.28). Such algebras are studied under their general form in Rosinger [1,2,3] and Chapters 2-7 in this volume. Their utility becomes apparent, among other, in connection with the possibility of increased *stability* of generalized solutions of nonlinear partial differential equations.

It should be pointed out that a different but not less convincing argument about the natural character of Colombeau's differential algebras is presented in Colombeau [1, pp 50-66]. For convenience we present it here in a summary version.

Let us remember that the linear space $\mathcal{D}'(\mathbb{R}^n)$ of the Schwartz distributions is the set of all *linear* and continuous mappings

$$(8.A1.30) \qquad T : \mathcal{D}(\mathbb{R}^n) \to \mathbb{C}$$

Therefore, if we look at distributions as complex valued functions defined on $\mathcal{D}(\mathbb{R}^n)$, it is natural to try to define the product of two distributions

$$(8.A1.31) \qquad T_1, T_2 : \mathcal{D}(\mathbb{R}^n) \to \mathbb{C}$$

as the *usual* product of complex valued functions, that is

$$(8.A1.32) \qquad T_1 \cdot T_2 = T : \mathcal{D}(\mathbb{R}^n) \to \mathbb{C}$$

where

(8.A1.33) $T(\phi) = T_1(\phi)\cdot T_2(\phi), \quad \phi \in \mathcal{D}(\mathbb{R}^n)$

In this way it is natural to try to embed $\mathcal{D}'(\mathbb{R}^n)$, that is the set of linear and continuous mappings (8.A1.30), into a differential algebra $\mathcal{C}^\infty(\mathcal{D}(\mathbb{R}^n))$ of $\mathcal{C}^\infty$-smooth complex valued functions on $\mathcal{D}(\mathbb{R}^n)$. However, an embedding

(8.A1.34) $\mathcal{D}'(\mathbb{R}^n) \subset \mathcal{C}^\infty(\mathcal{D}(\mathbb{R}^n))$

does raise two immediate difficulties.

First, one has to find a suitable concept of partial derivation for complex valued functions on $\mathcal{D}(\mathbb{R}^n)$, such that (8.A1.34) will hold and the partial derivation of functions in $\mathcal{C}^\infty(\mathcal{D}(\mathbb{R}^n))$ will extend that of distribution in $\mathcal{D}'(\mathbb{R}^n)$. This problem has been dealt with in Colombeau [5].

The second difficulty is more elementary and it is also more basic. Indeed, the multiplication in (8.A1.32) does *not* even generalized the usual multiplication of functions in $\mathcal{C}^\infty(\mathbb{R}^n)$. In order to see that, let us assume that it does, and let us take $f_1, f_2 \in \mathcal{C}^\infty(\mathbb{R}^n)$, and denote by

(8.A1.35) $f = f_1\cdot f_2 \in \mathcal{C}^\infty(\mathbb{R}^n)$

the usual product of functions. Let us denote by $T_1, T_2, T \in \mathcal{D}'(\mathbb{R}^n)$ the distributions generated by f_1, f_2 and f respectively, according to (5.4.2). Then in view of (8.A1.35), we should have

(8.A1.36) $T = T_1\cdot T_2$

in the sense of (8.A1.32). Hence according to (8.A1.33) we would obtain

(8.A1.37)
$$\int_{\mathbb{R}^n} f_1(x)f_2(x)\phi(x)dx = \left[\int_{\mathbb{R}^n} f_1(x)\phi(x)dx\right]\cdot\left(\int_{\mathbb{R}^n} f_2(x)\phi(x)dx\right), \quad \phi \in \mathcal{D}(\mathbb{R}^n)$$

which is obviously false for arbitrary $f_1, f_2 \in \mathcal{C}^\infty(\mathbb{R}^n)$.

However, this second difficulty need not be fatal: indeed, one can naturally think about using a suitable *quotient structure* on $\mathcal{C}^\infty(\mathcal{D}(\mathbb{R}^n))$ which factors out the difference between the right and left hand terms in

(8.A1.37). Fortunately, such a quotient structure - which because of the multiplication involved in (8.A1.37) should rather be a ring or an algebra - can easily be constructed.

Indeed, let us first notice that $\mathcal{D}(\mathbb{R}^n)$ is dense in $\mathcal{E}'(\mathbb{R}^n)$, Schwartz [1], and $\mathcal{E}'(\mathbb{R}^n)$ is a De Silva space, Colombeau [5,1], therefore the restriction mapping

(8.A1.38) $$\mathcal{C}^\infty(\mathcal{E}'(\mathbb{R}^n)) \ni f \longmapsto f\Big|_{\mathcal{D}(\mathbb{R}^n)} \in \mathcal{C}^\infty(\mathcal{D}(\mathbb{R}^n))$$

is *injective*. Hence we can consider the *embedding*

(8.A1.39) $$\mathcal{C}^\infty(\mathcal{E}'(\mathbb{R}^n)) \subset \mathcal{C}^\infty(\mathcal{D}(\mathbb{R}^n))$$

The idea is first to try to correct on the smaller space $\mathcal{C}^\infty(\mathcal{E}'(\mathbb{R}^n))$ the lack of identity in (8.A1.37). For that we recall the relation $\mathcal{E}''(\mathbb{R}^n) = \mathcal{C}^\infty(\mathbb{R}^n)$, Schwartz [1]. Hence, each linear and continuous functional $F \in \mathcal{E}''(\mathbb{R}^n)$ can be identified with the function $f \in \mathcal{C}^\infty(\mathbb{R}^n)$ defined by

(8.A1.40) $$f(x) = \langle F, \delta_x \rangle, \quad x \in \mathbb{R}^n$$

where δ_x is the Dirac δ distribution at x and $\langle\ ,\ \rangle$ is the bilinear form defined by duality.

The fact of interest which follows now is that in view of (8.A1.40), the elements of $\mathcal{E}''(\mathbb{R}^n)$ can easily be multiplied according to

(8.A1.41) $$F = F_1 \cdot F_2, \quad F_1, F_2 \in \mathcal{E}''(\mathbb{R}^n)$$

where we define F by

(8.A1.42) $$\langle F, \delta_x \rangle = \langle F_1, \delta_x \rangle \cdot \langle F_2, \delta_x \rangle, \quad x \in \mathbb{R}^n$$

which is nothing but the usual multiplication of the corresponding functions in $\mathcal{C}^\infty(\mathbb{R}^n)$.

Now we note that

(8.A1.43) $$\mathcal{E}''(\mathbb{R}^n) \subset \mathcal{C}^\infty(\mathcal{E}'(\mathbb{R}^n))$$

according, for instance, to the partial derivation on $\mathcal{E}'(\mathbb{R}^n)$ used in Colombeau [5,1]. Hence (8.A1.40) suggests the definition of an equivalence relation $\equiv$ on $\mathcal{C}^\infty(\mathcal{E}'(\mathbb{R}^n))$ as follows:

given $F_1, F_2 \in C^\infty(\mathcal{E}'(\mathbb{R}^n))$, then we define

$$(8.\text{A}1.44) \qquad F_1 \equiv F_2 \iff \left(\begin{array}{c} \forall \ x \in \mathbb{R}^n : \\ F_1(\delta_x) = F_2(\delta_x) \end{array} \right)$$

The utility of this equivaslence relation $\equiv$ is obvious. Indeed, let us define the linear mapping

$$(8.\text{A}1.45) \qquad \alpha : C^\infty(\mathcal{E}'(\mathbb{R}^n)) \to C^\infty(\mathbb{R}^n)$$

by

$$(8.\text{A}1.46) \qquad (\alpha(F))(x) = F(\delta_x), \quad F \in C^\infty(\mathcal{E}'(\mathbb{R}^n)), \quad x \in \mathbb{R}^n$$

which is thus an extension of the canonical mapping i in (8.A1.40). It is easy to see that α is well defined. Indeed, the mapping $\mathbb{R}^n \ni x \longmapsto \delta_x \in \mathcal{E}'(\mathbb{R}^n)$ is C^∞-smooth, Colombeau [5,1], hence $\alpha(F) \in C^\infty(\mathbb{R}^n)$, for $F \in C^\infty(\mathcal{E}'(\mathbb{R}^n))$.

Obviously, we obtain the following *commutative diagram*

$$(8.\text{A}1.47)$$

$$
\begin{array}{ccc}
C^\infty(\mathcal{E}'(\mathbb{R}^n)) & \longrightarrow & C^\infty(\mathbb{R}^n) \\
& \nwarrow \quad \nearrow & \\
c & \mathcal{E}''(\mathbb{R}^n) & i
\end{array}
$$

and furthermore

$$(8.\text{A}1.48) \qquad \ker \alpha = \{F \in C^\infty(\mathcal{E}'(\mathbb{R}^n)) \,|\, F \equiv 0\}$$

It follows that

$$(8.\text{A}1.49) \qquad C^\infty(\mathcal{E}'(\mathbb{R}^n))/\ker \alpha, \quad \text{and} \quad C^\infty(\mathbb{R}^n) \quad \text{are isomorphic algebras}$$

In this way we have found a suitable factorization on $C^\infty(\mathcal{E}'(\mathbb{R}^n))$ by the *ideal* $\ker \alpha$, which does indeed correct the difficulty mentioned in (8.A1.37).

Now it only remains to extend the ideal $\ker \alpha$ of $C^\infty(\mathcal{E}'(\mathbb{R}^n))$ to a similarly suitable ideal I in $C^\infty(\mathcal{D}(\mathbb{R}^n))$, or in an appropriate sub-

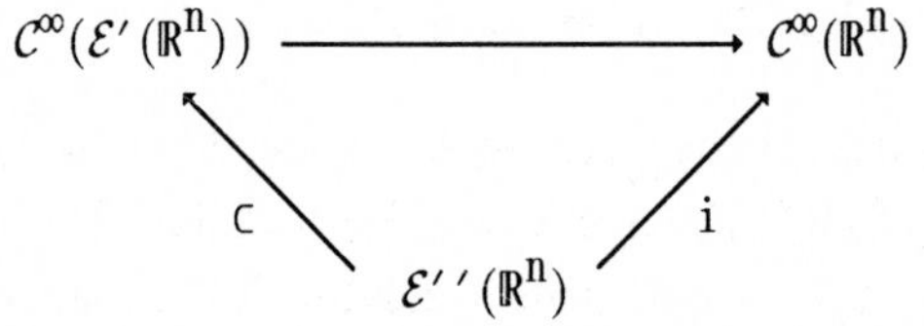

algebra $\mathcal{A}$ of $\mathcal{C}^\infty(\mathcal{D}(\mathbb{R}^n))$. This means that the following condition has to be satisfied

$$\mathcal{I} \cap \mathcal{C}^\infty(\mathcal{E}'(\mathbb{R}^n)) = \ker \alpha \tag{8.A1.50}$$

which is necessary and sufficient for the existence of a canonical embedding

$$\mathcal{C}^\infty(\mathcal{E}'(\mathbb{R}^n))/\ker \alpha \subset \mathcal{A}/\mathcal{I} \tag{8.A1.51}$$

The way to obtain that is indicated by the following basic result, Colombeau [1, pp 57-60].

Suppose given $F \in \mathcal{C}^\infty(\mathcal{E}'(\mathbb{R}^n))$, then $F \in \ker \alpha$ if and only if

$$\begin{aligned} &\forall \quad K \subset \mathbb{R}^n \text{ compact}, \quad q \in \mathbb{N}_+, \quad \phi \in \Phi_q : \\ &\exists \quad c, \eta > 0 : \\ &\forall \quad x \in K, \quad \epsilon \in (0, \eta) : \\ &\qquad |F(\phi_{\epsilon,x})| \leq c\epsilon^{q+1} \end{aligned} \tag{8.A1.52}$$

where $\phi_{\epsilon,x}(y) = \phi((y-x)/\epsilon)/\epsilon^n$, with $y \in \mathbb{R}^n$.

We are now nearing the end of our argument. Indeed, we only have to recall that for certain $F \in \mathcal{C}^\infty(\mathcal{D}(\mathbb{R}^n))$ and suitable $\phi \in \Phi$, $F(\phi_\epsilon)$ can exhibit a very fast growth in $1/\epsilon$, see the example following (8.1.25). Then (8.A1.52) obviously implies that the extended ideal $\mathcal{I}$ has to be an ideal in a strictly smaller subalgebra $\mathcal{A}$ of $\mathcal{C}^\infty(\mathcal{D}(\mathbb{R}^n))$, such that for $F \in \mathcal{A}$. $F(\phi_\epsilon)$ does not grow faster in $1/\epsilon$ than a polynomial. In this way one can easily arrive at the definitions (8.1.18) and (8.1.19), if one also remembers (8.1.22).

APPENDIX 2

ASYMPTOTICS WITHOUT A TOPOLOGY

The space $\mathcal{G}(\mathbb{R}^n)$ of generalized functions was constructed in (8.1.21) in a way which involved *both* some algebra and topology.

Indeed, on the one hand, (8.1.21) is a purely algebraic quotient construction, where $\mathcal{A}$ is an algebra and $\mathcal{I}$ is an ideal in $\mathcal{A}$. On the other hand, the definitions of $\mathcal{A}$ and $\mathcal{I}$ in (8.1.18) and (8.1.19) respectively, do obviously involve some kind of topology, owing to the respective asymptotic conditions when $m \rightarrow \infty$ and $\epsilon \rightarrow 0$.

In view of that, one may ask whether or not the construction of $\mathcal{G}(\mathbb{R}^n)$ could be seen for instance as a usual completion of $\mathcal{C}^\infty(\mathbb{R}^n)$ in a certain vector space topology $\mathcal{T}$ on $\mathcal{C}^\infty(\mathbb{R}^n)$, in which case $\mathcal{A}$ would be the set of Cauchy nets in $\mathcal{C}^\infty(\mathbb{R}^n)$, while $\mathcal{I}$ would be the set of the nets convergent to zero in $\mathcal{C}^\infty(\mathbb{R}^n)$ within the uniform topology $\mathcal{T}$.

We shall show that there is *no* such a uniform topology $\mathcal{T}$ on $\mathcal{C}^\infty(\mathbb{R}^n)$. The argument is quite simple and straightforward and is based on a well known result in general topology, Kelley, on the necessary and sufficient condition on a convergence class in order to be identical with the convergence generated by a topology. For convenience, we repeat that result here.

Suppose give a nonvoid set X and a class $\mathcal{C}$ of pairs (S,x), where S is a net in X and $x \in X$. Then there exists a topology $\mathcal{T}$ on X such that

(8.A2.1) $\quad (S,x) \in \mathcal{C} \Longleftrightarrow S$ converges to x in $\mathcal{T}$

if and only if $\mathcal{C}$ satisfies the following four conditions:

(8.A2.2) $\quad$ If S is the constant net x then $(S,x) \in \mathcal{C}$

(8.A2.3) $\quad$ If $(S,x) \in \mathcal{C}$ then $(S',x) \in \mathcal{C}$ for every subnet S' of S

(8.A2.4) $\quad$ If $(S,x) \notin \mathcal{C}$ then there xists a subnet S' of S such that for every subnet S'' of S', we have $(S'',x) \notin \mathcal{C}$

Finally, given a directed set D and a family of directed sets E_d with $d \in D$, let us consider the directed set

$$F = D \times \prod_{d \in D} E_d$$

and let us denote

$$G = \bigcup_{d \in D} (\{d\} \times E_d)$$

and then define

$$T : F \longrightarrow G$$

by $R(d,\eta) = (d,\eta(d))$. Further, for $S : G \longrightarrow X$ and $d \in D$ let us define $S_d : E_d \longrightarrow X$ by $S_d(e) = S(d,e)$. In that case we have

(8.A2.5)
If $(S_d, x_d) \in \mathcal{C}$, with $d \in D$,
and $(T,x) \in \mathcal{C}$, where $T(d) = x_d$,
with $d \in D$, then $(S_o R, x) \in \mathcal{C}$

Obviously, the above characterization of convergence classes does refer to a general, possibly nonuniform topology $\mathcal{T}$ on X. However, with a slight modification, the mentioned characterization can be applied to our case, when

(8.A2.6) $\quad X = \mathcal{C}^\infty(\mathbb{R}^n)$

Indeed, as is well known in the case of a vector space topology $\mathcal{T}$ on X, the class $\mathcal{Z}$ of nets convergent to zero determines in a unique way the class $\mathcal{C}$ of all convergent sequences, according to the relation

(8.A2.7) $$\mathcal{C} = \left\{ (S,x) \,\middle|\, \begin{array}{l} x \in X \\ (S-x,0) \in Z \end{array} \right\}$$

It follows that in our case when X is given by (8.A2.6) and $\mathcal{I}$ is assumed to be the class of nets convergent to zero, the class $\mathcal{C}$ of all convergent nets would consist of all

(8.A2.8) $\quad (S,\psi)$

where $S : \Phi \longrightarrow \mathcal{C}^\infty(\mathbb{R}^n)$, $\psi \in \mathcal{C}^\infty(\mathbb{R}^n)$, and if we define $f : \Phi \times \mathbb{R}^n \longrightarrow \mathbb{C}$ by $f(\phi,x) = (S(\phi))(x) - \psi(x)$, then

(8.A2.9) $\quad f \in \mathcal{I}$

In particular, it follows that all nets in $\mathcal{C}$ are defined on the index set Φ.

Then, as a first remark, it follows that Φ should be a directed set, although no such explicit provision is made or is even needed in Colombeau's theory. However, as mentioned above, the condition (8.1.19) defining $\mathcal{I}$ does involve an asymptotic property for $m \longrightarrow \infty$ and $\epsilon \longrightarrow 0$, which could eventually suggest a directed order on Φ.

Nevertheless, even if Φ could be made into a directed set, the class $\mathcal{C}$ defined in (8.A2.8) would still obviously fail to satisfy contion (8.A2.5)

Therefore, there is no topology on $\mathcal{C}^\infty(\mathbb{R}^n)$ in which $\mathcal{I}$ would be the class of nets convergent to zero. Consequently, the quotient structure in (8.1.21), according to which

(8.A2.10) $\qquad \mathcal{G}(\mathbb{R}^n) = \mathcal{A}/\mathcal{I}$

cannot be seen as a completion of $\mathcal{C}^\infty(\mathbb{R}^n)$ in any vectore topology.

We should however note that there may exist a vector space topology on $\mathcal{C}^\infty(\mathbb{R}^n)$ with Cauchy nets $\mathcal{B}$, and nets convergent to zero $\mathcal{J}$, such that

(8.A2.11) $\qquad \mathcal{G}(\mathbb{R}^n) = \mathcal{B}/\mathcal{J}$

and possibly

(8.A2.12) $\qquad \mathcal{I} \subset \mathcal{J}, \quad \mathcal{A} \subset \mathcal{B}, \quad \mathcal{J} \cap \mathcal{A} = \mathcal{I}$

The interesting thing however is that, even without such or any other topology on $\mathcal{C}^\infty(\mathbb{R}^n)$, the *direct and explicit, as well as natural asymptotics* in the definitions of $\mathcal{A}$ and $\mathcal{I}$, can offer Colombeau's method a surprising efficiency in solving large classes of linear and nonlinear partial differential equations.

See further Appendix 4.

APPENDIX 3

CONNECTIONS WITH PREVIOUS ATTEMPTS IN DISTRIBUTION MULTIPLICATION

There exists a considerable literature on a large varieity of attempts to define suitable distribution multiplications. This literature, published both before and after the so called Schwartz impossibility result, Schwartz [2], has mainly developed along rather independent, pure mathematical lines, and the results obtained could hardly be used in order to set up sufficiently general nonlinear theories dealing with nonlinear partial differential equations. An account of most of that literature can be found for instance in Rosinger [2,3]. Two recent papers with some of the most relevant results in that field are Ambrose and Oberguggenberger [1]. The latter paper is the best account so far of the essence of the mentioned literature and we shall present here shortly its main results which establish the relationship between four of the most important earlier distribution multiplications and the multiplication in Colombeau's algebra $\mathcal{G}(\mathbb{R}^n)$. Full details concerning proofs can be found in Oberguggenberger [1], as well as the references cited there.

Suppose given two arbitrary distributions $S,T \in \mathcal{D}'(\mathbb{R}^n)$.

The Ambrose product - which extends the product in Hörmander [2] - is denoted by $S\cdot T$, and exists by definition, if and only if

$\forall \quad x \in \mathbb{R}^n$;

$\exists \quad V$ open neighbourhood of x :

$\forall \quad \alpha,\beta \in \mathcal{D}(V)$:

(A1) $\mathcal{F}(\alpha S)\mathcal{F}^{-1}(\beta T) \in \mathcal{L}^1(\mathbb{R}^n)$

(A2) $\int_{\mathbb{R}^n} \mathcal{F}(\alpha S)\mathcal{F}^{-1}(\beta T)dx = \int_{\mathbb{R}^n} \mathcal{F}(\alpha T)\mathcal{F}^{-1}(\beta S)dx$

(A3) the linear mapping

$$\mathcal{D}(V) \ni \alpha \longmapsto \int_{\mathbb{R}^n} \mathcal{F}(\alpha S)\mathcal{F}^{-1}(\beta T)dx \in \mathbb{C} \quad \text{is continuous,}$$

where $\mathcal{F}$ and $\mathcal{F}^{-1}$ denote the direct and inverse Fourier transformas respectively. It is easy to see that, if

(8.A3.1) $\qquad \beta = 1$ on supp α

then the linear mapping in (A3) does not depend on β, hence it defines a distribution in $\mathcal{D}'(V)$. In this way, $S\cdot T$ is defined as the distribution in $\mathcal{D}'(\mathbb{R}^n)$ generated by (A3) and (8.A3.1).

The Mikusinski [3], Hirata-Ogata product $[S][T]$ exists, if and only if

(MHO) $$\lim_{\nu\to\infty} (\alpha_\nu * S)(\beta_\nu * T) \text{ exists in } \mathcal{D}'(\mathbb{R}^n)$$

for every δ-sequence $(\alpha_\nu | \nu \in \mathbb{N})$ *and* $(\beta_\nu | \nu \in \mathbb{N})$. We recall that $(\alpha_\nu | \nu \in \mathbb{N}) \in (\mathcal{D}(\mathbb{R}^n))^{\mathbb{N}}$ is called a δ-sequence, if and only if

(8.A3.2) $$\alpha_\nu \geq 0, \quad \nu \in \mathbb{N}$$

(8.A3.3) $$\text{supp } \alpha_\nu \to \{0\} \subset \mathbb{R}^n, \quad \text{when} \quad \nu \to \infty$$

(8.A3.4) $$\int_{\mathbb{R}^n} \alpha_\nu(x)dx = 1, \quad \nu \in \mathbb{N}$$

It follows easily that, if it exists, the limit in (MHO) does not depend on the δ-sequences $(\alpha_\nu | \nu \in \mathbb{N})$ and $(\beta_\nu | \nu \in \mathbb{N})$. Hence $[S][T]$ is defined as the limit in (MHO), whenever it exists.

A first result is the following implication

(8.A3.5) $$(A1) \Rightarrow (A2), (A3), (MHO)$$

and if (A1) holds, thus $S \cdot T$ exists, then

(8.A3.6) $$S \cdot T = [S][T]$$

It should however be noted that the existence of $[S][T]$ does *not* imply the existence of $S \cdot T$. For instance, if $f \in \mathcal{L}^\infty(\mathbb{R}^n)$ is continuous at $x = 0 \in \mathbb{R}^n$, without being continuous in a whole neighbourhood, then $[f][\delta]$ exists, but $f \cdot \delta$ does not exist.

The Vladimirov product $S_o T$ exists, if and only if

(VL)
$$\begin{aligned} &\forall \ x \in \mathbb{R}^n \\ &\exists \ V \text{ open neighbourhood of } x, \ \beta \in \mathcal{D}(\mathbb{R}^n) : \\ &\quad *) \ \beta = 1 \text{ on } V \\ &\quad **) \ \mathcal{F}(\beta S) * \mathcal{F}(\beta T) \in \mathcal{S}'(\mathbb{R}^n) \end{aligned}$$

It is easy to see that the linear mapping

(8.A3.7) $\mathcal{D}(V) \ni \alpha \longmapsto (\mathcal{F}(\beta S) * \mathcal{F}(\beta T))(\mathcal{F}^{-1}\alpha) \in \mathbb{C}$

is continuous. In this way, $S_o T \in \mathcal{D}'(\mathbb{R}^n)$ is defined by (8.A3.7).

It can be shown that

(8.A3.8) (A1) $\Longleftrightarrow$ (VL)

and whenever $S \cdot T$ exists, we have

(8.A3.9) $S \cdot T = S_o T$

We come now to the Kaminski Δ-product. We call $(\alpha_\nu | \nu \in \mathbb{N}) \in (\mathcal{D}(\mathbb{R}^n))^{\mathbb{N}}$ a Δ-sequence, if and only if it satisfies (8.A3.3), (8.A3.4), as well as

(8.A3.10)
$$\begin{aligned} &\forall \ p \in \mathbb{N}^n : \\ &\exists \ M_p > 0 : \\ &\forall \ \nu \in \mathbb{N} : \\ &\quad \epsilon_\nu^{|p|} \int_{\mathbb{R}^n} |D^p \alpha_\nu(x)| dx \leq M_p \end{aligned}$$

where $\operatorname{supp} \alpha_\nu \subset B(o, \epsilon_\nu) \subset \mathbb{R}^n$ and $\epsilon_\nu \longrightarrow 0$, when $\nu \longrightarrow \infty$, with $B(x,r)$ denoting the ball of radius $r > 0$ around $x \in \mathbb{R}^n$.

The Kaminski Δ-product $S \Delta T$ exists, if and only if

(KΔ) $\lim_{\nu \to \infty} (\alpha_\nu * S)(\beta_\nu * T)$ exists in $\mathcal{D}'(\mathbb{R}^n)$

for every Δ-sequences $(\alpha_\nu | \nu \in \mathbb{N})$ and $(\beta_\nu | \nu \in \mathbb{N})$. Again, it can be seen that when it exists, the limit in (KΔ) does not depend on the Δ-sequences involved, hence it is denoted by $S \Delta T$.

It can be shown that if $[S][T]$ exists, then so does $S \Delta T$ and we have

(8.A3.11) $S \Delta T = [S][T]$

The converse however is not true. Indeed, if we take

(8.A3.12) $S = \delta, \quad T = \sum_{1 < m < \infty} \frac{1}{m^r} \delta_{1/m} \in \mathcal{D}'(\mathbb{R})$

where $r > 2$, then $S \Delta T = 0$, but $[S][T]$ does not exist.

Finally, the relation with Colombeau's multiplication is obtained by taking particular Δ-sequences in (KΔ). Then one can show that if $S\Delta T \in \mathcal{D}'(\mathbb{R}^n)$ exists, we have

(8.A3.13) $\quad ST \approx S\Delta T$

where $ST \in \mathcal{G}(\mathbb{R}^n)$ denotes the product of S and T in Colombeau's algebra $\mathcal{G}(\mathbb{R}^n)$, and $\approx$ is the relation of association defined in Section 4.

We can recapitulate as follows:

(8.A3.14) $\quad S\cdot T$ exists $\Longleftrightarrow S_oT$ exists

in which case

(8.A3.15) $\quad S\cdot T = S_oT$

Further

(8.A3.16) $\quad S\cdot T$ exists $\Rightarrow [S][T]$ exists and $S\cdot T = [S][T]$

also

(8.A3.17) $\quad [S][T]$ exists $\Rightarrow S\Delta T$ exists and $[S][T] = S\Delta T$

finally

(8.A3.18) $\quad S\Delta T$ exists $\Rightarrow ST \approx S\Delta T$

In this way, the products $S\cdot T$ and S_oT are identical and also the least general. The product $[S][T]$ is a further generalization, while the product $S\Delta T$ is the most general among the four mentioned products, which all lead again to distributions in $\mathcal{D}'(\mathbb{R}^n)$.

The product in Colombeau's algebra $\mathcal{G}(\mathbb{R}^n)$ is related to all of the above four products by being related to the most general of them, that is the Kaminski Δ-product, as seen in (8.A3.18).

APPENDIX 4

AN INTUITIVE ILLUSTRATION OF THE STRUCTURE OF COLOMBEAU'S ALGEBRAS

It suffices to consider the one dimensional case of the algebra $\mathcal{G}(\mathbb{R})$.

As seen in Theorem 1, Section 2, the inclusion

(8.A4.1) $\mathcal{C}^\infty(\mathbb{R}) \subset \mathcal{G}(\mathbb{R})$

is an inclusion of differential algebras. In other words, the algebra structure and the derivatives on $\mathcal{G}(\mathbb{R})$, when restricted to functions in $\mathcal{C}^\infty(\mathbb{R})$, do perfectly coincide with the respective classical operations on $\mathcal{C}^\infty$-smooth functions.

Concerning the differential structure of $\mathcal{G}(\mathbb{R})$, that situation goes much further. Indeed, in view of Theorem 3, Section 2, the inclusion

(8.A4.2) $\mathcal{D}'(\mathbb{R}) \subset \mathcal{G}(\mathbb{R})$

still preserves the differential structure. That is, the derivatives on $\mathcal{G}(\mathbb{R})$, when restricted to $\mathcal{D}'(\mathbb{R})$, are precisely the usual distributional derivatives.

So that the peculiarity about the *structure* of $\mathcal{G}(\mathbb{R})$ only appears in connection with the way it extends the usual *multiplication* of *non* $\mathcal{C}^\infty$-smooth functions or distributions, whenever the latter are defined. Indeed, since a useful multiplication has to be compatible with differentiation - through the Leibnitz rule of product derivative, for instance - it all comes down to the possible relationships between multiplication and differentiation in the case of non $\mathcal{C}^\infty$-smooth functions or distributions. Here, the conflict between insufficient smoothness, multiplication and differentiation, in particular, the so called Schwartz impossibility result, come to set up some of the basic limitations on such possible relationships.

Fortunately, very simple examples can clearly illustrate the difficulties involved in extending the usual multiplication of non $\mathcal{C}^\infty$-smooth functions or distributions, see Appendix 1, Chapter 1. And then, a rather direct analysis of these difficulties can easily lead to an intuitive understanding of the *structure* of $\mathcal{G}(\mathbb{R}^n)$.

In view of the fact that - as mentioned above - the problem centers around the multiplication and differentiation of non $\mathcal{C}^\infty$-smooth functions and distributions, it means that a good illustration can be obtained if we consider products and derivatives of *discontinuous* functions. Indeed, applying differentiation to a non $\mathcal{C}^\infty$-smooth function for a suitable finite number of times, we must end up with a continuous function whose derivative does no longer exist as a continous function and may for instance exist as a distribution given by a discontinuous functions. And as $\mathcal{G}(\mathbb{R})$ has a sheaf structure, it suffices to consider all the above locally only, that is, in the neighbourhood of a discontinuity point of such a function.

Now obviously, the simplest nontrivial example is given by the continuous and non $\mathcal{C}^1$-smooth function

$$(8.A4.3) \qquad x_+ = \begin{vmatrix} 0 & \text{if} & x \leq 0 \\ x & \text{if} & x > 0 \end{vmatrix}$$

whose usual derivative no longer exists, and it only has the distributional derivative given by the Heaviside function

$$(8.A4.4) \qquad H(x) = \begin{vmatrix} 0 & \text{if} & x < 0 \\ 1 & \text{if} & x > 0 \end{vmatrix}$$

We note that

$$(8.A4.5) \qquad H \in \mathcal{L}^\infty(\mathbb{R}) \subset \mathcal{L}^1_{loc}(\mathbb{R}) \subset \mathcal{D}'(\mathbb{R}) \subset \mathcal{G}(\mathbb{R})$$

and $\mathcal{L}^\infty(\mathbb{R})$ is an algebra with the usual multiplication of functions. In this way, it suffices to study the relationship between the multiplication in $\mathcal{L}^\infty(\mathbb{R})$ and $\mathcal{G}(\mathbb{R})$, and do so with respect to the derivative on $\mathcal{G}(\mathbb{R})$, and in particular, on $\mathcal{D}'(\mathbb{R})$.

First, it is obvious that, for $m \in \mathbb{N}_+$, we have

$$(8.A4.6) \qquad H^m = H \quad \text{in} \quad \mathcal{L}^\infty(\mathbb{R})$$

However, as a power, H^m, with $m \in \mathbb{N}$, $m \geq 2$, is not defined in $\mathcal{D}'(\mathbb{R})$, since it involves the $m \geq 2$ factor product $H...H$. Yet, in view of (8.A4.6), (8.A4.5), for $m \in \mathbb{N}$, $m \geq 2$, we have the relation

$$(8.A4.7) \qquad H^m = H \quad \text{in} \quad \mathcal{D}'(\mathbb{R})$$

only via the relation (8.A4.6), that is, if H^m is computed in the algebra $\mathcal{L}^\infty(\mathbb{R})$.

As we also have $H \in \mathcal{G}(\mathbb{R})$, it is obvious that H^m, with $m \in \mathbb{N}_+$, is defined in $\mathcal{G}(\mathbb{R})$, but nevertheless, for $m \in \mathbb{N}$, $m \geq 2$, we have

$$(8.A4.8) \qquad H^m \neq H \quad \text{in} \quad \mathcal{G}(\mathbb{R})$$

Indeed, if we had equality in (8.A4.8), then we would obtain by differentiation that

$$mH^{m-1}DH = DH, \quad m \in \mathbb{N}, \quad m \geq 2$$

or

$$H^m DH = \frac{1}{m + 1} DH, \quad m \in \mathbb{N}_+$$

Let us take $m,p \in \mathbb{N}_+$, then we can compute

$$H^{m+p}DH$$

in two ways: first, we have

$$H^{m+p}DH = \frac{1}{m + p + 1} DH$$

or, we can also have

$$H^{m+p}DH = H^p(H^m DH) = H^p(\frac{1}{m + 1} DH) =$$

$$= \frac{1}{m + 1} H^p DH = \frac{1}{m + 1} \frac{1}{p + 1} DH$$

which means that

$$\frac{1}{m + p + 1} DH = \frac{1}{m + 1} \frac{1}{p + 1} DH, \quad m,p \in \mathbb{N}_+$$

But $DH \neq 0 \in \mathcal{G}(\mathbb{R})$. Thus we obtain the *absurd* result that

$$\frac{1}{m + p + 1} = \frac{1}{m + 1} \frac{1}{p + 1}, \quad m,p \in \mathbb{N}_+$$

which ends the proof of (8.A4.8).

The *difference* between (8.A4.7) and (8.A4.8) gives us a very simple example, and its following analysis can offer an intuitive insight into the structure of $\mathcal{G}(\mathbb{R})$.

In view of (8.A4.4), we have - among many other possible ones - the following *representation*

(8.A4.9) $\quad H = h + \mathcal{I} \in \mathcal{G}(\mathbb{R})$

where

$$(8.A4.10)\qquad h(\phi,x) = \int_{-x}^{\infty} \phi(y)dy, \quad \phi \in \Phi, \quad x \in \mathbb{R}$$

Then, in view of the way multiplication is defined in $\mathcal{G}(\mathbb{R})$, we obtain

$$(8.A4.11)\qquad H^m = h^m + \mathcal{I} \in \mathcal{G}(\mathbb{R}), \quad m \in \mathbb{N}_+$$

Now (8.A4.10) yields for $\phi \in \Phi$ the relation

$$(8.A4.12)\qquad \lim_{\epsilon \downarrow 0} h(\phi_\epsilon, x) = \begin{vmatrix} 0 & \text{if} \quad x < 0 \\ 1 & \text{if} \quad x > 0 \end{vmatrix}$$

while for $\phi \in \Phi$, $\epsilon > 0$ and $x \in \mathbb{R}$, we obtain

$$(8.A4.13)\qquad |h(\phi_\epsilon, x)| \le \int_{\mathbb{R}} |\phi(y)|dy < \infty$$

Hence, for $m \in \mathbb{N}_+$ and $\phi \in \Phi$, we have

$$(8.A4.14)\qquad \lim_{\epsilon \downarrow 0} h^m(\phi_\epsilon, x) = \begin{vmatrix} 0 & \text{if} \quad x < 0 \\ 1 & \text{if} \quad x > 0 \end{vmatrix}$$

and

$$(8.A4.15)\qquad |h^m(\phi_\epsilon, x)| < \left(\int_{\mathbb{R}} |\phi(y)|dy\right)^m < \infty$$

with $\epsilon > 0$ and $x \in \mathbb{R}$.

In view of (8.A4.12) and the continuity of $h(\phi_\epsilon, \cdot)$, we obviously have for $m \in \mathbb{N}$, $m \ge 2$, $\phi \in \Phi$ and $\epsilon > 0$

$$(8.A4.16)\qquad h^m(\phi_\epsilon, \cdot) \ne h(\phi_\epsilon, \cdot) \quad \text{in} \quad \mathcal{C}^\infty(\mathbb{R})$$

which is expected to happen, owing to (8.A4.8).

The crucial point of the analysis is the comparison of the relations (8.A4.7) and (8.A4.8), via the relations (8.A4.6).

Suppose given $\psi \in \mathcal{D}(\mathbb{R})$, then (8.A4.12)-(8.A4.15) and Lebesque's bounded convergence theorem give

$$(8.A4.17)\qquad \lim_{\epsilon \downarrow 0} \int_{\mathbb{R}} (h^m(\phi_\epsilon, x) - h(\phi_\epsilon, x))\psi(x)dx = 0$$

for $\phi \in \Phi$, $m \in \mathbb{N}_+$. This is in fact identical with

(8.A4.18) $$H^m \approx H \quad \text{in} \quad \mathcal{G}(\mathbb{R})$$

for $m \in \mathbb{N}_+$

But when seen in $\mathcal{D}'(\mathbb{R})$, the relation (8.A4.17) has the following different meaning: in view of (8.A4.12)-(8.A4.15) and Lebesque's bounded convergence theorem, it follows that

(8.A4.19) $$\lim_{\epsilon \downarrow 0} h(\phi_\epsilon, \cdot) = \lim_{\epsilon \downarrow 0} h^m(\phi_\epsilon, \cdot) = H \quad \text{in} \quad \mathcal{D}'(\mathbb{R})$$

and hence, as also follows from (8.A4.17), we have

(8.A4.20) $$\lim_{\epsilon \downarrow 0} (h^m(\phi_\epsilon, \cdot) - h(\phi_\epsilon, \cdot)) = 0 \quad \text{in} \quad \mathcal{D}'(\mathbb{R})$$

where all the three limits above are in the sense of the weak topology on $\mathcal{D}'(\mathbb{R})$, and hold for $m \in \mathbb{N}_+$ and $\phi \in \Phi$.

We can now conclude that, although H^m, with $m \in \mathbb{N}$, $m \geq 2$, is not definable as a power in $\mathcal{D}'(\mathbb{R})$ and is only defined via (8.A4.6), nevertheless, H^m and H are *indistinguishable* in $\mathcal{D}'(\mathbb{R})$, just as they are in $\mathcal{L}^\infty(\mathbb{R})$. In other words, $\mathcal{D}'(\mathbb{R})$ *cannot* retain any information on h or h^m in (8.A4.19), except to register their common *limit* H. It follows that the way *discontinuities* of functions such as in $\mathcal{L}^\infty(\mathbb{R})$, or in general $\mathcal{L}^1_{loc}(\mathbb{R})$ appear in $\mathcal{D}'(\mathbb{R})$, its *too simple* in order to allow for a suitable relation between multiplication and differentiation, such as given by the Leibnitz rule of product derivative. This excessive simplicity in dealing with discontinuities is apparent in the following general situation: given any distribution $T \in \mathcal{D}'(\mathbb{R})$, there exists families of functions $f_\epsilon \in \mathcal{C}^\infty(\mathbb{R})$, with $\epsilon > 0$, see (8.1.7), such that

(8.A4.21) $$\lim_{\epsilon \downarrow 0} f_\epsilon = T \quad \text{in} \quad \mathcal{D}'(\mathbb{R})$$

in the sense of the weak topology on $\mathcal{D}'(\mathbb{R})$, which means that

(8.A4.22) $$\lim_{\epsilon \downarrow 0} \int_{\mathbb{R}} f_\epsilon(x)\psi(x)dx = T(\psi), \quad \psi \in \mathcal{D}(\mathbb{R})$$

Now in view of (8.A4.22), it is obvious that, just as with (8.A4.18), the only thing retained in $\mathcal{D}'(\mathbb{R})$ from (8.A4.21) is the *limit* value given by the distribution T, all other information about f_ϵ, with $\epsilon > 0$, being lost. In particular, the *averaging* process (8.A4.22) involving arbitrary test functions $\psi \in \mathcal{D}(\mathbb{R})$, is *too coarse* in order to be able to accommodate the discrimination in (8.A4.16).

On the other hand, the picture in $\mathcal{G}(\mathbb{R})$, as given by (8.A4.9), (8.A4.11), is more sophisticated. Indeed, H and H^m are defined by h and h^m respectively, through the *quotient representations* in the mentioned two relations. Morevoer, the relation between h and H, as well as h^m and H^m is *not* through a limit or convergence process - see Appendix 2 - but through an *asymptotic interpretation*. And as seen in (8.A4.16), (8.A4.18), and of course (8.A4.8), that asymptotic interpretation can distinguish between H and H^m, precisely because it does retain *sufficient* information on h and h^m.

The above may serve as an instructive example in illustrating the fact that *asymptotic interpretations* can be more *sophisticated* - and thus useful - than limit, convergence or topological processes.

FINAL REMARKS

With this volume, the presentation of basic features of the 'algebra first' approach to a systematic and comprehensive nonlinear theory of generalized functions, needed for the solution of large classes of nonlinear partial differential equations, comes to a certain completion.

A first stage of this 'algebra first' approach was started in Rosinger [7,8], and developed later in Rosinger [1,2,3]. One of its particular, nevertheless quite natural and rather central cases, were presented in Colombeau [1,2].

This first stage focused on the 'near embedding'

(1) $\quad \mathcal{D}'(\mathbb{R}^n) \text{ '}\subset\text{' } (\mathcal{C}^\infty(\mathbb{R}^n))^{\mathbb{N}}$

mentioned in (8.1.8) for instance, which leads in a natural way to the idea of constructing embeddings, see (1.5.25)

(2) $\quad \mathcal{D}'(\mathbb{R}^n) \subset A = \mathcal{A}/\mathcal{I}$

where

(3) $\quad \mathcal{A}$ is a subalgebra in $(\mathcal{C}^\infty(\mathbb{R}^n))^\Lambda$

while

(4) $\quad \mathcal{I}$ is an ideal in $\mathcal{A}$

and Λ is a suitable, infinite index set.

The important point to note with (2)-(4) is that we have the obvious inclusions

(5) $\quad \mathcal{I} \subset \mathcal{A} \subset (\mathcal{C}^\infty(\mathbb{R}^n))^\Lambda \subset (\mathcal{C}^0(\mathbb{R}^n))^\Lambda = \mathcal{C}(\Lambda \times \mathbb{R}^n)$

if we consider Λ with the discrete topology. In that case however $\Lambda \times \mathbb{R}^n$ will become a completely regular topological space, and in view of (5), we have

(6) $\quad \mathcal{I} \subset \mathcal{A} \subset \mathcal{C}(\Lambda \times \mathbb{R}^n)$

in other words, $\mathcal{A}$ is a subalgebra in the algebra $\mathcal{C}(\Lambda \times \mathbb{R}^n)$ of continuous functions on $\Lambda \times \mathbb{R}^n$, while $\mathcal{I}$ is an ideal in $\mathcal{A}$.

Now we can recall the well known *rigidity* property between the *topological* properties of a completely regular space X and the *algebraic* properties of its ring of continuous functions $\mathcal{C}(X)$, Gillman & Jerison. For instance, two real compact spaces X and Y are topologically homeomorphic, if and only if $\mathcal{C}(X)$ and $\mathcal{C}(Y)$ are isomorphic algebras. It follows that a good deal of the algebraic properties of the quotient algebras $A = \mathcal{A}/\mathcal{I}$ in (2) may well depend on the simple topological properties of $\Lambda \times \mathbb{R}^n$. The extent to which that proved indeed to be the case is presented in Rosinger [1,2,3], Colombeau [1,2] and Chapters 2-8 of this volume.

In short, one may say that the essence of this 'algebra first' approach centers around the following:
Fortunate Inversion:
We have schematically the situation:

nonlinear PDEs
↑
linear algebra (semigroups, groups, vector spaces) —— functional analysis —— linear PDEs
↑
nonlinear algebra (rings, etc.)

where an arrow $\longrightarrow$ indicates the direction of increasing generality.

Nevertheless, in view of the mentioned *rigidity* property of *rings of continuous functions*, we can establish a rather powerful *inverse connection* :

nonlinear PDEs
/
nonlinear algebra

Concerning the second stage of that 'algebra first' approach, let us note the following.

The embedding (2) has of course *differential* aspects as well, which may go beyond the algebraic ones involved in (6) for instance. Indeed, one would like to have on $A = \mathcal{A}/\mathcal{I}$ partial derivatives which extend the distributional ones on $\mathcal{D}'(\mathbb{R}^n)$. Historically, that issue has led to a long lasting misundertanding, started with a misinterpretation of L. Schwartz's so called impossibility result of 1954.

However, as seen in Sections 1-4 and Appendix 1 in Chapter 1, the difficulties with the differential structure on $A = \mathcal{A}/\mathcal{I}$ happen to have a most simple *algebraic* nature and center around a *conflict* between discontinuity, multiplication and abstract differentiation.

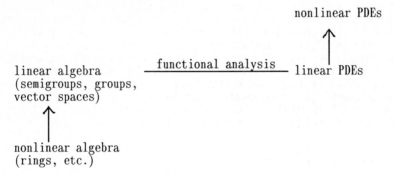

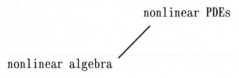

It is precisely the clarification in Chapter 1 of that second algebraic phenomenon which, together with the earlier dealt with algebraic aspects involved in (1)-(6), bring to a certain completion the mentioned 'algebra first' approach.

And now, where do we go from here?

Well, perhaps not so surprisingly, one most promising direction seems to be that along the lines of further *desescalation* in the sense mentioned in the Foreword.

Indeed, the 'algebra first' approach has already brought with it a significant desescalation from the functional analytic methods, so much customary in the study of partial differential equations during the last four or five decades, to the 'nonlinear algebra' of rings of continuous functions. And one of the aspects of that desescalation which is particularly important, yet it is seldom noticed, is that - unlike in functional analysis - the 'nonlinear algebra' of rings is in fact a *particular* and enriched case of the 'linear algebra' of vector spaces, groups or semigroups. In this way, the 'algebra first' approach in the study of *nonlinear* partial differential equations leads to *more particular* and not to more general algebraic structures!

But now, in view of the hierarchy

- set theory
- binary relations, order
- algebra
- topology
- functional analysis
- etc.

it is time for one *further desescalation*, namely, from 'algebra first' to 'order first'!

And the fact is that there exists a precedent for it, for nearly a decade by now, Browsowski.

Indeed, in that paper, the linear Dirichlet problem

$$(7)\qquad \begin{aligned} &\Delta U = 0 \quad \text{in} \quad \Omega \subset \mathbb{R}^n, \quad \Omega \text{ open, bounded} \\ &U = u \quad \text{on} \quad \partial\Omega \end{aligned}$$

is given a method of solution based on the Dedekind *order* completion of the Riesz space $\mathcal{C}(\partial\Omega)$ of continuous functions on the compact space $\partial\Omega \subset \mathbb{R}^n$.

Unfortunately, as it stands, that method is not applicable to *nonlinear* partial differential equations.

However, one can proceed along different lines in that 'order first' approach.

A promising direction is offered by an extension of the Cauchy-Kovalevskaia theorem to *continuous* nonlinear partial differential equations, using a Dedekind order completion of spaces of smooth functions, a joint result obtained in collaboration with M. Oberguggenberger, which is to be published elsewhere.

REFERENCES

Abbott, M.B. : Computational Hydraulics. Elements of the Theory of Free Surface Flows. Pitman, London, 1979

Abbott, M.B., Basco, D.R. : Computational Fluid Dynamics, An Introduction for Engineers, Longman, New York, 1989

Adamczewski, M. : Vectorisation, Analyse et Optimisation d'un Code Bidimensionnel Eulerien. Doctoral Thesis, University of Bordeaux 1, Talence, 1986

Ambrose, W. : Products of Distributions with Values in Distributions. J. Reine Angew. Math., 315(1980), 73-91

Aragona, J. : [1] Theoreme d'existence pour l'operateur $\bar{\partial}$ sur les formes differentielles generalisees. C.R. Acad. Sci Paris Ser I, Math, 300(1985), p. 239-242

Aragona, J. : [2] On existence theorema for the $\bar{\partial}$ operator for generalized differential forms. Proc. London Math. Soc. (3) 53, 1986, p. 474-488.

Aragona, J., Colombeau, J.F. : The $\bar{\partial}$ equation for generalized functions. J. Math. Anal. Appl. 110, 1(1985) p. 179-1990.

Ball, J.M. : Convexity conditions and existence theorems in nonlinear elasticity. Arch. Rat. Mech. Anal. 63 (1977) 337-403

Bell, J.L., Slomson, A.B. : Models and Ultraproducts. North Holland, Amsterdam, 1969

Biagioni, H.A. : [1] The Cauchy problem for semilinear hyperbolic systems with generalized functions as initial conditions, Resultate Math, 14(1988) 231-241

Biagioni, H.A. : [2] A Nonlinear Theory of Generalized Functions. Lecture Notes in Mathematics, vol. 1421, Springer, New York, 1990

Biagioni, H.A., Colombeau, J.F. : [1] Borel's theorem for generalized functions. Studia Math. 81(1985) p. 179-183.

Biagioni, H.A., Colombeau J.F. : [2] Whitney's extension theorem for generalized functions. J. Math. Anal. Appl. 114, 2(1986), p. 574-583

Braunss, G., Liese, R. : Canonical Products of Distributions and Causal Solutions of Nonlinear Wave Equations. J. Diff. Eq., 16(1974), 399-412

Brézis, H., Friedman, A. : Nonlinear Parabolic Equations Involving Measures as Initial Conditions. J. Math. Pures et Appl., 62(1983), 73-97

Brosowski, B. : An application of Korovkin's theorem to certain PDEs. In Lecture Notes in Mathematics, vol. 843, 1981, pp. 150-162, Springer, New York

Buck, R.C. : The solutions to a smooth PDE can be dense in $\mathcal{C}(\mathrm{I})$. J. Diff. Eq. 41(1981) 239-244

Carroll, R.W. : Abstract Methods in Partial Differential Equations. Harper & Row, New York, 1969

Cauret, J.J. : Analyse et Developpement d'un Code Bidimensionnel Elastoplastique. Doctoral Thesis, University of Bordeaux 1, Talence, 1986

Cauret, J.J., Colombeau, J.F., Le Roux, A.Y. : [1] Solutions generalisees discontinues de problemes hyperboliques non conservatifs. C.R. Acad Sci. Paris Serie I Math 302, (1986) p. 435-437.

Cauret, J.J., Colombeau, J.F., Le Roux, A.Y. : [2] - Discontinuous generalized solutions of nonlinear nonconservative hyperbolic equations. J. Math. Anal. Appl., 139(1989) 552-573

Colombeau, J.F. : [1] New Generalized Functions and Multiplication of Distributions. North Holland Mathematics Studies, vol. 84, 1984

Colombeau, J.F. : [2] Elementary Introduction to New Generalized Functions. North Holland Mathematics Studies, vol. 113, 1985

Colombeau, J.F. : [3] New General Existence Results for Partial Differential Equations in the $\mathcal{C}^\infty$ Case. University of Bordeaux, 1984

Colombeau, J.F. [4] A Mathematical Analysis Adapted to the Multiplication of Distributions. Springer Lecture Notes (to appear)

Colombeau, J.F. : [5] Differential Calculus and Holomorphy. Real and Complex Analysis in Locally Convex Spaces. North-Holland Mathematics Studies, vol. 64, 1982

Colombeau, J.F. : [6] A General Existence Result for Solutions of the Cauchy Problem for Nonlinear Partial Differential Equations. University of Bordeaux, 1985

Colombeau, J.F. : [7] A multiplication of distributions. J. Math. Anal. Appl. 94, 1(1983) p. 98-115.

Colombeau, J.F. : [8] New generalized functions, Multiplication of distributions. Physical applications. Portugal, Math. 41, 1-4(1982), p. 57-69.

Colombeau, J.F. : [9] Une multiplication generale des distributions. C.R. Acad. Sci. Paris Ser I Math. 296(1983), p. 357-360.

Colombeau, J.F. : [10] Some aspects of infinite dimensional holomorphy in mathematical physics. In "aspects of Mathematics and its Applications", editor J.A. Barroso, North-Holland Math. Library 34(1986) p. 253-263.

Colombeau, J.F. : [11] A new theory of generalized functions. In "Advances of Holomorphy and Approximation Theory", editor J. Mujica, North-Holland Math. Studies, 123, 1986, p. 57-66.

Colombeau, J.F. : [12] Nouvelles solutions d'equations aux derivees partielles. C.R. Acad. Sci. Paris Ser. I Math. 301(1985) p. 281-283.

Colombeau, J.F. : [13] Multiplication de distributisons et acoustique, Revue d' accoustiqaue, to appear.

Colombeau, J.F. : [14] Generalized functions, multiplication of distributions, applications to elasticity, elastoplasticity fluid dynamics and acoustics. Proceedings of the Congress of generalized functions, Dubrovnik, 1987, Plenum Pub. Comp., to appear.

Colombeau, J.F., Gale, J.E. : [1] Holomorphic generalized functions. J. Math. Anal. Appl. 103, 1(1984) p. 117-133

Colombeau, J.F., Gale J.E. : [2] The analytic continuation for generalized holomorphic functions. Acta Math. Hung., in press.

Colombeau, J.F., Langlais, M. : Existence et Unicité de Solutions d'Equations Paraboliques Nonlineaire avec Conditions Initiales Distributions. Comptes Rendus, 302(1986), 379-382.

Colombeau, J.F., le Roux, A.Y. : [1] Numerical techniques in elasto dynamics. Lecture Notes in Math. 1270, Springer (1987) p. 103-114

Colombeau, J.F., Le Roux, A.Y. : [2] Generalized functions and products appearing in equations of physics, preprint.

Colton, D.L. : Analytic Theory of Partial Differential Equations. Pitman Advanced Publishing Program, Boston, 1980

Dacorogna, B. : Weak Continuity and Weak Lower Semicontinuity of Nonlinear Functionals. Lecture Notes in Math., vol. 922, Springer, New York, 1982

Di Perna, R.J. : Compensated Compactness and General Systems of Conservation Laws. Trans. AMS, vol 292, no 2, Dec 1985, 383-420

Ehrenpreis, L. : Solutions of Some Problems of Division I. Amer. J. Math., 76(1954), 883-903

Eringen, C. (Ed) : Continuum Physics. Vol. II. Acad. Press, New York, 1975

Evans, L.C. : Weak Convergence Methods for Nonlinear PDEs. Conference Board of the Mathematical Sciences, no. 74. Providence, 1990

Fischer, B. : [1] The Neutrix Distribution Product $x_+^{-r}\delta^{(r-1)}(x)$. Stud. Sci. Math. Hung., 9(1974), 439-441

Fischer, B : [2] Distributions and the Change of Variables. Bull. Math. Soc. Sci. Math. Rom., 19(1975), 11-20.

Folland, G.B. : Introduction to Partial Differential Equations. Princetown Univ. Press, 1976

Friedlander, F.G. : Introduction to the Theory of Distributions. Cambridge Univ. Press, Cambridge, 1982

Fung, Y.C. : A First Course in Continuum Mechanics. Prentice-Hall, New Jersey, 1969

Gillman, L., Jerison, M. : Rings of Continuous Functions. Van Nostrand, New York, 1960

Golubitsky, M., Schaeffer, D.G. : Stability of shock waves for a single conservation law. Adv. Math., 15 , 1975, 65-71

Grushin, V.V. : A Certain Example of a Differential Equation Without Solutions. Math. Notes, 10(1971), 449-501

Gutterman, M. : An Operational Method in Partial Differential Equations. SIAM J Appl. Math, vol 17, no 2, March 1969, 468-493

Hatcher, W.S. : Calculus is Algebra. Amer. Math. Month., (1982), 362-370

Hirata, Y., Ogata, H. : On the Exchange Formula for Distributions. J. Sci. Hiroshima Univ., Ser. A, 22(1958), 147-152

Hörmander, L. : [1] Linear Partial Differential Operators, (fourth printing) Springer, New York, 1976

Hörmander, L. : [2] Fourier Integral Operators. Acta Math, 127(1971), 79-183

Kamiński, A. : Convolution, product and Fourier transform of distributions. Studia Math., 74(1982), 83-96

Kelley, J.L. : General Topology. Van Nostrand, New York, 1955

Köthe, G. : Topologische lineare Raume, vol. 1, Springer, Berlin, 1960

Kranz, S.G. : Function Theory of Several Complex Variables. J. Wiley, New York, 1982

Kuo, H. H. : Differential Calculus for Measures on Banach Spaces, pp. 270-285 in Springer Lecture Notes in Mathematics, vol. 644, New York, 1978

Lax, P.D. : The Formation and Decay of Shock Waves. Amer. Math. Month., (1972). 227-241

Lewy, H. : An Example of a Smooth Linear Partial Differential Equation without Solution. Ann. Math., vol. 66, no. 2 (1957), 155-158

Lions, J.L. : [1] Une Remarque sur les Problemes D'evolution Nonlineaires dans les Domaines Noncylindrique. Rev. Romaine Math. Pure Appl., 9(1964), 11-18

Lions, J.L. : [2] Quelques Methods de Resolution des Problemes aux Limites Nonlineaires. Dunod, Paris, 1969

Majda, A. : Compressible Fluid Flow and Systems of Conservation Laws in Several Space Variables. Spinger, New York, 1984

Malgrange, B. : Existence et Approximation des Solutions des Equations aux Derivees Partielles et des Equations de Convolutions. Ann. Inst. Fourier, Grenoble, 6(1955-56), 271-355

Mikusinski, J. : [1] Irregular Operations on Distributions. Stud. Math., 20(1960), 163-169

Mikusinski, J. : [2] On the Square of the Dirac Delta Distribution. Bull. Acad. Pol. Sc., vol. 14, no. 9(1966), 511-513

Mikusinski, J. : [3] Criteria for the Existence and Associativity of the Product of Distributions. Studia Math., 21(1962), 253-259

Murat, F. : Compacite par compensation : Condition necessaire et sufficante de continuite faible sous une hypotheses de rang constant. Ann. Scuola Norm. Sup. 8(1981) 69-102.

Narasimhan, R. : Analysis on Real and Complex Manifolds. Masson & Cie, Paris, 1973

Nirenberg, L. & Trèves, F. : Solvability of a First Order Linear Partial Differential Equation. Comm. Pure Appl. Math., 16(1963), 331-351

Oberguggenberger, M. : [1] Products of Distributions. J. Reine Angew. Math, 365(1986), 1-11

Oberguggenberger, M. : [2] Weak Limits of Solutions to Semilinear Hyperbolic Systems. Math. Ann., 274(1986), 599-607

Oberguggenberger, M. : [3] Multiplication of Distributions in the Colombeau Algebra $\mathcal{G}(\Omega)$. Boll. Unione Mat. Ital. (6)5-A(1986)

Oberguggenberger, M. : [4] Generalized Solutions to Semilinear Hyperbolic Systems. Monatsch. Math. 103(1987) 133-144

Oberguggenberger, M : [5] Private Communication, 1986

Oberguggenberger, M. : [6] Semilinear wave equations with rough initial data : generalized solutions. In Antosik, P., Kaminski, A. (Eds.) Generalized Functions and Convergence. World Scientific Publishing, London, 1990.

Oberguggenberger, M. : [7] Propagation of singularities for semilinear hyperbolic initial-boundary value problems in one space dimension. J. Diff. Eq. 61(1986), 1-39

Oberguggenberger, M. : [8] Propagation and reflection of regularity for semilinear hyperbolic (2×2) systems in one space dimension. Nonlinear Anal. 10(1986), 965-981

Oberguggenberger, M. : [9] Weak limits of solutions to semilinear hyperbolic systems. In: Hyperbolic Equations, Ed. F. Colombini, M.K.V. Morthy, Pitman Research Notes in Math. Longman 1987, 278-281.

Oberguggenberger, M. : [10] Solutions generalisees de systemes hyperbolic semilineaires. Computes Rendus Acad. Sci. Paris Ser. I, 305(1987), 17-18

Oberguggenberger, M. : [11] Hyperbolic systems with discontinuous coefficients: examples. In: B. Stankovic, E. Pap, S. Philipovic, V.S. Vladimirov (Ed.), Generalized functions, Convergence Structures, and Their Applications. Plenum Press, New York 1988, 257-266

Oberguggenberger, M. : [12] Products of distributions: Nonstandard methods. Z. Anal. Anw. 7(1988), 347-365

Oberguggenberger, M : [13] Systemes hyperboliques a coefficients discontinus: solutions generalisees et une application a l'acoustique lineaire. C.R. Math. Rep. Acad. Sci. Canada 10(1988), 143-148

Oberguggenberger, M. : [14] Hyperbolic systems with discontinuous coefficients: Generalized solutions and a transmission problem in acoustics. J. Math. Anal. Appl. 142(1989), 452-467

Oberguggenberger, M. : [15] Multiplications of Distributions and Applications to PDEs. Technical Report UPWT 90/3, Department of Mathematics and Applied Mathematics, University of Pretoria, South Africa, 1990

Oleinik, O.A. : The Analyticity of Solutions of PDEs and its Applications. Trends in Applications of Pure Mathematics to Mechanics (Ed. Fichera, G.), Pitman, London, 1976

Oxtoby, J.C. : Measure and Category. Springer, New York, 1971

Pathak, R.S. : [1] Orthogonal series representations for generalized functions. J. Math. Anal. Appl. 13(1988) 316-333

Pathak, R.S. : [2] Ultradistributions as boundary values of analytic functions. Trans. Amer. Math. Soc. 286(2)(1984) 536-566

Peyret, R., Taylor, T.D. : Computational Methods for Fluid Flow. Springer, New York, 1983

Raju, C.K. : Products and compositions with the Dirac dela function. J. Phys. A : Math. Gen. 15(1982) 381-396

Rauch, J., Reed, M. : Nonlinear superposition and absorption of delta waves in one space dimension. J. Funct. Anal. 73(1987) 152-178)

Reed, M.C. : [1] Abstract Nonlinear Wave Equations. Springer Lecture Notes in Mathematics. vol. 507. 1976

Reed, M.C. : [2] Propagation of singularities for nonlinear wave equations in one dimension. Comm. Part Diff. Eq., 3, 1978, 153-199

Reed, M.C. : [3] Singularities in nonlinear waves of Klein-Gordon type. Springer Lecture Notes in Mathematics, vol. 648, 1978, 145-161

Reed, M.C., Berning, J.A. : Reflection of singularities of one dimensional semilinear wave equations at boundaries. J. Math. Anal. Appl., 72, 1979, 635-653

Richtmyer, R.D. : Principles of Advanced Mathematical Physics, vol. 2, Springer, New York, 1981

Rosinger, E.E. : [1] Distributions and Nonlinear Partial Differential Equations. Springer Lectures Notes in Mathematics, vol. 684, 1978

Rosinger, E.E. : [2] Nonlinear Partial Differential Equations, Sequential and Weak Solutions. North Holland Mathematics Studies, vol. 44, 1980

Rosinger, E.E. : [3] Generalized Solutions of Nonlinear Partial Differential Equations. North Holland Mathematics Studies, vol. 146, 1987

Rosinger, R.E. : [4] Nonlinear Equivalence, Reduction of PDEs to ODEs and Fast Convergent Numerical Methos. Research Notes in Mathematics, vol. 77. Pitman, Boston, 1982

Rosinger, E.E. : [5] Propagation of round off errors and the role of stability in numerical methods for linear and nonlinear PDEs. Appl. Math. Modelling, 1985, 9, 331-336

Rosinger, E.E. : [6] Convergence paradox in numerical methods for linear and nonlinear PDEs. In R. Vichnevetsky & R.S. Stepleman (Eds) Advances in Compuer Methods for PDEs, vol. VI, pp. 431-435, IMACS, New Brunswick, 1987

Rosinger, E.E. : [7] Embedding of the $\mathcal{D}'$ Distributions into Pseudo-topological Algebras. Stud. Cerc. Mat. vol. 18, no. 5, 1966, 687-729

Rosinger, E.E. : [8] Pseudotoplogical Spaces. The embedding of the $\mathcal{D}'$ Distributions into Algebras. Stud. Cerc. Math. vol. 20, no. 4, 1968, 553-582

Rubel, L.A. : A universal differential equation. Bull. AMS, vol. 4, no. 3(1981) 345-349

Rudin, W. : Functional Analysis. McGraw-Hill, New york, 1973

Sato, M., Kawai, T., Kashiwara, M. : Hyperfunctions and Pseudodifferential Equations. Springer Lecture Notes in Mathematics, vol. 287, 1973

Schmieden, C., Laugwitz, D. : Eine Erweiterung der Infinitesimalrechung. Math. Zeitschr., 69(1985), 1-39

Schwartz, L. : [1] Theorie des Distributions I, II. Hermann, Paris 1950, 1951

Schwartz, L. : [2] Sur L'impossibilite de la Multiplication des Distributions. C.R. Acad. Sci. Paris, 239(1954), 847-848

Seebach, J.A. Jr., Seebach, L.A., Steen, L.A. : What Is a Sheaf? Amer. Math. Month. (1970), 681-703

Shapira, P. : Une Equation aux Derivees Partielles Sans Solution dans L'espace des Hyperfunctions. C.R. Acad. Sci. Paris, 265(1967), 665-667

Slemrod, M. : Interrelationships among mechanics, numerical analysis, compensated compactness and oscillation theory. In Oscillation Theory, Computation and Methods of Compensated Compactness (Dafermos, C., et. al, eds) Springer, New York, 1986

Smoller, J. : Shock Waves and Reaction-Diffusion Equations. Springer, New York, 1983

Sobolev, S.L. : [1] Le Probleme de Cauchy dans L'espace des Functionelles. Dokl. Acad. Sci. URSS, vol. 7, no. 3(1935), 291-294

Sobolev, S.L. : [2] Methode Nouvelle a Resondre le Probleme de Cauchy pour les Equations Lineaires Hyperboliques Normales. Mat. Sbor., vol. 1, no. 43(1936), 39-72

Stroyan, K.D., Luxemburg, W.A.J. : Introduction to the Theory of Infinitesimals. Acad. Press, New York, 1976

Struble, R.A. : Operator Homomorphisms. Math. Z., 130(1973), 275-285

Tartar, L. : Compensated compactness and applications to PDEs. In Nonlinear Analysis and Mechanics: Herriot Watt Symposium 4 (Knops, R.J. ed.) Pitman, 1979

Temam, R. : Infinite Dimensional Dynamical Systems in Mechanics and Physics. Applied Mathematical Sciences, vol. 68, Springer, New York, 1988

Todorov, T.D. : [1] Colombeau's new generalized functions and Nonstandard Analysis. Proceedings of the Congress of generalized functions, Dubrovnik 1987, Plenum Pub. Comp.

Todorov, T.D. : [2] Sequential approach to Colombeau's theory of generalized functions. Publications IC/87/128. International Center for Theoretical Physics, Trieste

Treves, F. : [1] Linear Partial Differential Equations, Notes on Mathematics and its Applications. Gordon and Breach, New York, 1971

Treves, F. : [2] Basic Linear Partial Differential Equations. Acad Press, New York, 1978

Treves, F. : [3] Introduction to Pseudodifferential and Fourier Integral Operators I, II. Plenum Press, New York, 1980

Treves, F. : [4] On Local Solvability of Linear Partial Differential Equations. Bull. AMS, 76(1970), 552-571

Truesdell, C. : A First Course in Rational Continuum Mechanics. Vol. 1. Acad. Press, New York, 1977

Van der Corput, J. G. : Introduction to Neutrix Calculus. J. D'Analyse Math., 7(1959), 281-398

Van Rootselaar, B. : Bolzano's Theory of Real Numbers. Arch. Hist. Exact Sc., 2(1964), 168-180

Vladimirov, V.S. : Generalized functions in mathematical physics. Mir Publishers, Moscow, 1979

Waelbroeck, L. : The Category of Quotient Bornological Spaces. Aspects of Mathematics and its Applications (Ed. Barroso, J. A.). North Holland, Amsterdam, 1984

Walker, R.C. : The Stone-Cech compactification. Springer, Berlin, 1974

Walter, W. : An Elementary Proof of the Cauchy-Kovalevskaia Theorem. Amer. Math. Monthly, Feb. 1985, 115-126

Whitney, H. : Analytic Extensions of Differentiable Functions Defined on Closed Sets. Trans. AMS, 36(1934), 63-89

Yoshida, K. : Functional Analysis. Springer, New York, 1965

Zabuski, N.J. : Computational Synergetics and Mathematical Innovation. J. Comp. Physics, 1981, 43, 195-249